Dychweler y llyfr hwn [illegible]
neu pan e[illegible]ir am[illegible]

EXPERIENTIA SUPPLEMENTUM 26

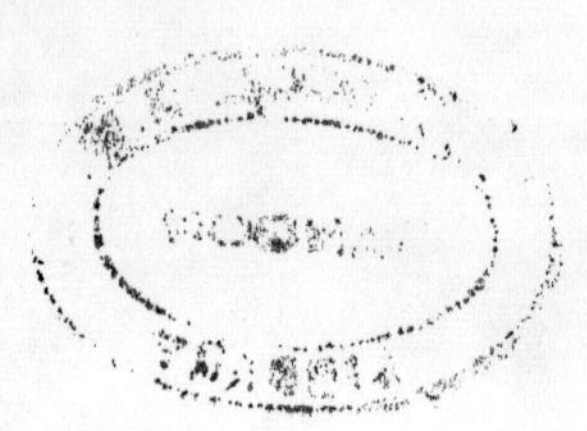

Enzymes and Proteins from Thermophilic Microorganisms

Structure and Function

Proceedings of the International Symposium
Zürich, July 28 to August 1, 1975

Edited by Herbert Zuber

1976 Birkhäuser Verlag, Basel und Stuttgart

761739

SCIENCE
QR90.S95
1975

UNIV. COLL. N.W.
BANGOR
LIBRARY

CIP-Kurztitelaufnahme der Deutschen Bibliothek

Enzymes and proteins from thermophilic microorganisms: structure and function; proceedings of the internat. Symposium Zürich, July 28 to August 1, 1975/ ed. by Herbert Zuber. - 1. Aufl. - Basel, Stuttgart: Birkhäuser, 1976.
(Experientia: Suppl.; 26)
NE: Zuber, Herbert [Hrsg.]

All rights reserved
No part of this book may be reproduced in any form, by photostat, microfilm, or any other means, without written permission from the publishers.

© Birkhäuser Verlag Basel, 1976
Printed in Switzerland

ISBN 3-7643-0827-3

PREFACE

The papers collected in this volume were presented at the "Symposium on Enzymes and Proteins from Thermophilic Microorganisms" held in Zürich, Switzerland, from July 28 - August 1, 1975.

The phenomenon of life at higher temperatures, thermophily, has stimulated natural scientists for over twothousand years. Concern with this biological problem has experienced recurring renaissances over the centuries. Therefore, it is not surprising to find in our time when molecular biology radiates its influence into many disciplines of natural sciences and medicine, that the study of thermophily is again a central subject of interest and now from the molecular aspects. Together with nucleic acids, enzymes and other proteins are important molecules for the processes of life and they should in their thermophilic forms have an immediate relation to the phenomenon of thermophily.

The structure and function of thermophilic enzymes and proteins has fascinated us for a long time in the laboratory and it occurred to us, about two years ago, their importance warranted discussion within a larger group. Thus we decided to organize a symposium. A preliminary inquiry that was circulated indicated to our pleasure great and spontaneous interest in the idea of holding such a symposium. Obviously there was a gap here that had not been filled by symposia already held on thermophilic microorganisms.

The major purpose of this international meeting was to bring together for the first time scientists of several countries, joined by their common interest in the structure and properties of thermophilic enzymes and proteins and their importance to the complex processes of the thermophilic cell. Therefore, the object of attention was not restricted to the problem of thermostability, which is of course of considerable practical interest.

A further purpose of this conference served to focus attention to this expanding area and it is hoped that this monograph will not only highlight the present frontiers and the accompanying problems, but will also serve as a useful reference source and as an entry into the field in general.

Various aspects of the structure and function of thermophilic enzymes and proteins were discussed during the first 6 sessions, and in a subsequent round table discussion centred on this subject. The last 2 sessions and a second round table discussion were devoted to thermophilic metabolism, especially to the thermophilic protein synthesizing system and the nucleic acids, and to the problem of thermophily in general.

The present volume records most but not all of the lectures and the remarks made during the two round table discussions.

The meeting could not have been held without the generous help of the following firms: Boehringer Mannheim GmbH, Ciba-Geigy AG, Hoechst Aktiengesellschaft, Hoffmann-LaRoche AG, Nestlé Alimentana SA, Novo Industri AS, Procter & Gamble Company, Sandoz AG. We wish to express our thanks to them and to the Eidgenössische Technische Hochschule, Zürich for financial support. The chairman is also heavily indebted to the organizing committee of the symposium: Dr. G. Frank, Dr. H.-U. Haberstich, Dr. L. Jung and Dr. W. Sidler, to the symposium-secretary Miss A. Schumacher and to the scientific and administrative staff of Eidgenössische Technische Hochschule Zürich-Hönggerberg. Special acknowledgement is made to Dr. G. Frank, Miss A. Schumacher and the Birkhäuser Verlag, Basel, for their great assistance in preparing this publication.

Zürich
August 1975 H. Zuber

Contents

LIST OF PARTICIPANTS

Andresen, O.	Novo Industri AS DK-2880 Bagsvaerd, Denmark
Artavanis, S.	Biozentrum der Universität CH-4056 Basel, Switzerland
Atkinson, A.	Microbiol. Research Establishment Porton, Near Salisbury, Wilts, England
Beaucamp, K.	Boehringer Mannheim GmbH D-8132 Tutzing, Germany
Biffen, J.H.F.	London Hospital Medical College London E1 2AD, England
Blomeyer, K.F.	Procter & Gamble E.T.C. B-1820 Strombeek-Bever, Belgium
Boccù, E.	Institute of Pharmaceutical Chemistry I-35100 Padova, Italy
Born, W.	Eidgenössische Technische Hochschule CH-8049 Zürich, Switzerland
Boyer, E.	Miles Laboratories Inc. Elkhart, Indiana 46514, USA
Brewer, J.	University of Georgia Athens, Georgia 30602, USA
Buswell, J.A.	Paisley College of Technology Paisley, Renfrewshire PA1 2BE, England
Campbell, L.L.	University of Delaware Newark, Delaware 19711, USA
Charlier, Josée	Université Libre de Bruxelles B-1640 Rhode-Saint-Genèse, Belgium
Cuendet, P.	Eidgenössische Technische Hochschule CH-8049 Zürich, Switzerland
Degen, L.	Snamprogetti S.p.A. I-00015 Monterotondo (Roma) Italy
Degryse, E.	Laboratorium voor Genetica en Microbiologie van de V.U.B. B-1070 Brussels, Belgium
Dhala, S.A.	Bhavan's College Andheri, Bombay 400058, India

Downey, R.J.	Ohio University Athens, Ohio 45701, USA
Dreher, Renate	Institut für Biologie II der Universität D-74 Tübingen, Germany
Feder, J.	Monsanto Company, NED S213 St. Louis, Missouri 63166, USA
Fiechter, A.	Eidgenössische Technische Hochschule CH-8006 Zürich, Switzerland
Fontana, A.	Institute of Organic Chemistry I-35100 Padova, Italy
Frank, G.	Eidgenössische Technische Hochschule CH-8049 Zürich, Switzerland
Friedman, S.M.	Hunter College of the City University New York, N.Y. 10021, USA
Gambacorta, Agata	Laboratorio per la Chimica di Molecole di Interesse Biologico del C.N.R. Arco Felice, Naples, Italy
Gloger, M.	Boehringer Mannheim GmbH D-8132 Tutzing, Germany
Grosjean, H.	Université Libre de Bruxelles B-1640 Rhode-Saint-Genèse, Belgium
Gysi, J.	Eidgenössische Technische Hochschule 8049 Zürich, Switzerland
Haberstich, H.U.	Eidgenössische Technische Hochschule CH-8049 Zürich, Switzerland
Harris, J.I.	University of Cambridge Cambridge CB2 2QH, England
Heinen, W.	University of Nijmegen, Faculty of Science Nijmegen, The Netherlands
Hengartner, H.	University of Cambridge Cambridge CB2 2QH, England
Hollaus, F.	Zuckerforschungs-Institut Fuchsenbigl A-2286 Haringsee, Austria
Holz, G.	Boehringer Mannheim GmbH D-8132 Tutzing, Germany
Imahori, K.	University of Tokyo, Faculty of Agriculture J-113 Tokyo, Japan
Isemura, T.	Kinki University School of Medicine Osaka-Prefecture, Japan

Jansonius, J.	Biozentrum der Universität CH-4056 Basel, Switzerland
Julian, G.	Biol. Res. Center, Hungarian Academy of Sciences Szeged, Hungary
Jung, L.	Eidgenössische Technische Hochschule CH-8049 Zürich, Switzerland
Kao, O.H.W.	New York State Department of Health Albany, N.Y. 12201, USA
Kaye, J.W.	Clara Maass Professional Building Belleville, New Jersey 07109, USA
Kuhn, H.J.	Eidgenössische Technische Hochschule CH-8006 Zürich, Switzerland
Ljungdahl, L.G.	Institut für Mikrobiologie der Universität D-34 Göttingen, Germany
Lungershausen, R.	Institut für Mikrobiologie der Universität D-34 Göttingen, Germany
Mammi, M.	Institute of Organic Chemistry I-35100 Padova, Italy
Matthews, B.W.	University of Oregon Eugene, Oregon 97403, USA
Mizusawa, K.	Kikkoman Shoyu Company Ltd. J-399 Noda, Japan
Nesemann, G.	Hoechst AG D-6230 Frankfurt/M-Höchst, Germany
Neurath, H.	University of Washington Seattle, Washington 98195, USA
Nosoh, Y.	Tokyo Institute of Technology Tokyo 152, Japan
Ogasahara, Kyoko	Nagasaki University School of Medicine Nagasaki-Shi 852, Japan
Outtrup, Helle	Novo Industri AS DK-2880 Bagsvaerd, Denmark
Pinsky, A.	Bar Ilan University Ramat Gan, Israel
Pollach, G.	Zuckerforschungs-Institut Fuchsenbigl A-2286 Haringsee, Austria
Poralla, K.	Institut für Biologie II der Universität D-74 Tübingen, Germany
Ramaley, R.	University of Nebraska School of Medicine Omaha, Nebraska 68105, USA

Rao, K.	King's College, Dept. of Plant Sciences London SE 24 9JF, England
Reigel, F.	Eidgenössische Technische Hochschule 8049 Zürich, Switzerland
Reiling, H.E.	Zentralinstitut für Biochemie und Biophysik FUB D-1 Berlin 33, Germany
Roeder, A.	Boehringer Mannheim GmbH D-8132 Tutzing, Germany
De Rosa, M.	Laboratorio per la Chimica di Molecole di Interesse Biologico del C.N.R. Arco Felice, Naples, Italy
Saiki, T.	University of Tokyo, Dept. of Agricult. Chemistry Bunkyo-ku, Tokyo, Japan
Schaer, H.P.	Eidgenössische Technische Hochschule CH-8049 Zürich, Switzerland
Schindler, J.	Henkel & Cie GmbH D-4000 Düsseldorf, Germany
Schmid, R.	Henkel & Cie GmbH D-4000 Düsseldorf, Germany
Schmitt, H.	Eidgenössische Technische Hochschule CH-8049 Zürich, Switzerland
Sidler, W.	Eidgenössische Technische Hochschule CH-8049 Zürich, Switzerland
Singleton, R., Jr.	University of Delaware, Dept. of Biol. Sciences Newark, Delaware, USA
Stellwagen, E.	University of Iowa, Dept. of Biochemistry Iowa City, Iowa 52242, USA
Sund, H.	Universität Konstanz, Fachbereich Biologie D-7750 Konstanz, Germany
Sundaram, T.K.	University of Manchester Manchester M60 1QD, England
Tairo, O.	Mitsubishi-Kasei Inst. of Life Sciences Machida, Tokyo 194, Japan
Thein, Iris	Institut für Biologie II der Universität D-74 Tübingen, Germany
Tratschin, J.D.	Eidgenössische Technische Hochschule CH-8049 Zürich, Switzerland
Veronese, F.	Institute of Pharmaceutical Chemistry I-35100 Padova, Italy

Vroemen, A.J.	Gist-Brocades N.V., Research & Development Delft, The Netherlands
Walker, J.	University of Cambridge Cambridge CB2 2QH, England
Wedler, F.C.	Rensselaer Polytechnic Institute Troy, N.Y. 12181, USA
Widmer, H.	Eidgenössische Technische Hochschule CH-8049 Zürich, Switzerland
Williams, R.A.D.	London Hospital Medical College London E1 2AD, England
Yaguchi, M.	Max-Planck-Institut für Molekulare Genetik D-1000 Berlin 33, Germany
Yoshida, M.	Jichi Medical School, Dept. of Biochemistry Tochigi-Ken, Japan 329-04
Yutani, K.	Osaka University, Inst. for Protein Research Suita-city, Osaka, Japan
Zuber, H.	Eidgenössische Technische Hochschule CH-8049 Zürich, Switzerland
Heimer, Y.M.	Nuclear Research Centre - Negev Beer Sheva, Israel
Kolb, Edith	Kantonsspital Zürich, Chirurgie A CH-8006 Zürich, Switzerland
Lindsay, G.	Australian National University Canberra, Australia

THERMOPHILIC HYDROLYTIC ENZYMES (PROTEINASES, AMYLASES)

UNIV. COLL. N.W.
BANGOR
LIBRARY
19

THERMAL STABILITY OF HOMOLOGOUS NEUTRAL METALLOENDOPEPTIDASES IN THERMOPHILIC AND MESOPHILIC BACTERIA: STRUCTURAL CONSIDERATIONS

M.K. PANGBURN, P.L. LEVY, K.A. WALSH and H. NEURATH
Department of Biochemistry, University of Washington, Seattle, WA 98195

ABSTRACT

Thermolysin and neutral protease A are neutral metalloendopeptidases having similar specificity, molecular weight, metal content, and amino acid composition. Thermolysin, derived from the thermophilic organism *Bacillus thermoproteolyticus*, is heat inactivated at about 84° whereas neutral protease A, derived from the mesophilic organism *Bacillus subtilis*, is inactivated at about 59°. Structural analyses reveal that the two enzymes are homologous. Of the 326 residues of neutral protease A, 171 have been placed in sequence and 49% of these have been found in identical loci in thermolysin. These include many of the residues corresponding to the active site of thermolysin. The sensitivity of both enzymes to thermal inactivation is dependent upon the presence of calcium and neutral protease appears to bind less calcium than thermolysin. Structural data indicate that many of the ligands associated with calcium sites 1 and 2 (double site of thermolysin) are present in neutral protease and that calcium site 4 cannot exist in neutral protease. The structural homology and functional analogy of these two proteins support the concept that they have similar conformations. The known structure of thermolysin is used as a model to discuss structural differences which might be related to thermal stability.

INTRODUCTION

One approach to explore the chemical basis of the thermal stability of proteins is to define sets of homologous proteins which differ in thermal stability but are similar in amino acid sequence and three-dimensional structure. Differences in stability can then be related to dissimilarities of structural details. Two neutral metalloendopeptidases, thermolysin from the thermophilic organism *B. thermoproteolyticus* and neutral protease A from the mesophilic organism *B. subtilis*, appear to meet these criteria[1,2]. The present report summarizes current evidence of their structural homology, functional analogy and differences in thermal stability. Attempts are made to identify components of structure which may be important in explaining the differing thermal stability of these two enzymes.

MATERIALS AND METHODS

Crystalline thermolysin was purchased from Daiwa Kasei K.K., Osaka, Japan; crude neutral protease (*Bacillus subtilis* strain NRRLB3411) was obtained from the Monsanto Company, St. Louis, MO through the courtesy of Dr. J. Feder. Two forms of neutral protease (A and B) were purified by affinity chromatography as described by Pangburn *et al.*[2]. A specific adsorbent for affinity chromatography was prepared by coupling BrCN-activated Sepharose 4B through a spacer of triethylenetetramine to chloroacetyl-D-phenylalanine[3].

The concentration of active enzyme was determined spectrophotometrically at 25° by following the hydrolysis of furylacryloyl-Gly-Leu-NH_2 (FAGLA, ΔEm of 317 cm^{-1} at 345 nm) using the conditions described by Pangburn et al.[2]. Protein concentrations and specific activities were determined using the extinction coefficients and kinetic constants reported by the same authors. Caseinolytic activity was assayed according to Keay and Wildi[4].

Amino acid analyses were performed on a Durrum (model D-500) amino acid analyzer. Sequence analysis of neutral protease A is described in a separate communication[5].

The inhibition and modification of neutral protease A by ethoxyformic anhydride followed the procedures described by Burstein et al.[6] for the analogous reaction with thermolysin.

RESULTS

General Similarity and Functional Analogy

Neutral protease from Bacillus subtilis and thermolysin from Bacillus thermoproteolyticus have been grouped as neutral metalloendopeptidases in accord with structural and functional similarities[1,5,6,7]. The enzymes have similar molecular weights and amino acid compositions (Tables 1, 2). Each enzyme requires one mole of zinc per mole of protein for catalytic activity.

Table 1: A Comparison of Thermolysin and the Neutral Proteases.

	Thermolysin	Neutral Protease A	Neutral Protease B
Molecular Weight	34,400[a]	37,000[b]	37,000[b]
Thermal Inactivation Temperature[c]	84°	59°	58°
Metal Content (At./mol)			
Zn^{++} [d]	1.01	1.05	0.95
Ca^{++}	4.0	2.0[f]	N.D.
Activity (FAGLA) k_{cat}, sec^{-1}	47	26	68
K_m, mM	2.1	4.7	4.3
k_{cat}/K_m	21	5.6	16
(casein)[g]	1.0	1.05	0.95

[a] Calculated from the complete amino acid sequence[8].
[b] Determined by SDS-gel electrophoresis[9].
[c] Temperature at which 50% of the enzyme lost activity in 15 min. Conditions: 0.1 mg enzyme/ml, 10 mM HEPES, 2 mM $CaCl_2$, pH 7.2.
[d] Measured by atomic absorption spectroscopy.
[e] Voordouw and Roche[10].
[f] R. Roche, personal communication.
[g] Relative rate of casein hydrolysis.

Table 2: Amino Acid Compositions of Thermolysin and Neutral Protease A

	Thermolysin[a]	Neutral Protease A		Thermolysin	Neutral Protease A
LYS	11	17.0	ALA	28	28
HIS	8	5.6	CYS	0	0
ARG	10	8.7	VAL	22	18.2
ASX	44	48.0	MET	2	3.0[c]
THR	25	29.4[b]	ILE	18	13.9
SER	26	28.2[b]	LEU	16	20.5
GLX	21	25.9	TYR	28	23.9
PRO	8	11.2	PHE	10	11.3
GLY	36	30.5	TRP	3	3.0[d]
			Total	316	326

[a] According to the sequence analyses of Titani <u>et al</u>.[8].
[b] Extrapolated to zero time of hydrolysis.
[c] Determined as methionine sulfone after performic acid oxidation.
[d] Determined after base hydrolysis[11].

They catalyze the hydrolysis of peptide bonds to which hydrophobic amino acids donate the amino groups and bind to immobilized peptides containing D-Leu or D-Phe residues[2]. Their catalytic activities are similarly dependent upon pH (Fig. 1). All three enzymes exhibit comparable proteolytic activity towards casein (Table 1) but they show four-fold differences in maximum activity with the synthetic dipeptide substrate.

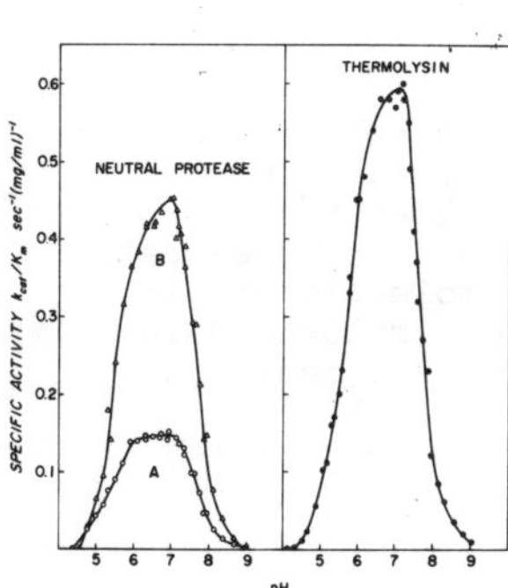

Fig. 1
pH dependence of FAGLA hydrolysis by neutral proteases A and B and by thermolysin. All solutions contained 1 mM FAGLA, 50 mM buffer, 10^{-5} M $ZnCl_2$ and were brought to an ionic strength of 0.1 with NaCl. The buffers employed were NaAc (pH 4.38-5.21), PIPES (pH 5.28-7.07), HEPES (pH 6.14-7.93) and Tris (pH 7.01-8.99). The initial rate of hydrolysis was determined with a Cary 16 recording spectrophotometer at 345 nm (slit width 1 mm). The sample chamber was thermostated at 25°.

Inhibitors of thermolysin also inhibit neutral protease. Both enzymes are sensitive to ethoxyformic anhydride (Fig. 2) and this inhibition appears to be due to the reversible modification of a single histidyl residue in each[6,7]. Furthermore, both enzymes are inactivated by low concentrations of metal ions such as Ag^{+}, Hg^{++} and o-chloromercuriphenol [7,12,13,14].

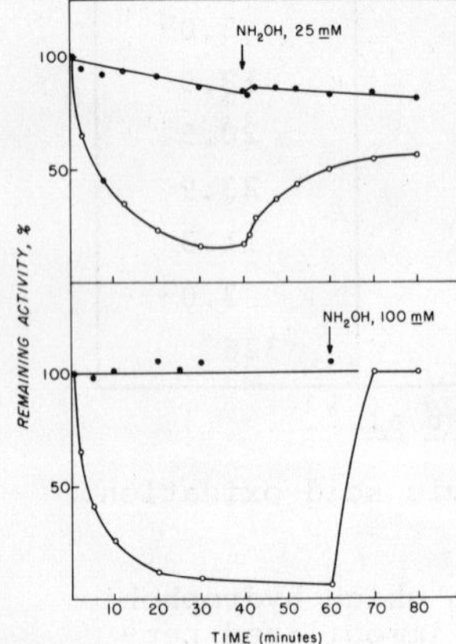

Fig. 2
Upper: Inactivation of neutral protease A by 10 mM ethoxyformic anhydride at 25°, pH 6.5, in the presence (closed circles) and the absence (open circles) of 50 mM Cbz-L-phenylalanine. Reactivation was accomplished by the addition of 1 M hydroxylamine (pH 6.5) to a final concentration of 25 mM. Lower: Inactivation of thermolysin by 3 mM ethoxyformic anhydride at 25°, pH 7.2, in the presence (closed circles) and the absence (open circles) of 5 mM Cbz-L-phenylalanine[6]. Reactivation was accomplished by the addition of hydroxylamine (pH 7.2) to a final concentration of 0.1 M.

Structural Comparison

Although only 54% of the amino acid sequence of neutral protease A has been determined[5], it is clear that this enzyme and thermolysin are homologous, (Fig. 3), indicating that these two proteins have evolved from a common ancestral protein. The greatest similarity is found in the central region of the molecule (residues 110 through 177) while a less similar portion is located in the amino terminal region (residues 1-55). Of the 171 residues placed in sequence in neutral protease, 49% appear to be in identical loci in thermolysin. The identity in the central region is 59%, and in the amino terminal region 27%.

Fig. 3
Comparison of the partial sequence of neutral protease A[5] with portions of the complete sequence of thermolysin. The residue numbering is that of thermolysin[8]. Four gaps are inserted (dashed lines) to optimize apparent homology. Unidentified residues are indicated by question marks; tentative assignments are underlined.

Matthews and coworkers[15,16] identified a number of functional residues in the active site of the three-dimensional structure of thermolysin. Table 3 compares these residues with those found in corresponding regions of the neutral protease sequence. All three zinc ligands in thermolysin occur at identical loci in neutral protease. Of the six active site residues identified by Colman _et al._[15], three are found in identical locations in neutral protease whereas the other three have not yet been conclusively identified. The seven residues defining the substrate binding pocket of thermolysin are found in corresponding sequences in neutral protease. Six are identical and the other (Val/Ile_{192}) is indicative of a structurally conservative interchange.

Table 3: Comparison of Residues Implicated in Function and Stability of Thermolysin (TL) with the Corresponding Residues in Neutral Protease A (NPA).

Active Site		
Residue Number	TL	NPA
143	Glu	Glu
157	Tyr	Tyr
170	Asp	Asx
203	Arg	Arg
226	Asp	?
231	His	?
Binding Pocket		
130	Phe	Phe
133	Leu	Leu
139	Val	Val
188	Ile	Ile
189	Gly	Gly
192	Val	Ile
202	Leu	Leu

Metal Ligands			
Metal	Residue Number	TL	NPA
Zn	142	His	His
	146	His	His
	166	Glu	Glu
Ca(1)	138	Asp	Asp
	187	Glu	Asp
	174	Thr	Tyr
Ca (1&2)	177	Glu	Asp
	185	Asp	?
	190	Glu	Glu
Ca(2)	183	Asn	?
	191	Asp	Asp
Ca(3)	57	Asp	?
	59	Asp	?
	61	Gln	?
Ca(4)	190	Glu	Glu
	193	Tyr	Thr
	194	Thr	-
	197	Ile	Val
	200	Asp	Pro

The numbering of residues is that for thermolysin (Fig. 3). Symbols in the neutral protease column indicate that the residue is not yet identified (?) or deleted from neutral protease (-). Numbers in parenthesis identify calcium binding sites in thermolysin[16].

The degree of similarity in the primary structure of these two proteases and the conservation of functionally crucial residues not only indicate that the two molecules are homologous but also suggest that they have similar tertiary structures. If this suggestion is correct, the extent and location of the known sequences of neutral protease can be diagramatically illustrated as in Fig. 4. It can be seen that the residues in the central portion of thermolysin which surround and define the active site, with its bound zinc atom, are largely identified in neutral protease. The similarity of the two enzymes in enzymatic specificity[17], kinetic parameters and susceptibility to inhibitors[7] is consistent with a hypothesis of similar conformations in the region of their active sites.

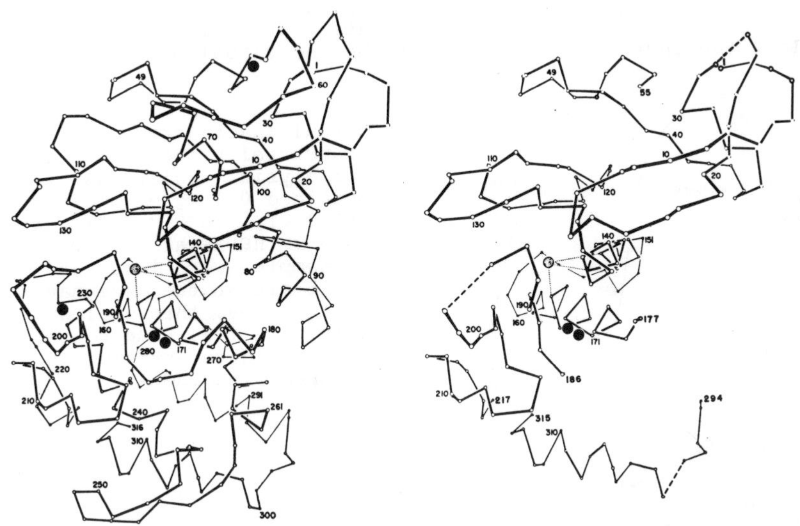

Fig. 4
The backbone conformation of thermolysin[15] is illustrated on the left. Zn^{++} and Ca^{++} are indicated by stippled and solid circles respectively. The Zn atom is located in the active site cleft. On the right are illustrated only those portions of thermolysin corresponding to the sequence data for neutral protease given in Fig. 3. Presumed deletions in neutral protease are indicated by dashed lines (residues 26-27, 194-196, 299-300). The actual conformation of neutral protease is not known.

Thermal Stability

The large difference in thermal stability between neutral protease A and thermolysin is illustrated in Fig. 5. Under these particular conditions neutral protease has a half-life of 15 min at 59°, whereas thermolysin is quite stable at that temperature and a temperature of 84° is required to inactivate 50% in 15 min. Autolysis does not appear to be the rate-limiting step in the denaturation of neutral protease, since the thermal inactivation curve is not significantly affected by a 10-fold increase in enzyme concentration (Fig. 6) and since the time course of loss of activity (not shown) approximates an exponential decay. The biphasic temperature dependency of the inactivation of thermolysin at the higher concentration (1.0 mg/ml) requires a more complex interpretation than the denaturation at concentrations of 0.1 mg/ml. At the lower concentration, the denaturation profiles are assumed to reflect the intrinsic stability of the individual molecules.

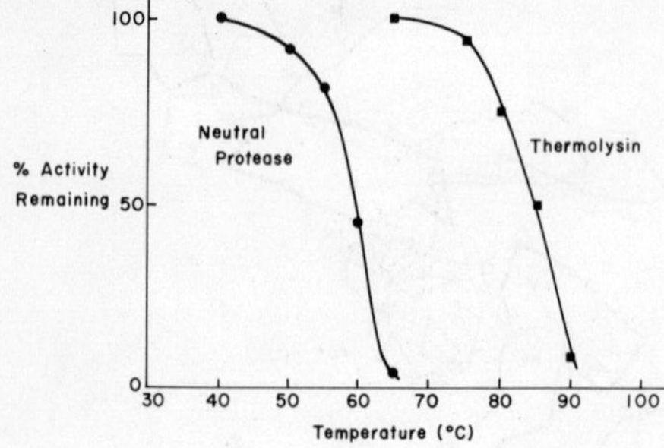

Fig. 5
Thermal inactivation curves for neutral protease A and thermolysin. Solutions contained 0.1 mg enzyme/ml, 2 mM $CaCl_2$, 10^{-5} M $ZnCl_2$ and 5 mM HEPES at pH 7.0. The samples were incubated for 15 min at the designated temperatures, then cooled rapidly in ice. The residual peptidase activity was measured with FAGLA.

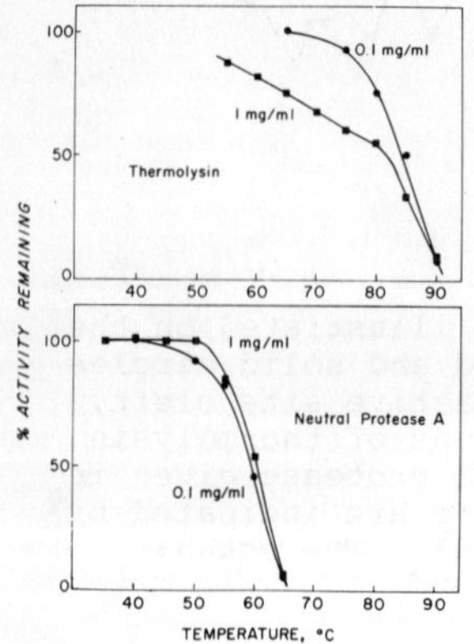

Fig. 6
The effect of enzyme concentration on thermal inactivation. Conditions and procedures were the same as in Fig. 5 except for enzyme concentrations.

The effect of calcium ions on the thermal stability was explored by determining the thermal denaturation temperature (T_m) at various calcium concentrations. T_m is defined as the temperature at which 50% of the enzyme is irreversibly inactivated in 15 min. Control experiments demonstrated that differences in ionic strength alone do not have a significant effect on T_m of either enzyme (T_m varied less than 2° between zero and 2 M NaCl). Since both $CaCl_2$ and KSCN have been shown to be denaturants of other proteins at high concentrations[18], the effect of KSCN concentration was also examined as a control. Fig. 7

shows that calcium ions in the range 10^{-4} M to 10^{-1} M, enhance the thermal stability of thermolysin as much as 12° but have little effect on neutral protease. Higher concentrations of $CaCl_2$ appear to mimic the denaturing action of KSCN. The controls demonstrate that the enhancement was not due to chloride ions, ionic strength or nonspecific denaturation.

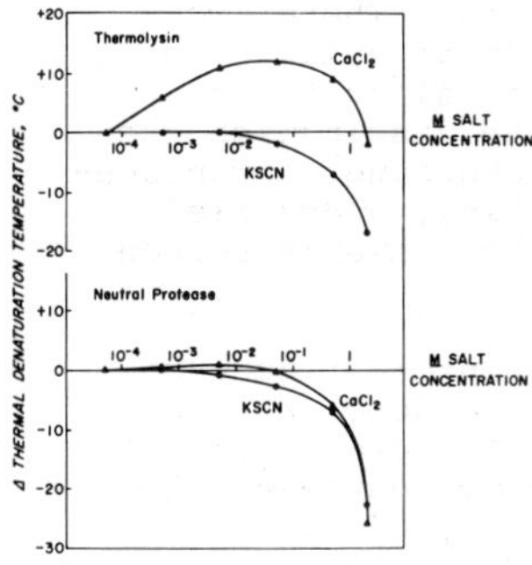

Fig. 7
The effect of $CaCl_2$ and KSCN on the thermal denaturation temperature (T_m) of thermolysin and neutral protease A. T_m was measured as described in Fig. 5 and each point represents a complete denaturation curve as shown there. The buffer was 5 mM HEPES, 10^{-5} M $ZnCl_2$, (pH 7.0) with the indicated concentration of $CaCl_2$ or KSCN (KSCN solutions contained also 2 mM $CaCl_2$). Protein concentration was 0.1 mg/ml.

DISCUSSION

Evidence of Homology

Chemical, physical and enzymatic data indicate that thermolysin and neutral protease A are homologous enzymes which have preserved virtually identical catalytic functions while diverging from a common ancestral protein. The similarities are most evident among the 16 proposed components of the active site of thermolysin[15,16] where 12 of the 13 residues placed so far in neutral protease are found to be identical to their counterparts in thermolysin (Table 3). The only established difference among these residues is a conservative interchange of valine and isoleucine at position 192. Comparison of the segments of sequence from residues 110 to 177 (Fig. 3) reveals 40 pairs of identical residues (59%) in the two molecules. Analysis of all of the data in Fig. 3 demonstrates 49% identity.

The three dimensional model of thermolysin can be examined for compatability with the partial sequence of neutral protease A. While the illustration in Fig. 4 is only intended as a diagramatic guide to the regions of sequence which could be considered in neutral protease, the indications of homology in Fig. 3 coupled with the analogies in catalytic mechanisms[7] support the idea of conformational similarity. Crystallographic evidence is needed to test this proposal.

Thermal Stability

The most obvious difference between thermolysin and neutral protease A lies in their thermal stability. Neutral protease is heat inactivated at a temperature 25° below that of thermolysin. The invasion of B. subtilis and B. thermoproteolyticus into their respective biological niches may well have been facilitated by the generation of proteins with differing thermal stabilities, allowing migration into new thermal environments. These two proteins may thus provide a model pair for seeking evidence of the modulations of structure which are the basis of alterations in thermal stability. Although it is difficult to distinguish the significant amino acid replacements from a background of replacements which may be unrelated to thermal stability, a detailed structural comparison may ultimately define the elements responsible for differential thermal stability.

The loci of calcium binding are of particular interest because past studies of both enzymes have documented the dependence of stability on calcium ions[13,19]. In the case of thermolysin, it has been shown that the removal of bound calcium decreases the thermal stability, but not the catalytic activity[20,21]. This suggests a stabilizing role for the four bound calcium ions in the thermolysin molecule. Neutral protease, on the other hand, binds fewer calcium ions (Table 1) and is less thermostable.

Comparison of the calcium binding regions of thermolysin to the corresponding regions of the sequence of neutral protease reveals that one of the binding sites (Ca 4 in Table 3) cannot exist in neutral protease. The exchange of Asp 200 in thermolysin with proline in neutral protease eliminates one of the counter ions in neutral protease and three residues of the chelating loop (Glu 190-Asp 200, Figs. 3, 4) in thermolysin have been completely deleted. Alterations in the regions corresponding to calcium sites 1 and 2 raise some question about their integrity in neutral protease. These two sites, however, have maintained their anionic residues and it is likely that more structural information will place these ions in these loci in neutral protease. In spite of the existence of a number of studies of stabilization by calcium[1], including those reported here, there is a need for accurate binding constants and data relating structural stabilization in both enzymes to the number of bound calcium ions.

Matthews and coworkers identified a network of salt bridges in the three-dimensional structure of thermolysin[15] and suggested that these ionic interactions contribute to the structural stability of this molecule. In neutral protease six of the corresponding fifteen sites have been examined and five have lost the residues necessary to form salt bridges. This contrast supports the suggestion of a stabilizing role for these interactions but two points suggest caution. It is possible that a different network of salt bridges stabilizes neutral protease. It is also possible that the salt bridges are not important factors in the thermal stability of either molecule. This uncertainty is raised by the observation that changes in the ionic strength from 0.01 to 2.0 had no significant effect on the thermal inactivation temperatures of thermolysin or neutral protease. It is not clear, however, to what extent the ionic strength of the solvent would affect intramolecular ionic interactions.

Comparison of the amino acid compositions of neutral protease and thermolysin (Table 2) reveals that thermolysin has the higher proportion of hydrophobic residues (7 more Phe, Tyr, Trp, Ile, Leu, Val and Ala) and the lower proportion of hydrophilic residues (20 less residues of Asx, Glx, Lys, Arg, Ser, and Thr). It might be suggested that the relatively more hydrophobic nature of thermolysin promotes a more stable protein core and that the larger number of calcium binding sites and/or salt bridges increases the activation energy for the transition to the denatured state.

ACKNOWLEDGMENT

This work was supported by a grant from the National Institutes of Health (GM-15731) and the American Cancer Society (BC-91P).

REFERENCES

1. H. Matsubara and J. Feder, The Enzymes, Third edition, ed. by P.D. Boyer (Academic Press, N.Y.) Vol. 3, p. 721 (1971).
2. M.K. Pangburn, Y. Burstein, P.H. Morgan, K.A. Walsh and H. Neurath. Biochem. Biophys. Res. Commun. 54, 371 (1973).
3. K.A. Walsh, Y. Burstein and M.K. Pangburn. Methods Enzymol. 34, 435 (1974).
4. L. Keay and B.S. Wildi. Biotechnol. Bioeng. 12, 179 (1970).
5. P.L. Levy, M.K. Pangburn, Y. Burstein, L.H. Ericsson, H. Neurath and K.A. Walsh. Proc. Nat. Acad. Sci. USA, in press.
6. Y. Burstein, K.A. Walsh and H. Neurath. Biochemistry 13, 205 (1974).
7. M.K. Pangburn and K.A. Walsh. Biochemistry, in press.
8. K. Titani, M.A. Hermodson, L.H. Ericsson, K.A. Walsh and H. Neurath. Nature New Biology 238, 35 (1972).
9. K. Weber and M. Osborn. J. Biol. Chem. 244, 4406 (1969).
10. G. Voordouw and R.S. Roche. Biochemistry 13, 5017 (1974).
11. T.E. Hugli and S. Moore. J. Biol. Chem. 247, 2828 (1972).
12. B. Holmquist and B.L. Vallee. J. Biol. Chem. 249, 4601 (1974).
13. S. Endo. Hakko Kagaku Zasshi 40, 346 (1962).
14. J.D. McConn, D. Tsuru and K.T. Yasunobu. Arch. Biochem. Biophys. 120, 479 (1967).
15. P.M. Colman, J.N. Jansonius and B.W. Matthews. J. Mol. Biol. 70, 701 (1972).
16. B.W. Matthews, L.H. Weaver and W.R. Kester. J. Biol. Chem. 249, 8030 (1974).
17. J. Feder and J.M. Schuck. Biochemistry 9, 2784 (1970).
18. P.H. von Hippel and K.Y. Wong. J. Biol. Chem. 240, 3909 (1965).

19. J.D. McConn, D. Tsuru and K.T. Yasunobu. J. Biol. Chem. 239, 3706 (1964).
20. S.A. Latt, B. Holmquist and B.L. Vallee. Biochem. Biophys. Res. Commun. 37, 333 (1969).
21. J. Feder, L.R. Garrett and B.S. Wildi. Biochemistry 10, 4552 (1971).

THE STRUCTURE AND STABILITY OF THERMOLYSIN

L.H. WEAVER, W.R. KESTER, L.F. TEN EYCK and B.W. MATTHEWS
Institute of Molecular Biology, University of Oregon
Eugene, Oregon 97403, U.S.A.

INTRODUCTION

Thermolysin is an extracellular proteolytic enzyme isolated from *Bacillus thermoproteolyticus* (Endo, 1962). The enzyme is quite thermostable, retaining over half of its activity after being heated in an aqueous solution for an hour at 80°C, while at 65°C practically no inactivation occurs (Endo, 1962; Matsubara, 1967).

Because of its unusual stability, the enzyme was an interesting candidate for structural analysis, and its amino acid sequence (Titani *et al.*, 1972) and three-dimensional structure have been determined (Matthews *et al.*, 1972a, 1972b; Colman *et al.*, 1972a; Matthews *et al.*, 1974), this being the first structure determination of a protein from a thermophilic source.

The protein, illustrated in Fig. 1, consists of two distinct lobes, one of which contains predominantly β-structure, while the other consists primarily of helices. The active site of the molecule is formed by a deep cleft between the two lobes, with the zinc ion, essential for activity, at the bottom of the cleft. On removal of the zinc ion, only very slight conformational changes are observed, suggesting that the zinc ion is not essential for the stability of the molecule.

In the context of this meeting, one of the most important conclusions to be drawn from the study of the structure of thermolysin is, in a sense, a negative one: viz., that the conformation of thermolysin does not include any unusual structural elements not observed in non-thermostable proteins. With the possible exception of the calcium binding sites, discussed in the following section, the conformation of thermolysin resembles that of a "typical" protein. As we shall discuss in more detail in the final section, we suggest that the structures of thermostable proteins will in general be very similar to their mesophilic counterparts, and that the additional energy of stabilization can come from rather subtle differences in hydrophobic, ionic or hydrogen bonding interactions (Matthews *et al.*, 1974).

CALCIUM STABILIZATION OF THERMOLYSIN

It has long been recognized that calcium stabilizes thermolysin (Endo, 1962). Feder *et al.* (1971) showed that in solution four atoms of calcium are bound to the enzyme, and that removal of three to four of these ions results in a loss of thermostability, but not activity. The X-ray results confirm this conclusion. In the three-dimensional structure (Fig. 1) two calcium ions, Ca(1) and Ca(2), occupy a double site 13 Å from the essential zinc ion; Ca(4) occupies a site about 11 Å from Ca(2), and Ca(3) binds on the surface of the molecule about 30 Å from the double site.

The role of calcium in the stabilization of thermolysin is complicated by the fact that loss of enzymatic activity on calcium removal can be due *either* to destabilization of the protein leading to conformational changes and/or unfolding or to autodigestion and degradation of the molecule. Furthermore, the analysis of calcium binding is complicated by the presence of four binding sites.

In order to try to determine the relative binding constants of the four calcium sites, we have attempted to sequentially remove calcium from crystalline thermolysin and monitor calcium occupancy by calculating difference Fourier syntheses. It is well known that in solution calcium-depleted thermolysin is rapidly degraded through autodigestion. Also, crystals of native thermolysin in the presence of EDTA are unstable, exhibiting cracking and being dissolved away, presumably due to autodigestion by enzyme molecules in solution. We have, however, been able to stabilize the crystals somewhat by performing experiments in the presence of the potent thermolysin inhibitor phosphoramidon ($K_I \sim 10^{-6}$M) (Suda et al., 1973). Crystals of thermolysin soaked in solutions of 1.0×10^{-4} M phosphoramidon bind the inhibitor with essentially 100% occupancy (Weaver and Matthews, unpublished observations). On exposing crystals of inhibited thermolysin to increasing concentrations of a calcium chelating agent such as EDTA, the crystals are unaffected up to about 0.1 mM EDTA, and, as judged by their diffraction patterns and by difference Fourier syntheses, show no loss of calcium from any of the four binding sites. As the concentration of EDTA is increased, there is a rather sharp transition, at about 0.5 - 1.0 mM, beyond which the crystals tend to crack and dissolve over a period of a few hours. It is, however, possible to take photographs of crystals in the transition region, and a difference map showing the projected difference in electron density between phosphoramidon-inhibited crystal in the presence of 0.5 mM EDTA and an inhibited crystal in the presence of 0.01 M Ca^{++} is shown in Fig. 2. In the thermolysin space group, this projection map provides three independent views of each binding site. The negative density, notably at Site 2, shows that calcium has been at least partially removed from this site, whereas there is no indication of calcium depletion at Sites 1 and 3.

A quantitative estimate of the loss of calcium at each site was obtained by refining the respective calcium occupancies and other parameters (Hart, 1961). In order to calibrate the refinement results, which are given in Table 1, we have included the results for the difference in occupancy between gadolinium-substituted thermolysin and native thermolysin (Colman et al., 1972b; Matthews and Weaver, 1974). Exposure of thermolysin to gadolinium, or any other of the lanthanides, results in lanthanide replacement of calcium at Sites 1, 3 and 4, reflected by increased occupancy, whereas Ca(2) is displaced but not replaced. The occupancy of -25.2 for Site 2 in Table 1 therefore corresponds to loss of a full calcium ion. Comparison with the occupancies in the second column of Table 1 suggests that exposure of thermolysin to 0.5 mM EDTA for about one day results in approximately 50% loss of calcium from Sites 2 and 4, but no loss from Sites 1 and 3.

We have also attempted to remove calcium from thermolysin by using solutions containing both EDTA and free calcium, and have found that under these conditions higher concentrations of EDTA can be used, even though the molar excess of EDTA over calcium exceeds that of EDTA alone which would rapidly destroy the crystals. For example, we have taken photographs of phosphoramidon-inhibited thermolysin crystals after soaking for 24 hrs in a mixture of 8 mM EDTA and 2 mM Ca^{++}. Under these conditions calcium is partly removed from Sites 2, 3 and 4 (Fig. 2; Table 1). Equal concentrations of chelator and of calcium (e.g., 5 mM EDTA, 5 mM Ca^{++}) also result in partial calcium removal (Table 1). Such an equimolar solution has a low concentration of free calcium, estimated as not more than 10^{-5} M, which remains essentially constant as calcium is removed from the crystals.

The results given in Table 1 suggest that calcium is removed most easily from Site 2, then from Site 4, followed by Site 3, but never from Site 1 under the conditions used here. Assuming that the sites which exhibit the greatest loss of calcium have the weakest binding constants, the results given in Table 1 suggest that Site 1 binds calcium most strongly, and that Sites 2 and 4 have the lowest affinity for calcium. The ordering of the relative affinity constants of the four thermolysin calcium-binding sites suggested by these experiments is as follows:

$$\text{Site 1} \gg \text{Site 3} > \text{Site 4} \gtrsim \text{Site 2}$$

This ordering is also consistent with what one might expect from inspection of the thermolysin structure. Ca(1) occupies an enclosed binding site, its ligands including four acid groups; Ca(3) is more exposed, and has two aspartic acids among its ligands; Ca(4) is also somewhat exposed, and although it is coordinated by five protein oxygens, only one of these is that of an acid (Asp 200); Ca(2) occupies the outer position of the double binding site, and is coordinated by three acids, but these are shared with Ca(1). In addition, Ca(2) has an unfavorable electrostatic interaction with Ca(1) which is 3.8 Å away, and it is not surprising that Ca(2) is relatively weakly bound.

These results may be compared with those of Voordouw and Roche (1974) who studied the binding of calcium ions to thermolysin by gel filtration, and found that at free calcium concentrations of about 5 x 10^{-5} M, two calcium ions dissociated simultaneously from the enzyme, while the remaining two calciums remained bound to the enzyme in the presence of 10^{-6} M free calcium. Voordouw and Roche found the binding of the two displaced ions to be cooperative, and argued that they were Ca(1) and Ca(2), the ions bound at the double site.

The present results are not compatible with the suggestion that Ca(1) is one of the less-tightly bound calcium ions, and suggest, instead, that the two ions displaced first are Ca(2) and Ca(4). Inspection of the thermolysin structure suggests that there may be some interaction between Sites 2 and 4. Figure 6 of Matthews _et al._ (1974) shows that Glu 190 and Asp 191 (via a water molecule) are both ligands of Ca(2), and that the carbonyl oxygen of the peptide linking these two residues is, via another water molecule, liganded to Ca(4). Furthermore, the Ca(2) ligands Glu 190 and Asp 191 are directly linked via the polypeptide backbone to Tyr 193 and Thr 194, the start of the Ca(4) binding loop. Therefore if, say, Ca(2) were displaced from the enzyme, it would be expected that a conformational change might occur at residues 190-191, and that this would in turn modify the conformation, and perhaps weaken calcium binding, at Site 4. On the other hand, it is not obvious that such an effect would be of sufficient magnitude to cause full cooperativity, as suggested by Voordouw and Roche, particularly since we have shown previously (Matthews and Weaver, 1974) that displacement of Ca(2) by lanthanides causes only slight conformational adjustments in the calcium ligands.

The results which we have described suggest the following role for the respective calcium binding sites in the stabilization of thermolysin. At high calcium concentrations all calcium sites are fully occupied, both stabilizing the enzyme and protecting it against autolysis. As the exogenous calcium concentration is lowered, calcium ion begins to dissociate initially from Site 2, and also in part from Site 4, possibly somewhat cooperatively. Dahlquist _et al._ (1975) have shown that below 50°C one calcium ion can be removed from active thermolysin, without loss of enzymatic activity, and suggest that this ion is Ca(2). Our results are consistent with this suggestion. Dahlquist _et al._ (1975) have also shown that removal of additional calcium ions results in an irreversible loss in enzymatic activity. We suggest that the second calcium ion removed is Ca(4), and that its removal permits the polypeptide loop including residues 193-200 (Fig. 3) to adopt a conformation which permits autolysis, very likely at the Gly 196 - Ile 197 bond. This "nicked" molecule would not necessarily unfold or dissociate at low temperatures. Further loss of calcium, from Site 3 and finally Site 1, would ultimately lead to more extensive autolysis and complete degradation of the enzyme. Inspection of the three dimensional structure of thermolysin suggests that the function of Ca(3) is to stabilize a loop on the surface of the molecule, whereas the role of tightly-bound Ca(1) is essentially structural, serving to link together the two molecular lobes.

THERMOSTABILITY OF OTHER PROTEINS

We have remarked above that the thermolysin molecule contains no structural elements which do not have counterparts in other less stable proteins, and that there is nothing particularly unusual in the conformation of thermolysin to which one might attribute thermostability (Matthews _et al_., 1974).

Furthermore, as we have pointed out previously (Matthews _et al_., 1974), there is now convincing evidence that the three dimensional structures of related proteins from thermophilic and mesophilic sources are often very similar. For example, the primary structure of the heat-stable ferredoxin from the thermophile _Clostridium tartarivorum_ is obviously similar to those of four mesophilic bacterial ferredoxins (Tanaka _et al_., 1971), suggesting that the three dimensional structures are also closely related. The same conclusion was reached from a comparison of the amino acid sequences of glyceraldehyde-3-phosphate dehydrogenases from _Bacillus stearothermophilus_ and yeast, where an overall homology of approximately 70% was observed (Bridgen and Harris, 1973). Also, preliminary results indicate that a neutral protease from _Bacillus subtilis_, much less stable than thermolysin, has a homologous amino acid sequence (Pangburn _et al_, 1973; Pangburn, 1974). These findings all indicate that the three dimensional structures of these related enzymes must be quite similar, and led us to suggest (Matthews _et al_., 1974) that the difference between thermolability and thermostability for a given protein will in general be due to subtle changes in the protein structure which result in more favorable hydrophobic interactions, hydrogen bonding, metal binding, ionic interactions, or a combination of all of these. Although one or another of these types of interactions may be more important than the others in a given instance, we believe that thermal stabilization in general cannot be explained by a single determinant such as enhanced hydrophobic bonding or ionic interactions.

These ideas are also consistent with the thermodynamic description of the factors influencing the stability of proteins developed by Kauzmann, Tanford, Brandts and others (e.g., see Kauzmann, 1959; Tanford, 1962; Brandts, 1964a, 1964b, 1967; Schellman, 1955; Scheraga, 1963). These authors have shown that the net free energy of stabilization of a protein can be written as a sum of contributions arising from hydrogen bonds, hydrophobic bonds, electrostatics, chain entropy and so on. The important result to emerge is that while the sum of the positive contributions to the stability of the folded protein (dominated by entropy-driven hydrophobic bonding) is large (on the order of 100 kcal/mole for a small protein), there is a comparable negative contribution arising from the increase in conformational entropy on unfolding the protein, so that the _net_ free energy of stabilization is small, usually on the order of 10 kcal or less. Because of the delicate balance between stabilizing and destabilizing factors, slight modifications in the structure of a protein can radically alter its folding behavior.

Thermal stability of proteins has two aspects. Normally it is measured as a kinetic phenomenon--that is, the time required for 50% denaturation is measured at some temperature. Thermal stability in the kinetic sense implies a higher activation energy for denaturation than normal, which usually (but not necessarily) implies that the native structure is more stable than normal. The fact that denaturation occurs shows that the inactivated form of the protein is more stable than the active form at this temperature. Equilibrium thermal stability requires that the native enzyme be thermodynamically stable at elevated temperatures. In the case of thermolysin, its stability at 65°C shows that it is thermostable in the equilibrium sense of the term.

Brandts (1967) has discussed the effect of heat on the stability of proteins, and has shown that as temperature increases, the free energy contribution favoring the folded state increases for hydrophobic bonding, but remains constant for hydrogen bonding. Up to about 50-60°C the free energy increase arising from hydrophobic bonding roughly offsets the increasingly unfavorable contribution to the free energy arising from the conformational entropy of folding. Therefore, in the temperature range 10-50°C, the small margin of stability described above tends to be maintained. At higher temperatures the conformational entropy term will increasingly favor the unfolded state.

From this thermodynamic description of the stability if proteins, it is not surprising to find that the structures of thermophilic proteins are very similar to their mesophilic counterparts. Because the net free energy difference between a folded and unfolded protein is typically only 5-10 kcal/mole; i.e., equivalent to several hydrogen bonds or to the hydrophobic interaction energy generated by a single leucine side chain, subtle changes in the structure of a protein can dramatically modify its stability in solution.

A specific example of the stabilization of bacterial ferredoxins by salt bridges has been recently described by Perutz (1975) who suggests that in monomeric enzymes the extra energy of stabilization can be provided without disturbance of the tertiary structure by a few extra salt bridges on the molecular surface, and in oligomeric proteins such as hemoglobin by extra salt bridges, hydrogen bonds or non-polar bonds at the subunit interfaces.

We have suggested (Matthews et al., 1974) that the energy of stabilization in a given instance can come from slight changes in hydrophobic bonding, hydrogen bonding, metal binding, and ionic interactions, although not necessarily exclusively from ionic interactions as in the case of the ferredoxins.

ACKNOWLEDGMENTS

We thank Drs. T. Aoyagi and H. Neurath for a gift of phosphoramidon, and Dr. F.W. Dahlquist for numerous discussions on the stability of thermolysin. This work was supported in part by grants from the National Science Foundation (BMS74-18407) and the National Institutes of Health (GM20066, GM15423, GM00715) and by the award to B.W.M. of an Alfred P. Sloan Research Fellowship and a Public Health Service Career Development Award (GM70585) from the Institute of General Medical Sciences.

REFERENCES

Brandts, J.F. (1964a). J. Amer. Chem. Soc. 86, 4291.

Brandts, J.F. (1964b). J. Amer. Chem. Soc. 86, 4302.

Brandts, J.F. (1967). In "Thermobiology". A.H. Rose, ed., p. 25. Academic Press, New York.

Bridgen, J. and Harris, J.I. (1973). Ninth Int. Congr. Biochem. Stockholm, July.

Colman, P.M., Jansonius, J.N. and Matthews, B.W. (1972a). J. Mol. Biol. 70, 701.

Colman, P.M., Weaver, L.H. and Matthews, B.W. (1972b). Biochem. Biophys. Res. Commun. 46, 1999.

Dahlquist, F.W., Long, J.W. and Bigbee, W.L. (1975). Biochemistry, In Press.

Endo, S. (1962). J. Fermentation Tech. 40, 346.

Feder, J., Garrett, L.R. and Wildi, B.W. (1971). Biochemistry, 10, 4552.

Hart, R.G. (1961). Acta Cryst. 14, 1188.

Kauzmann, W. (1959). Advan. Protein. Chem. 14, 1.

Matthews, B.W., Colman, P.M., Jansonius, J.N., Titani, K., Walsh, K.A. and Neurath, H. (1972a). Nature New Biol. 238, 41.

Matthews, B.W., Jansonius, J.N., Colman, P.M., Schoenborn, B.P. and Dupourque, D. (1972b). Nature New Biol. 238, 37.

Matthews, B.W. and Weaver, L.H. (1974). Biochemistry, 13, 1719.

Matthews, B.W., Weaver, L.H. and Kester, W.R. (1974). J. Biol. Chem. 249, 8030.

Matsubara, H. (1967). In "Molecular Mechanisms of Temperature Adaption", C.L. Prosser, ed., p. 283. Washington, D.C.: American Association for the Advancement of Science.

Pangburn, M.K. (1974). Ph.D. Thesis, University of Washington.

Pangburn, M.K., Burstein, Y., Morgen, P.H., Walsh, K.A. and Neurath, H. (1973). Biochem. Biophys. Res. Commun. 54, 371.

Perutz, M.F. (1975). Nature 255, 256.

Schellman, J.A. (1955). Compt. Rend. Trav. Lab. Carlsberg Ser. Chim. 29, 223.

Scheraga, H.A. (1953). In "The Proteins", Vol. II, 2nd Edit., H. Neurath, ed. Academic Press, New York.

Suda, H., Aoyagi, T., Takeuchi, T. and Umezawa, H. (1973). J. Antibiotics 26, 621.

Tanford, C. (1962). J. Amer. Chem. Soc. 84, 4240.

Titani, K., Hermodson, M.A., Ericsson, L.H., Walsh, K.A. and Neurath, H. (1972). Nature New Biol. 238, 35.

Voordouw, G. and Roche, R.S. (1974). Biochemistry, 13, 5017.

TABLE 1. Refined Parameters for Calcium-Depleted Thermolysin Crystals.

Crystal Soaking Solution	10 mM Gd^{+++}	5 mM EDTA 5 mM Ca^{++}	0.5 mM EDTA	8 mM EDTA 2 mM Ca^{++}
Site 1				
Z	30.4	3.3	3.9	10.1
x	0.755	0.746	0.744	0.753
y	0.617	0.614	0.617	0.619
z	0.038	0.043	0.043	0.045
B	17.3	14.5	14.2	14.2
Site 2				
Z	-25.2	-13.2	-12.0	-13.2
x	0.760	0.756	0.753	0.757
y	0.624	0.624	0.621	0.619
z	0.066	0.064	0.066	0.064
B	37.6	12.9	14.5	12.6
Site 3				
Z	37.4	-1.2	2.3	-12.2
x	0.432	0.434	0.430	0.441
y	0.875	0.876	0.875	0.880
z	0.042	0.042	0.037	0.044
B	29.1	15.2	15.1	14.4
Site 4				
Z	32.6	-7.1	-13.5	-16.5
x	0.700	0.698	0.701	0.701
y	0.491	0.488	0.494	0.495
z	0.083	0.078	0.080	0.080
B	26.9	15.9	13.6	11.8

In the table, Z is the occupancy, approximately in electrons, relative to native thermolysin; B is the thermal parameter in $Å^2$; and x, y and z are the fractional coordinates. A negative occupancy indicates that electron density has been lost from the site. The refinements are based on 4.0 Å resolution data except for Gd^{+++} where 2.4 Å data were used.

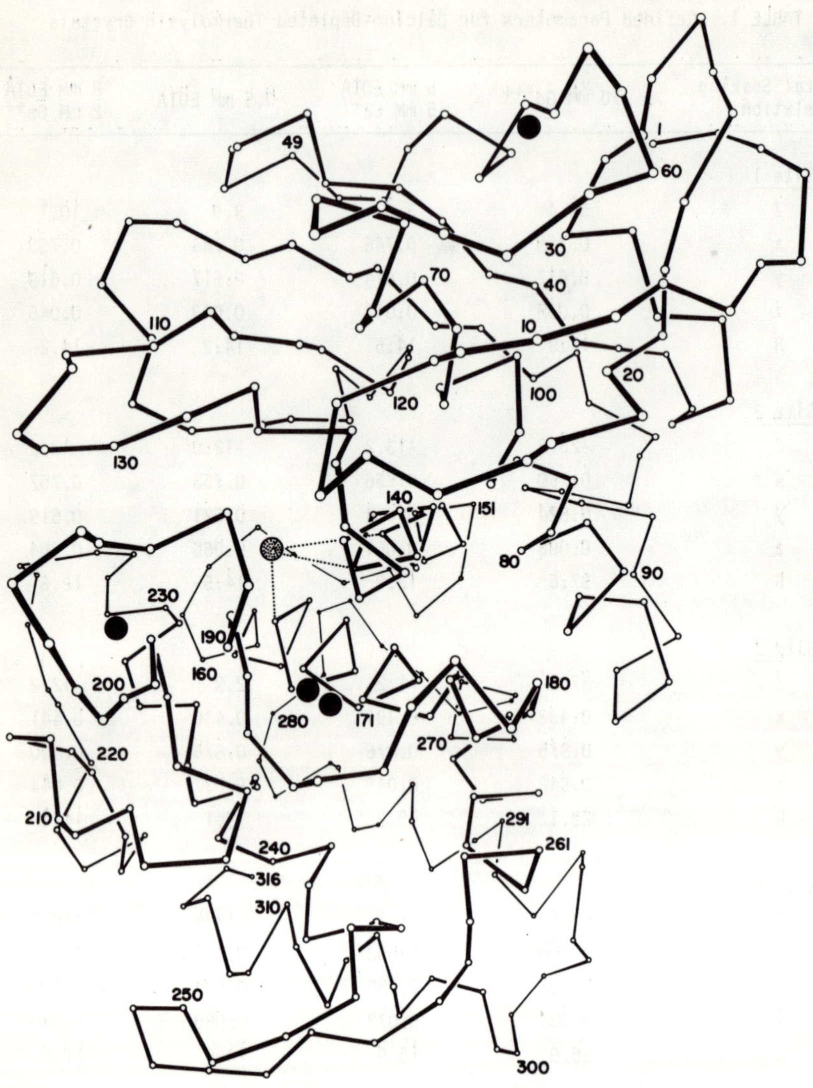

Figure 1 Perspective drawing illustrating the backbone conformation of thermolysin. The zinc atom is drawn stippled, and the calcium ions as solid circles.

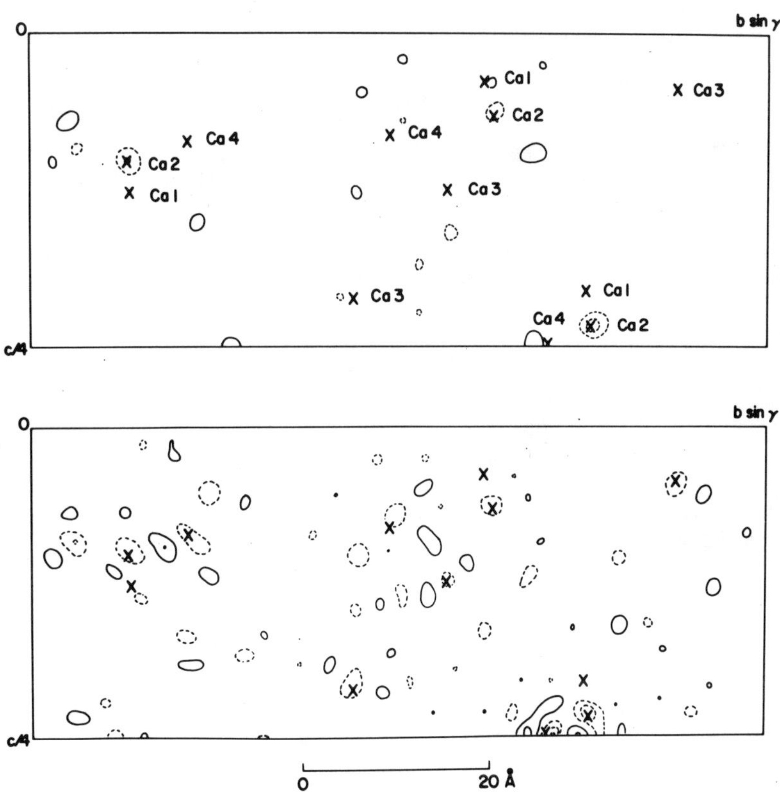

Figure 2 Difference maps showing at 4.0 Å resolution the projected difference in electron density between calcium-depleted and native crystals of thermolysin. The locations of the calcium binding sites, each of which appears in three places, are indicated. The positive contours (drawn solid) indicate an increase in density relative to native thermolysin, while the negative contours (broken) indicate a loss of density. In one case (upper drawing) the crystals were soaked for 16 hrs in 0.5 mM EDTA, pH 7.2; and in the other case (lower drawing) in 8 mM EDTA, 2 mM Ca^{++}, pH 7.2 for 24 hrs before X-ray photography.

STUDIES ON THE INHIBITION OF THERMOLYSIN

J. FEDER, N. AUFDERHEIDE AND B.S. WILDI
Monsanto Company, New Enterprise Division, St. Louis, Missouri (USA)

The neutral proteases are a group of metallo endopeptidases primarily of microbial origin which have a similar substrate specificity and have neutral pH activity optima. These enzymes have been isolated from Bacilli, Streptomyces, Aspergilli, and Pseudomonas among others.[1] All of these enzymes catalyze the hydrolysis of peptide bonds of which the imino group is contributed by a hydrophobic amino acid, particularly leucine and phenylalanine. The neutral proteases isolated from B. subtilis,[2] B. megaterium,[3-5] B. cereus[6,7] and B. thermoproteolyticus[8,9,10] contain a single zinc atom and are stabilized by calcium and other divalent cations. These enzymes generally lack cysteine in their structures and Pangburn et al.[11] recently reported similarity in

[1] A. Matsubara and J. Feder, in The Enzymes (Ed. P. Boyer; Academic Press, New York 1971), vol. 3, p. 721.

[2] J. D. McConn, D. Tsuru and K. T. Yasunobu, J. biol. Chem. 239, 3706 1964).

[3] J. Chaloupka and P. Kreckova, Biochem. biophys, Res. Commun. 8, 120 (1962).

[4] J. Millet and R. Acher, Biochim. biophys. Acta 151, 302 (1968).

[5] L. Keay, J. Feder, L. R. Garrett, M. H. Moseley and N. Cirulis, Biochim. biophys. Acta 251, 74 (1971).

[6] D. N. Salter, Biochem. J. 72, 23 p. (1959).

[7] J. Feder, L. Keay, L. R. Garrett, N. Cirulis and M. H. Moseley, Biochim. biophys. Acta 251, 74 (1971).

[8] Y. Ohta, Y. Ogura and A. Wada, J. biol. Chem. 241, 5919 (1966).

[9] S. A. Latt, B. Holmquist and B. L. Vallee, Biochem. biophys. Res. Commun. 37, 333 (1969).

[10] J. Feder, L. R. Garrett and B. S. Wildi, Biochemistry 10, 4552 (1971).

[11] M. K. Pangburn, Y. Bürstein, P. H. Morgan, K. A. Walsh and H. Neurath, Biochem. biophys. Res. Commun. 54, 371 (1973).

sequences of the B. subtilis neutral protease and thermolysin. In spite of the great similarity in catalytic and structural properties these enzymes differ markedly in thermal stability.

Inhibitors that are structurally related to the substrates of the neutral proteases have been employed to investigate the active site of thermolysin for comparison with others in this class. The work on thermolysin is reported here. Since a hydrophobic amino acid is required for substrate specificity, a variety of peptides involving phenylalanine and leucine was examined with respect to inhibition of thermolysin. Based upon the effects of substitutent groups, charge and position, a model for the binding of charged substrates and inhibitors is proposed.

METHODS AND MATERIALS

Materials and Methods. Crystalline thermolysin (B. thermoproteolyticus Rokko) was purchased from Daiwa Kasei Co., Ltd., Osaka, and used without further purification. FA-Gly-Leu-NH_2[12] and FA-Gly-Phe-NH_2 were synthesized as described (Feder and Schuck, 1970).[13] All peptides used as inhibitors were available from commercial sources. Z-Leu-Gly, Leu-Gly-NH_2, Leu-Gly-Gly, Leu-NH_2, Bz-Leu, Ac-Leu, Ac-norLeu, Z-Leu, Z-Phe-Gly, Z-Phe, Ac-Phe, Bz-Phe, L-Phe, D-Phe, Ac-D-Phe, Gly-D-Phe, Ac-Gly, Z-Gly, Bz-Gly, Bz-Gly-OMe, Ac-Gly-Leu and Z-Gly-Leu were purchased from Cyclo Research Corporation. Z-Leu-Gly-NH_2, Leu-Gly, Gly-D-Leu, Z-Phe-Gly-NH_2, Phe-Gly-NH_2, Phe-Gly, Z-Gly-Phe, Bz-Gly-Phe, Gly-Phe-NH_2, Gly-Phe and Phe-Phe were purchased from Mann Research Laboratories. Reagent grade salts and deionized water were used in these studies.

[12]Abbreviations used: FA, furylacryloyl; Z, carbobenzoxyl; Bz, benzoyl; Ac, acetyl.

The thermolysin catalyzed hydrolysis of FA-Gly-Leu-NH_2 was monitored spectrophotometrically at 345 mμ on a Cary 14PM recording spectrophotometer as previously described (Feder and Schuck, 1970[13], Feder, 1968.[14]) The FA-Gly-Leu-NH_2 was prepared in pH 7.2 N-2-hydroxyethylpiperazine-N'-2-ethanesulfonic acid (Hepes) buffer (0.1M) at a concentration of 9.6×10^{-4} M with and without appropriate inhibitor concentration. All the inhibition studies were carried out at 25.0 ± 0.1° unless otherwise indicated.

When S << K_m competitive inhibition is described by the rate equation

$$dp/dt = k_{cat}(E)(S)/K_m(1 + (I)/K_I) ,$$

the complete reactions yield a first-order rate constant $k = k_{cat}(E)/K_m (I + (I)/K_I)$. The ratio of the first order constants in the presence and absence of inhibitor, π, is equal to $1 + I/K_I$

$$\pi = k_o/k_i = 1 + (I)/K_I$$

A plot of π versus the inhibitor concentration is linear with the slope equal to the reciprocal of the K_I. This same rate dependence also is obtained for noncompetitive inhibition under pseudo-first-order conditions and only the competitive effect is measured.

The enzyme molarity was determined from the psuedo-first-order rate constant for the hydrolysis of FA-Gly-Leu-NH_2 ($k = k_{cat}(E)/K_m$) divided by k_{cat}/K_m ($19.11 \times 10^3\ M^{-1}\ sec^{-1}$).

[13] J. Feder and J. M. Schuck, Biochemistry 9, 2784 (1970).
[14] J. Feder, Biochem. biophys. Res. Commun. 32, 326 (1968).

RESULTS AND DISCUSSION

The inhibition of the thermolysin catalyzed hydrolysis of FA-Gly-Leu-NH_2 by phenylanine and leucine-containing dipeptides is shown in Table I and II.

Table I. Inhibition of Thermolysin by Phenylalanine Peptides[a]

Inhibitor	conc. x 10^3 (M)	K_I x 10^3 (M)
Z-Phe-Gly-NH_2	0.32 - 3.18	0.35
Z-Phe-Gly	2.72 - 27.20	4.50
Phe-Gly-NH_2	11.20 - 56.10	10.90
Phe-Gly	2.01 - 20.10	10.30
Z-Gly-Phe-NH_2		K_m = 14.70[b]
Z-Gly-Phe	2.45 - 24.50	14.70
Bz-Gly-Phe	2.11 -- 21.12	38.50
Gly-Phe-NH_2	3.45 - 17.20	53.00
Gly-Phe	13.20 - 87.40	141.00
Gly-D-Phe	13.20 - 65.40	16.00
Phe-Phe	0.57 - 5.69	2.13

[a]All reactions carried out in pH 7.2 Hepes buffer (0.1M), 25.0 ± 0.1. $(E_o) = 1.16 \times 10^{-6}$M; $(S_o) = 0.76 \times 10^{-4}$M;, additional details in text.

[b]From Morihara and Tsuzuki (1970).[14a]

Generally, greater inhibition was observed when the carboxyl and amino groups were blocked. The presence of a free charge reduced the inhibition in almost all cases. A comparison of the inhibition by Z-Leu-Gly-NH_2 with Z-Leu-Gly, Leu-Gly-NH_2 and Leu-Gly yielded ratios of the K_I values of 1.3, 2.7, and 7.7, respectively. Similarly the ratio of K_I values for Z-Phe-Gly-NH_2 compared with Z-Phe-Gly, Phe-Gly-NH_2 and Phe-Gly were 12.9, 31.1, and 29.4, respectively. A similar comparison of dipeptides with free amino groups with and without free carboxyl groups yielded K_I ratios of 4.1, 2.8

[14a]Morihara and H. Tsuzuki, Eru. J. biochem. 15, 374 (1970).

Table II. Inhibition of Thermolysin by Leucine Peptides[a]

Inhibitor	conc. x 10^3 (M)	K_I x 10^3 (M)
Z-Leu-Gly-NH_2	0.95 - 9.55	3.07
Z-Leu-Gly	2.25 - 45.20	4.03
Leu-Gly-NH_2	8.70 - 43.40	8.30
Leu-Gly	3.65 - 21.90	23.50
Leu-Gly-Gly	3.36 - 33.60	27.40
D-Leu-Gly	20.20 - 67.50	108.00
Z-Gly-Leu-NH_2		K_m = 20.6[b]
Z-Gly-Leu	1.65 - 16.48	5.65
Ac-Gly-Leu	2.82 - 16.94	36.00
Gly-Leu-NH_2	4.71 - 47.07	11.60
Gly-Leu	9.51 - 95.10	47.50
Gly-D-Leu	77.4	No inhibition
Leu-Leu	0.51 - 20.20	0.84

[a]All reactions carried out in pH 7.2 Hepes buffer (0.1M), 25.0 ± 0.1°, $(E_o) = 1.16 \times 10^{-6}$M; $(S_o) = 9.76 \times 10^{-4}$M; additional details in text.

[b]From Morihara and Tsuzuki (1970).[14a]

and 2.7 for Gly-Leu/Gly-Leu-NH_2, Leu-Gly/Leu-Gly NH_2 and Gly-Phe/Gly-Phe-NH_2 respectively.

It was previously reported[15] that no inhibition was observed with Bz-Gly-Lys and Z-Gly-Arg even at 3-4 x 10^{-2} M. These effects, however, depended upon the relative position of the phenylalanine or leucine in the dipeptide. In almost all cases the presence of a free amino group on the hydrophobic amino acid enhanced inhibition. Values of 4.9, 13.7, 1.4 and 2.0 were obtained for the K_I ratios for Gly-Phe-NH_2/Phe-Gly-NH_2, Gly-Phe/Phe-Gly, Gly-Leu-NH_2/Leu-Gly-NH_2 and Gly-Leu/Leu-Gly, respectively. Similarly, greater inhibition was observed with leucine and phenylalanine with free amino groups than with the N-acyl derivatives (Table III).

[15]J. Feder, L. R. Brougham and B. S. Wildi, Biochemistry 13, 1186 (1974).

Table III. Inhibition of Thermolysin by Leucine and Phenylalanine[a]

Inhibitor	conc. x 10^3 M	K_I x 10^3 M
L-Phe	5.86 - 29.40	15.60
Z-Phe	0.36 - 3.56	0.51
Bz-Phe	3.28 - 19.66	4.50
Ac-Phe	3.29 - 19.73	150.00
Ac-Phe-NH_2	3.52 - 21.23	83.00
D-Phe	1.96 - 19.60	4.98
Ac-D-Phe	3.37 - 20.20	15.80
L-Leu	2.29 - 22.87	13.33
Z-Leu	0.21 - 2.08	0.32
Bz-Leu	1.97 - 17.00	0.88
Ac-Leu	6.72 - 20.17	20.88
Ac-norLeu	4.19 - 20.96	25.50
Leu-NH_2	4.86 - 29.20	11.80
D-Leu	2.29 - 22.87	16.50
Ac-Gly	8.25 - 82.80	208.00
Z-Gly	3.86 - 15.40	11.70
Bz-Gly	18.00 -- 54.00	120.00
Bz-Gly-OMe	5.01 - 21.14	40.40

[a]All reactions carried out in pH 7.2 Hepes buffer (0.1M) 25.0 ± 0.1. (E_o) = 1.0 -1.2 x 10^{-6}M; (S_o) = 9.76 x 10^{-4}M; additional details in text.

K_I ratios of 9.6, 3.2, and 1.6 were obtained for Ac-Phe/Phe, Ac-D-Phe/D-Phe and Ac-Leu/Leu respectively. These effects were more pronounced for phenylalanine peptides than leucine peptides. The data also showed greater inhibition if the free carboxyl group were removed by the length of an amino acid from the hydrophobic amino acid. Ratios of K_I values of 3.3, 1.4, 8.0 and 3.9 were obtained for Z-Gly-Phe/Z-Phe-Gly, Z-Gly-Leu/Z-Leu-Gly, Z-Gly-Tyr/Z-Tyr-Gly and Z-Gly-Trp/Z-Trp-Gly[15] respectively.

In a previous report[15] it was suggested that the carbobenzoxyl group contributed to the binding of some amino acids and peptides such as Z-Gly without which no inhibition was observed. Table IV compares the inhibition of the thermolysin catalyzed hydrolysis of FA-Gly-Leu-NH_2 by carbobenzoxyl and acetyl blocked amino acids and peptides.

Table IV. Effect of Blocking Group on Inhibition of Thermolysin[a]

Inhibition	$K_I \times 10^3$ M	K_IAc-K_IZ
Ac-Phe	150.00	294.1
Z-Phe	0.51	
Ac-Leu	20.88	65.3
Z-Leu	0.32	
Ac-Gly-Leu	36.00	6.4
Z-Gly-Leu	5.65	
Ac-Gly	208.00	17.8
Z-Gly	11.70	

[a] All reactions in pH 7.2 Hepes buffer (0.1M) 25.0 ± 0.1°. $(E_o) \cong 1.0 \times 10^{-6}$ M; $(S_o) = 9.76 \times 10^{-4}$M; additional details in text.

Considerable enhancement in inhibition was observed with the carbobenzoxyl blocked compounds over the N-acetyl derivatives. This strongly suggests binding of the carbobenzoxyl group at the substrate binding site. Based on crystallographic studies Matthews, Weaver and Kester[16] suggested that Z-Phe was bound backwards with the acyl group in the hydrophobic pocket. Bigbee and Dahlquist[17] also proposed different modes of binding of this inhibitor based on nuclear magnetic resonance studies. The present report supports these suggestions.

[16]B. W. Matthews, L. H. Weaver and W. R. Kester, J. biol. Chem. 249, 8030 (1974).

[17] W. L. Bigbee and F. W. Dahlquist, Biochemistry 13, 3532 (1974).

A comparison of the inhibition by the leucine and phenylalanine peptides and amino acids suggested slightly different modes of binding for these amino acids. The effects of charge and position of the hydrophobic amino acid were more pronounced for phenylalanine than leucine derivatives. It was also interesting to note that comparison of the D and L amino acids yielded K_I ratios of 3.1 and 0.7 for D-Phe/L-Phe and D-Leu/L-Leu respectively. The D-Phe yielded almost three-fold the inhibition of the L-Phe, whereas the D-Leu was less inhibitory than the L-Leu. This suggested that there might be subtle differences in the mode of binding of FA-Gly-Leu-NH_2 and FA-Gly-Phe-NH_2 which would be reflected in the competitive inhibition by various peptides.

The inhibition of the thermolysin catalyzed hydrolysis of FA-Gly-Leu-NH_2 and FA-Gly-Phe-NH_2 was studied and the K_I values obtained from each substrate were compared (Table V, Figure 1).

Table V. Comparison of Inhibition of Thermolysin Catalyzed Hydrolysis of FA-Gly-Leu-NH_2 and FA-Gly-Phe-NH_2[a]

Inhibitor	$K_I \times 10^3$ (M) FA-Gly-Leu-NH_2	FA-Gly-Phe-NH_2
Z-Leu-Gly-NH_2	5.21	5.39
L-Leu	13.33	13.33
D-Leu	16.50	18.80
Bz-Leu	1.09	0.98
Z-Gly	11.56	14.19
Z-Phe	0.62	0.75

[a] All reactions carried out in pH 7.2 Hepes buffer (0.1M) 25.0 ± 0.1°. $(E_o) = 1 \times 10^{-6}$ M, $(S_o) = 9.76 \times 10^{-4}$M for FA-Gly-Leu-$NH_2$, 8.93×10^{-4} M for FA-Gly-Phe-NH_2; additional details in text.

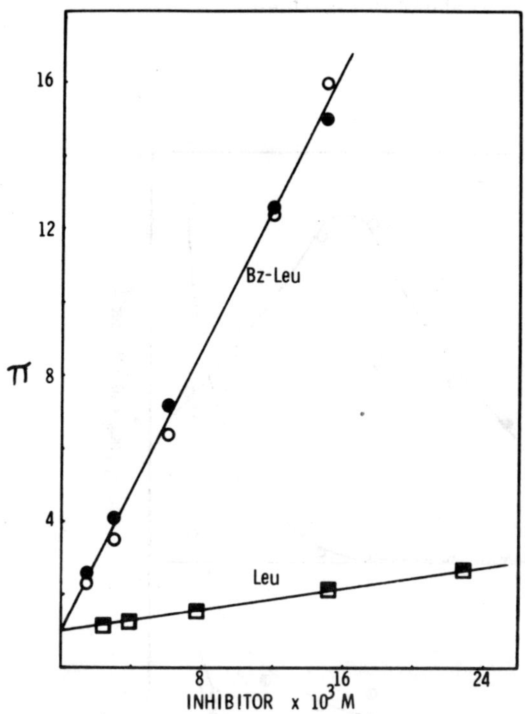

Fig. 1. The inhibition of the thermolysin catalyzed hydrolysis of FA-Gly-Leu-NH_2 (○, □) and FA-Gly-Phe-NH_2 (●■) by Bz-Leu and Leu, π plots. pH 7.2 Hepes buffer (0.1M), 25.0 ± 0.1°, $(S_o) = 9.76 \times 10^{-4}$M; $(E_o) = 1.0 \times 10^{-6}$M. Additional details in text.

The identical K_I values obtained for each inhibitor with both substrates suggest that FA-Gly-Leu-NH_2 and FA-Gly-Phe-NH_2 have an almost identical mode of productive binding at the active site.

A comparison of the pH dependence of the thermolysin catalyzed hydrolysis of FA-Gly-Leu-NH_2 and the K_I values for Z-Phe-Gly-NH_2, Z-Leu-Gly-NH_2, Bz-Leu and Z-Phe are shown in Figure 2.

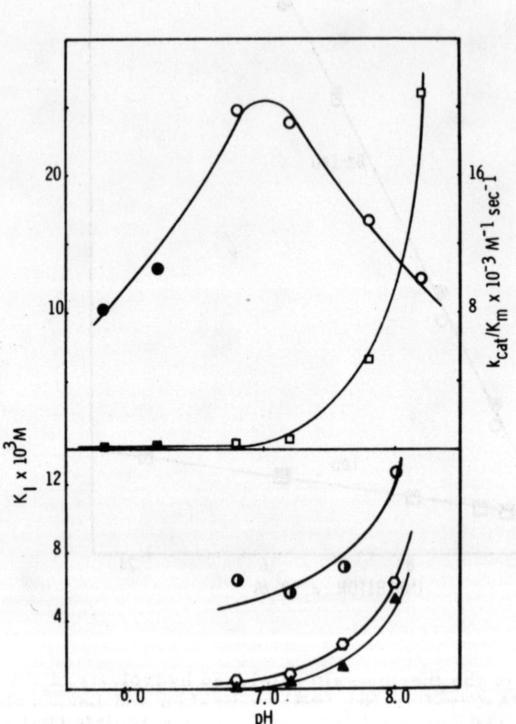

Fig. 2. Comparison of the pH dependence of the thermolysin catalyzed hydrolysis of FA-Gly-Leu-NH_2 (●○) and the pH dependence of the K_I for Z-Phe-Gly-NH_2 (■□), Z-Leu-Gly-NH_2 (◐), Bz-Leu (⬡) and Z-Phe (▲); cacodylate buffer 0.1M (●■); Hepes buffer, 0.1 (○□◐⬡▲); (E_o) = 0.6-1.0 x 10^{-6} M; (S_o) = 9.76 x 10^{-4} M.

A dependence upon a group with a pK near 8 was observed for the inhibition. Previously,[15] it was suggested that this might reflect the importance of salt bridges involving histidine at the active site required for maintenance of an active conformation.

The results reported here suggest a model not only for the binding of inhibitor peptides but perhaps also for the binding of substrate. The primary structure and conformation of thermolysin have been determined.[18-21] Colman, Jansonius and Matthews[21] found that the molecule consisted of two lobes with a cleft containing the essential zinc atom. A hydrophobic pocket was described about 8 Å from the zinc in which the phenylalanine side chain of the inhibitor Ala-Phe was found.[16] This pocket therefore would be the best candidate for the specificity binding site for the hydrophobic sidechain of the substrate. Any molecule which would bind at this site would competitively inhibit the enzyme. Colman et al.[21] also pointed out the presence of an Arg-203 in the active site which appeared to be an analogue to Arg-145 in carboxypeptidase. In carboxypeptidase Arg-145 is thought to bind the carboxyl group of the substrate. Also present near the zinc is Glu-143 in a position almost identical to Glu-270 in carboxypeptidase.[21] These two groups, Arg-203 and Glu-143, are located at opposite sides of the opening of the hydrophobic pocket, and although no role has been assigned to them they might be involved in the interaction of charged substrates and inhibitors. The difference

[18]K. Titani, M. A. Hermodson, L. H. Ericsson, K. A. Walsh and H. Neurath, Nature (London), New biol. 238, 35 (1972).

[19]B. W. Matthews, P. M. Colman, J. N. Jansonius, K. Titani, K. A. Walsh and H. Neurath, ibid. 238, 41 (1972).

[20]B. W. Matthews, J. N. Jansonius and P. M. Colman, ibid. 238, 37 (1972).

[21]P. M. Colman, J. N. Jansonius and B. M. Matthews, J. mol. Biol. 70, 701 (1972).

in inhibition constants observed relative to the position of the hydrophobic amino acid to the free amino or carboxyl group suggests multiple binding involving the hydrophobic pocket, Glu-143 and Arg-203. An inhibitor peptide such as Phe-Gly-NH_2 might bind both in the hydrophobic site and via an ionic interaction between the free α-amino group and Glu-143. However Gly-Phe would not fit into this mode of multiple binding. The presence of the Glu-143 might be so positioned that during catalysis it would interact with the amino group of the phenylalanine or leucine released on hydrolysis. Arg-203, however, might be displaced sufficiently from the hydrophobic pocket so that both binding of the hydrophobic side chain of the inhibitor and interaction with its free carboxyl group would not be possible. The displacement of a single amino acid length between the hydrophobic side chain and the free carboxyl, however, would allow multiple interaction. The presence of a free amino or carboxyl group almost completely prevents catalysis. Matsubara[22] reported that Gly-Phe and Gly-Phe-NH_2 were not cleaved by thermolysin while Z-Gly-Phe and Z-Gly-Leu were very slowly hydrolyzed. Likewise, Morihara and co-workers[23] reported that Gly-Leu-NH_2 was not hydrolyzed. We have observed that the thermolysin catalyzed hydrolysis of FA-Gly-Leu yielded a k_{cat}/K_m of about 1/1000 that for FA-Gly-Leu-NH_2. All of these studies employed dipeptides in which the free carboxyl was on the hydrophobic amino acid. Morihara and Tsuzuki,[14a] however, found that tripeptides with the general structure Z-X-Phe-Ala were good substrates in spite of the free carboxyl group. In these cases the free carboxyl is displaced by an amino acid length from the hydrophobic side chain and this, according to our model, would permit the multiple

[22]H. Matsubara, Biochem. biophy. Res. Commun. 24, 427 (1966).

[23]K. Morihara, H. Tsuzuki and T. Oka, Arch. biochem. biophys. 123, 572 (1968).

interaction of binding of the side chain in the hydrophobic pocket and interaction of the carboxyl group with Arg-203 (Figure 3).

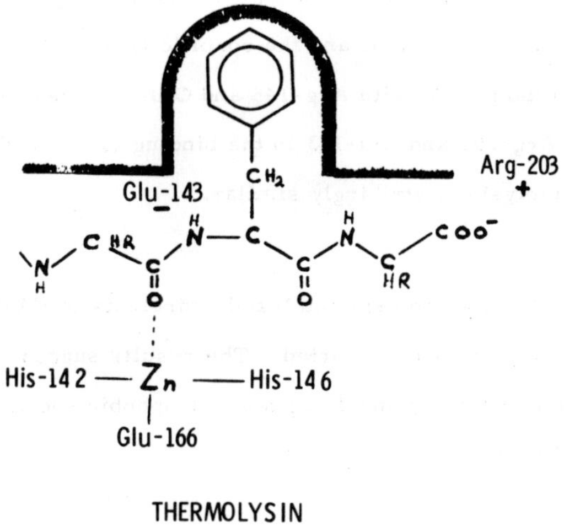

Fig. 3. Diagram of proposed orientation of charged inhibitors and substrates at the active site of thermolysin.

Based upon this model for inhibitor binding one might propose a similar orientation for a substrate molecule. The side chain of leucine or phenylalanine would bind in the hydrophobic specificity pocket and the extended peptide would be orientated with the carboxyl end of the molecule toward the Arg-203 side of the pocket. The imino group of the peptide bond to be cleaved would be near Glu-143. The presence of a free amino or carboxyl group, however, would probably distort the exact orientation relative to the zinc and essential histidines. Bigbee and Dahlquist[16] have suggested a similar model for the binding of a polypeptide substrate.

It is interesting to note that Hartsuck and Lipscomb[24] have proposed that the unusual stability of the (Gly-Tyr)-carboxypeptidase A complex involves, in addition to interaction with the zinc and the specificity pocket, interaction of the carboxyl and amino groups with Arg-145 and Glu-270, respectively. The proposed role of Arg-203 and Glu-143 in the binding of charged substrates and inhibitors to thermolysin is strikingly similar.

SUMMARY

The inhibition of the thermolysin catalyzed hydrolysis of FA-Gly-Leu-NH_2 and FA-Gly-Phe-NH_2 has been reported. The results suggest a model for substrate and inhibitor binding involving the hydrophobic specificity pocket, Arg-203 and Glu-143.

[24] J. A. Hartsuck and W. N. Lipscomb, in The Enzymes (Ed. P. Boyer: Academic Press, New York 1971), vol. 3, p. 1.

EFFECT OF EDTA ON THE CONFORMATIONAL STABILITY OF THERMOLYSIN

A. Fontana, E. Boccù and F.M. Veronese

Institute of Organic Chemistry and of Pharmaceutical Chemistry, University of Padova, Padova, Italy.

1. INTRODUCTION

Thermolysin is a remarkably thermostable metallo endopeptidase produced by Bacillus Thermoproteolyticus Rokko [1]. The enzyme contains one catalytically essential zinc as well as four calcium atoms per molecule [2]. Calcium ions are not directly involved with the catalytic activity, but the nature of interactions in which they partecipate, as obtained by X-ray analysis, suggest that these ions may contribute to the thermostability of the molecule [3,4]. The present communication describes the effect of the chelating agent EDTA which is an effective inhibitor of the enzyme [5,6], on the stability against heat and protein denaturants of the secondary structure of thermolysin, as assessed by circular dichroism and emission fluorescence measurements.

2. MATERIALS AND METHODS

Crystalline thermolysin (from Bacillus thermoproteolyticus, Rokko) was purchased from Daiwa Kasei K.K., Osaka, Japan. The enzyme was purified on a Sephadex G-75 column (2x75 cm) equilibrated with 1M NaCl-0.01M sodium acetate -2mM $CaCl_2$, pH 7.2. The fractions containing the enzyme were combined and the resulting solution extensively dialyzed at 4°C against 20mM Tris-HCl buffer, pH7.0, containing 1mM $CaCl_2$. This enzyme solution has been used in the present study. Protein concentration was estimated from the absorbance at 280nm, using a molar extinction (ε) of 52,400 [6].

Circular dichroism measurements were performed on a Cary Model 61 recording spectropolarimeter. Temperature effects were studied in jacketed cells through which an ethylene glycol/water mixture was circulated from a thermostatically controlled bath. Temperatures were measured inside the cells with a thermistor probe. The circular dichroic signal at 220 nm was continuously recorded, while the temperature of the sample in the cuvette was raised at a constant rate of 2 degrees/ min.

Fluorescence emission spectra were measured with a Hitachi MPF-2 spectrofluorimeter, equipped with a thermostated cuvette holder and connected to a Hitachi QPD_{33} recorder. The excitation wavelength was 290 nm. The contribution to the intensity of fluorescence by the solvent buffer or contaminants was routinely measured by a fluorescence scan from 300 to 380 nm upon excitation at 290 nm.

3. RESULTS

The exceptional thermostability of thermolysin has been verified by circular dichroism measurements. At neutral pH, the far uv spectrum shows shoulders near 220 and 210 nm, indicative of α-helix. Determination of the ellipticity at 220 nm as a function of temperature over the range 20-95°C in different solvent conditions and in presence of protein denaturants should be informative on the structural stability of the protein. The results previously reported by Ohta[7] on the effect of organic solvents and other unfolding agents have been confirmed by the present studies.

As shown in Fig. 1, thermolysin is stable up to about 75°C when heated in presence of Ca^{2+}. The thermostability of the enzyme is also dependent upon the

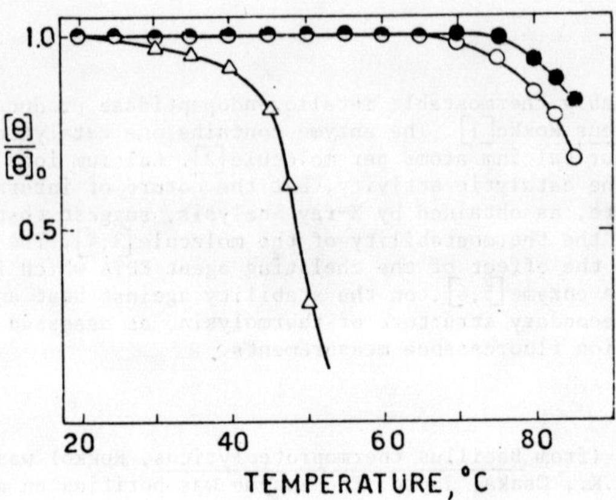

FIG. 1. Effect of temperature on the structural stability of thermolysin in different solvent conditions as assessed by circular dichroism measurements. The ellipticity value [θ] was measured at 220 nm as a function of temperature; $[\theta]_o$ represents the value at room temperature. The enzyme was dissolved in 20 mM Tris-HCl buffer, pH7.0, at a concentration of 0.1 mg/ml. ●---●, 10 mM $CaCl_2$; O---O, 1 mM $CaCl_2$; Δ---Δ, 10 mM EDTA.

concentration of calcium ion, since stability is enhanced in a 10mM in respect to 1mM concentration of the ion.

EDTA was found to have a pronouncedeffect on the stability of thermolysin. In presence of this sequestrating ion agent the structure of the enzyme is changed slightly, as shown by reduction of negative ellipticity values both in the far and near uv region, negative difference absorption measurements and fluorescence emission properties. In presence of 10mM EDTA the intensity of fluorescence of the enzyme is slightly reduced (about 10%) and the wavelength of maximum emission is red shifted upon excitation at 290 nm from 333 to 338nm. If we take the emission maximum of tryptophan residues in a protein as an indication of hydrophobicity of their environment [8], these residues seen to be more exposed to the solvent when the enzyme is treated with EDTA. In conclusion, the physical methods mentioned would indicate that a conformational transition occurs in thermolysin by the action of EDTA, all observed effects being in harmony with chain unfolding.

The effect of EDTA on the thermostability of the protein is dramatic, since the enzyme is thermolabile already at about 40°C(Fig.1). On the other hand, conformational studies carried out on thermolysin in presence of specific zinc chelating agents such as tetraethylenepentamine[6], have indicated that this catalytically essential ion does not influence the conformation and stability of the protein.

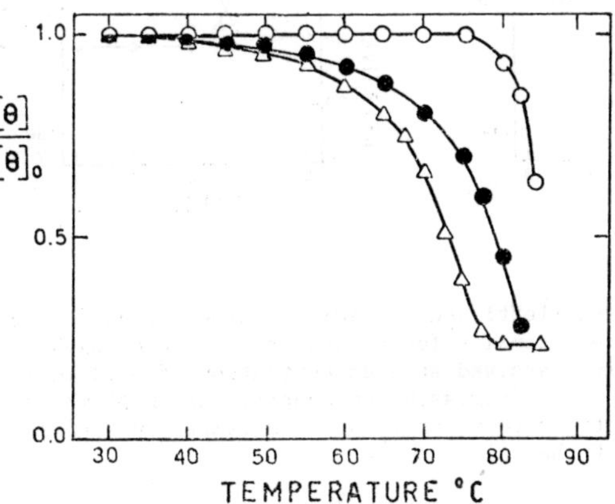

FIG. 2. Effect of urea and guanidine hydrochloride on heating denaturation curve of thermolysin. Conditions are as described in Fig. 1 and in the Experimental. The enzyme bas been dissolved in 20 mM Tris-HCl buffer, pH 7.0, containing 1 mM $CaCl_2$. o——o, Tris buffer;●——●,8M urea in Tris buffer;Δ——Δ, 6M guanidine hydrochloride in Tris buffer.

Fig. 2 shows the results on the thermal stability of thermolysin in presence of 8M urea and 6M guanidine hydrochloride. These strong denaturing agents for proteins do not show any effect at room temperature, whereas unfolding is accelerated at higher temperatures. On the other hand, in presence of these denaturants and 10mM EDTA thermolysin is immediately denatured, as ascertained by circular dichroism measurements.

The unfolding of thermolysin by the action of heat and protein denaturants has been also studied by emission fluorescence measurements. The intensity of fluorescence of thermolysin in 1-10mM Ca^{2+} solution at neutral pH upon heating up to 85°C decreases in an almost linear way due to thermal quenching[9]. A similar monotonic curve of relative intensity <u>versus</u> temperature was obtained with the enzyme as with a standard mixture of tyrosine and tryptophan in the same ratio as contained in the protein molecule. On the other hand a clear transition at about 47°C was observed when heating of the enzyme was performed in 1-10mM EDTA.

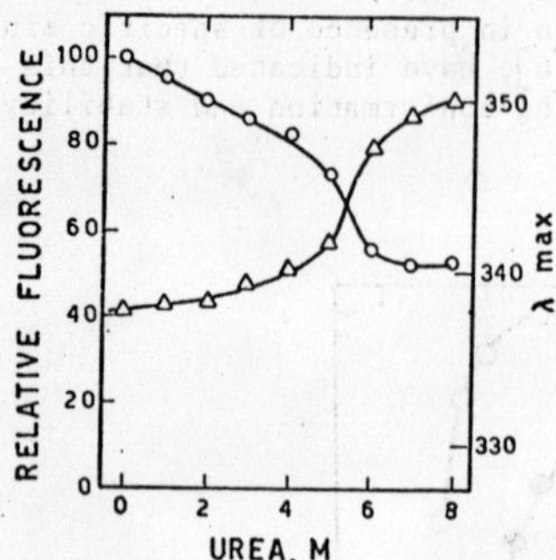

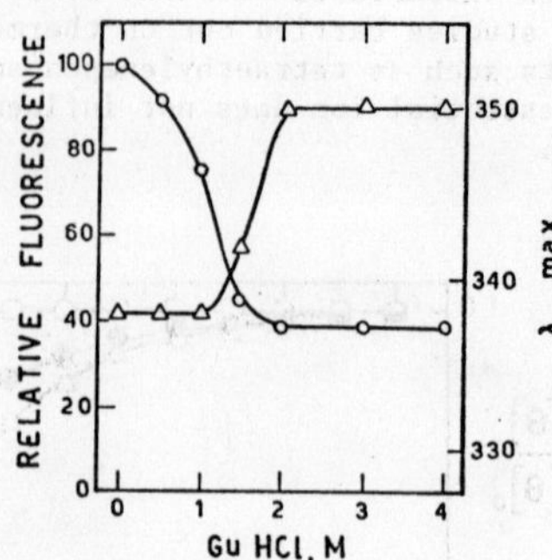

FIG. 3. Effect of urea (left) and guanidine hydrochloride (right) on the relative fluorescence intensity (o—o) and maximum wavelength of emission (Δ—Δ) of thermolysin dissolved at a concentration of 0.03 mg/ml in urea or guanidine solutions of increasing concentrations in 20 mM Tris-HCl buffer, pH7.0. The spectra were recorded at room temperature after 5 minutes of the preparation of the solution.

Fig. 3 shows the change in intensity and maximum of the emission of fluorescence of thermolysin dissolved in urea and guanidine hydrochloride in 20mM Tris-HCl buffer, pH 7.0, containing 10mM EDTA.
With both protein denaturants a gradual decrease of fluorescence intensity and red shift of maximum of emission of the enzyme was observed upon increasing the perturbant concentration. The changes in the fluorescence characteristics clearly indicate the unfolding of the enzyme. In 8 M urea and 6M guanidine the maximum wavelength of emission is at 350 nm, which is the maximum of emission in aqueous solvents of free tryptophan or simple tryptophyl peptides [8]. A midpoint transition can be evaluated at about 5M and 1.2M concentration of urea and guanidine hydrochloride respectively. These concentrations of denaturants are usually observed to be sufficient for the unfolding of most globular proteins from mesophilic sources.

4. DISCUSSION

The role of calcium in maintaining protein stability has been reported for a number of different enzymes, including trypsin [10], α-amylase [11], *Escherichia coli* glutamine synthetase [12] and several neutral proteases [13]. The results here reported suggest that the calcium ions play a major role in thermolysin in maintaining the integrity of its structure and its stability towards heat and protein denaturants.

Ohta [7] has made a thorough study of the effect of temperature on the structural stability of thermolysin and it was suggested that stability is provided by both hydrogen and hydrophobic bonding between phenolic groups. However, on the basis of the X-ray structure elucidation [3,4] no stabilizing effect could be ascribed to the tyrosine residues. Considering that the thermolysin molecule is a single polypeptide chain without disulfide bridges, the possibility exists that calcium might perform this bridging function within the molecule.

ACKNOWLEDGEMENT

The authors wish to thank Mr. M. Zambonin and Mr. F. Miozzo for skilful technical assistance and Dr. L. Callegaro for performing some experiments here mentioned. This work has been supported by the Italian C.N.R.

References

[1] S. Endo, J. Fermentation Techn. 40, 346(1962).

[2] J. Feder, L.R. Garrett and B.S. Wildi, Biochemistry 10, 4552(1971).

[3] P.M. Colman, J.N. Jansonius and B.W. Matthews, J. Mol. Biol. 70, 701(1972).

[4] B.W. Matthews, L.H. Weaver and W.R. Kester, J. Biol. Chem. 249, 8030(1974).

[5] K. Morihara and H. Tsuzuki, Biochim. Biophys. Acta 118, 215(1966).

[6] G. Woordouw and R.S. Roche, Biochemistry 13, 5017(1974)

[7] Y. Ohta, J. Biol. Chem. 242, 509(1967).

[8] L. Brand and B. Witholt, Methods in Enzymology 25, 776(1967).

[9] J.A. Gally and G.H. Edelman, Biochim. Biophys Acta 60, 499(1962).

[10] J.P. Abita, M. Delaage and M. Lazdunski, Eur. J. Biochem. 8, 314(1969).

[11] J. Hsiu, E.H. Fischer and E.A. Stein, Biochemistry 3, 61 (1964).

[12] B.M. Shapiro and A. Ginsburg, Biochemistry 7, 2153(1968).

[13] D. Tsuru, H. Kira, T. Yamamoto and J. Fukumoto, Agr. Biol. Chem. 30, 856(1966).

ROLE OF A SULFHYDRYL GROUP IN THE STRUCTURE AND FUNCTION OF ALKALINE PROTEASES FROM A THERMOPHILIC ACTINOMYCETE, *Streptomyces rectus* var. *proteolyticus*

K. MIZUSAWA and F. YOSHIDA
Central Research Laboratories, Kikkoman Shoyu Co., Ltd.
399 Noda, Noda-Shi, Chiba-Ken, Japan

INTRODUCTION

Thermophilic actinomycetes are generally defined as a group of actinomycetes which grow well above 50°C (1). It is well known that they are widely distributed in nature, inhabiting soils, composts, hay, peat, etc., and play an important role in the decomposition of organic matters.
In the past decade, a considerable attention has been paid to them as active producers of thermostable extracellular enzymes such as amylase, protease, cellulase, and cell wall lytic enzymes. As for proteolytic enzymes, most of them so far studied are alkaline proteases which are produced by several species belonging to the genera *Streptomyces* (2-4), *Thermoactinomyces* (5-14), *Thermomonospora* (10), and *Micropolyspora* (15, 16).
These alkaline proteases have been found to possess some common features. First, they are highly thermostable irrespective of the degree of purity. Second, they all seem to be inhibited by *p*-chloromercuribenzoate. This property suggests a participation of sulfhydryl groups in enzyme function, and is in striking contrast to mesophilic alkaline proteases which have no sulfhydryl groups. And third, it appears that they have a similar molecular weight with mesophilic ones.
Considering these facts, the alkaline proteases of thermophilic actinomycetes may have a different structure at the active site from that of mesophilic ones, and the sulfhydryl group(s) may bear a common role in supporting their thermostability.
As the first clue to elucidate these problems, we will summarize the results of investigation on a sulfhydryl group of alkaline proteases that have been purified from the culture filtrate of thermophilic *Streptomyces rectus* var. *proteolyticus*.

GENERAL FEATURES OF PROTEASE B

This organism produces two similar proteases, designated as proteases A and B according to the order of elution on SE-Sephadex chromatography (17). All the following results are of protease B, because the A-fraction had almost the same properties with the B-fraction except for a slight difference in amide content.
The enzyme is a typical serine protease inhibited by diisopropyl phosphofluoridate (DFP), active in alkaline pH range (2). Although the enzymatic properties are similar to those of subtilisins, it is characterized by high thermal stability and sensitivity to *p*-chloromercuribenzoate (*p*-MB).
The enzyme is a compactly folded globular protein with a molecular weight, 21,500, and an isoelectric point, 9.5, and it is composed of one polypeptide chain with 215 amino acid residues (17). Both amino- and carboxy-terminal residues are tyrosines. The amino acid composition is not unusual as a whole in comparison with mesophilic alkaline proteases such as subtilisin BPN' and *Aspergillus soyae* alkaline protease, but it is noteworthy that the thermostable enzyme has one cysteine residue whereas the mesophilic ones have neither cysteine nor cystine. Thus it has been verified that

the sensitivity to p-MB is based on the single cysteine residue. The results of ORD, CD, and IR-spectra (18) have shown that the enzyme protein consists of 15 to 20% α-helix, random and some anti-parallel β-structures. Accordingly, the thermostable enzyme does not possess any unusual structure in gross conformation. Nevertheless, it has a higher thermal stability and is far more resistant against protein-denaturing reagents such as urea, guanidine hydrochloride, and sodium dodecyl sulfate than mesophilic alkaline proteases like subtilisin BPN' and *Aspergillus* alkaline protease (19).

RELATION OF THE SULFHYDRYL GROUP TO ENZYME FUNCTION

The assumption that the single sulfhydryl (-SH) group is intimately concerned with active site has been drawn from the following experimental results (19).

1. Inactivation by p-MB is linearly related to the molar ratio of p-MB to enzyme. At the ratio of 0.7 to 0.8, complete loss of both proteinase and esterase activities occured.
2. The inactivation was reversible. Full activity was restored by cysteine, although the rate of reactivation was rather slow requiring 3 hours for p-MB-enzyme and 1 hour for Hg-enzyme in the presence of a large excess of cysteine.
3. The rate of p-MB inhibition decreased largely in the presence of substrates of low molecular weight.
4. The rate of reaction of enzyme -SH group with p-MB was almost instantaneous, whereas diisopropyl phosphoryl enzyme (DIP-enzyme) reacted very slowly with it.
5. Native enzyme incorporated about 0.7 μmoles of phosphorus per μmoles of enzyme with DFP-treatment. On the other hand, p-MB- and Hg-enzymes showed no incorporation of phosphorus.

It has been shown that the -SH group is partly buried in a hydrophobic region near active site, since 5,5'-dithiobis(2-nitrobenzoic acid) did not react with the -SH group of native enzyme, but it reacted instantly with the one of heat-denatured enzyme protein.

THERMAL INSTABILITY OF p-MB-ENZYME

An interesting phenomenon accompanying the modification of the -SH group is a marked change of thermal stability. Fig. 1 shows the thermal stability of the enzyme whose -SH group has been blocked with phenylmercuric acetate, p-MB, or mercuric chloride. From these thermal inactivation curves, temperatures causing 50% inactivation have been calculated to be 65°, 68°, 82°, 85° for phenylmercuric, p-MB-, native, and Hg-enzymes, respectively. Accordingly, the former two enzymes are more thermolabile as much as 20° than Hg-enzyme. The native enzyme has a slightly lower stability than Hg-enzyme, but the thermal stability of both enzymes is regarded essentially as the same, because they showed the same degree of thermostability at more dilute enzyme concentration under which autolysis of the native enzyme was negligible.
The marked change of thermostability was reflected in thermal denaturation of enzyme protein monitored by ORD or CD spectra (19). Temperature dependence curves of Moffitt-Yang parameters of ORD have shown that the p-MB-enzyme is stable below 50° and the Hg-enzyme is entirely stable up to 70°. The Tm values (i.e., temperatures of half-maximal change of b_0 values) were

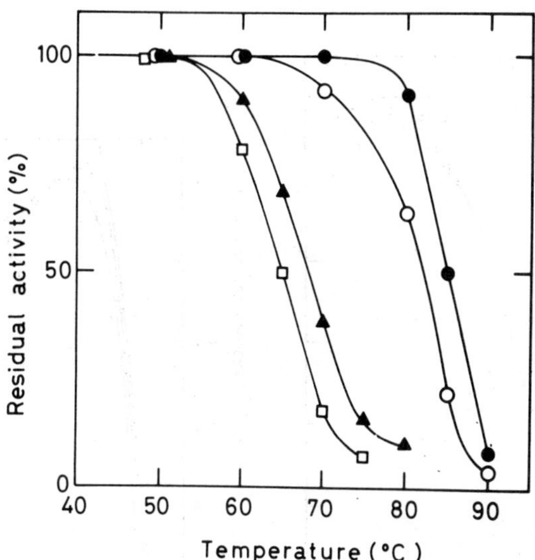

Fig. 1. Effect of -SH group modification on the thermostability of thermophilic *Streptomyces* alkaline protease. Heat treatment was conducted for 30 min in 0.05 M sodium acetate buffer (pH 5.6) containing 0.01 M calcium acetate. Enzyme concentration was 0.05 mg per ml. Residual activity was determined after reactivation by 0.05 M cysteine. ▲, p-MB-enzyme; □, phenylmercuric enzyme; ●, Hg-enzyme; ○, native enzyme.

55° for the former and 75° for the latter. Potential activity decreased in accordance with b_0 with both enzymes. These heat denaturations were completely irreversible, since any restoration of conformation has not been observed on cooling. Thus it has been verified that the decreased thermal stability of p-MB-enzyme is ascribed to increased thermal sensitivity of its protein structure.

CONFORMATIONAL CHANGE ACCOMPANYING SULFHYDRYL MODIFICATION

The results mentioned above suggest that some conformational change may occur at or around the active site by blocking the -SH group. In fact, this has been confirmed by CD spectra as indicated in Fig. 2.
Native enzyme had a large negative trough at 222 nm ($[\theta] = -10.69 \times 10^3$ deg cm^2 per dmole), a positive peak at 255 nm ($[\theta]$ = +28 to 29), a negative trough at 282 nm ($[\theta]$ = -255), and a shoulder at 292 nm. *p*-MB-enzyme had a little smaller value of mean residue ellipticity at 222 nm ($[\theta] = -10.13 \times 10^3$) and showed a slight change around 255 nm. Hg-enzyme had almost the same spectra with native enzyme except around 222 nm ($[\theta] = -10.34 \times 10^3$). Thus it has been indicated that the -SH modification with *p*-MB induces a slight conformational alteration around the -SH group, which results in a drastic loss of thermostability.

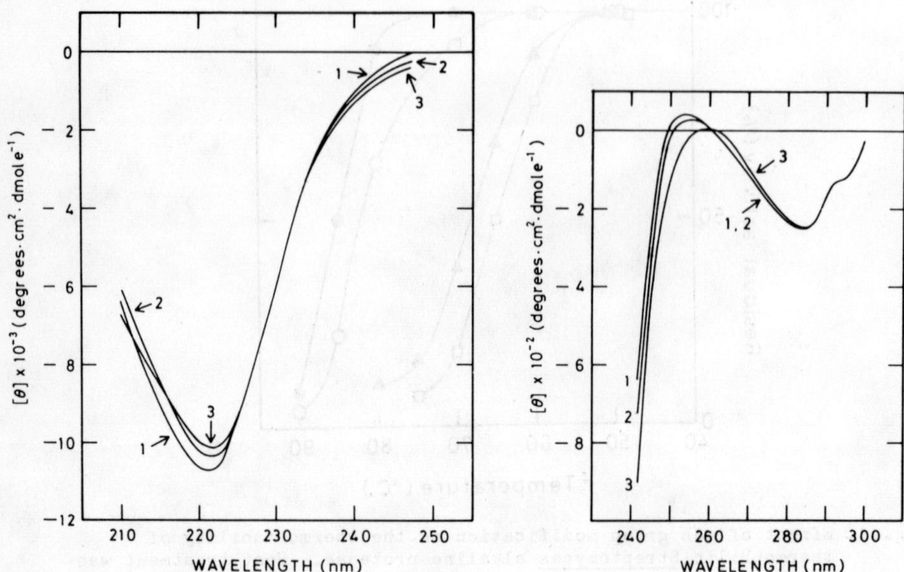

Fig. 2. Far-ultraviolet (left) and near-ultraviolet (right) CD spectra of native, p-MB-, and Hg-enzymes of thermophilic Streptomyces alkaline protease. Enzyme concentration was 0.30 mg per ml (far-UV range) or 0.50 mg per ml (near-UV range) in 0.05 M sodium acetate buffer (pH 5.6) containing 0.01 M calcium acetate. Curve 1, native enzyme; Curve 2, Hg-enzyme; Curve 3, p-MB-enzyme.

CONSIDERATION ON ROLE OF THE -SH GROUP

The assumption that the -SH group takes part directly in enzyme function as one of the catalytic sites is unlikely. It is probable that the enzyme has a homologous primary structure and the same charge relay system with subtilisin. Recent report by Borgia and Campbell (20) that the amino acid sequence around the active serine residue of this protease was identical with that of subtilisin supports further the above assumption.

Since this -SH group has been shown to be located at or near the active site, the slight conformational change caused by introducing a p-MB residue must have occured around the active site region to loosen its tight organization. Therefore, it is suggested strongly that the tight organization of active site is one of the decisive factors for supporting high thermal stability of this protease.

Fig. 3 is a schematic drawing of active site to interpret the role of the -SH group based on the data presented here. Supposing a section cut transversely at active site, the single cysteine residue will be present in the inner part near the position of catalytic sites or binding sites. The -SH group will be bound to another residue (A) via hydrogen bond to tighten the active site conformation just like a prop or a hinge.

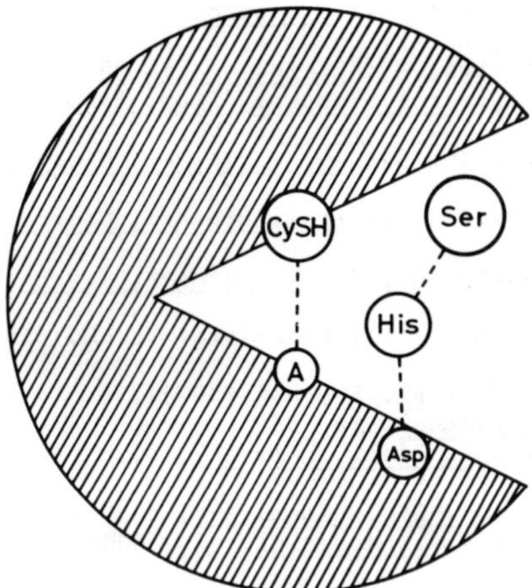

Fig. 3. Schematic drawing of active site. The area covered with oblique lines represents enzyme protein, and dashed lines indicate hydrogen bonds. See text for the details.

When a p-MB residue has been introduced to the -SH group, the bulky residue will provoke a slight conformational change which will either destroy the charge relay system or raise a steric hindrance. At the same time, it will cause a loosening of the rigid structure around the region, resulting in thermal instability of whole molecule.

On the other hand, in case of Hg-binding, conformational change will not be so large as to affect thermostability, or it may also be that mercury has been chelated to another residue to tighten the active site conformation. Thus it will be reasonable to consider that the -SH group participates in a tight organization of the active site through which a high thermal stability of the molecule is expressed.

Whether such a sulfhydryl participation in the structure and function of this alkaline protease is common to the ones from all genera of thermophilic actinomycetes, and whether it represents an event to gain thermostability during the evolutional pathway of alkaline protease molecules are interesting problems, but must await a future study.

Acknowledgments - We wish to express our sincere gratitude to Professors K. Arima and K. Imahori of Tokyo University for their valuable advice.

REFERENCES

1) Cross, T.: J. Appl. Bacteriol. 31, 36 (1968).
2) Mizusawa, K., Ichishima, E., Yoshida, F.: Agr. Biol. Chem. 28, 884 (1964).
3) Mizusawa, K., Ichishima, E., Yoshida, F.: Agr. Biol. Chem. 30, 35 (1966).
4) Mizusawa, K., Ichishima, E., Yoshida, F.: Appl. Microbiol. 17, 366 (1969).

5) Ovcharov, A. K.: Appl. Biochem. Microbiol. **2**, 272 (1966).
6) Konovalov, S. A., Dorokhov, V. A.: Appl. Biochem. Microbiol. **4**, 103 (1969).
7) Loginova, L. G., Negru-Vode, V. V.: Mikrobiologia **38**, 36 (1969).
8) Loginova, L. G., Tsaplina, L. A., Guzhova, E. P., Boltyanskaya, E. V., Kovalevskaya, I. D., Seregina, M.: Mikrobiologia **39**, 784 (1970).
9) Orlowska, B., Szewczuk, A.: Arch. Immunol. Ther. Exp. **20**, 543 (1972).
10) Desai, A. J., Dhala, S. A.: J. Bacteriol. **100**, 149 (1969).
11) Ovcharov, A. K., Rassulin, Yu. A.: Appl. Biochem. Microbiol. **4**, 485 (1968).
12) Ovcharov, A. K., Konovalov, S. A.: Appl. Biochem. Microbiol. **4**, 373 (1968).
13) Matsuo, M., Yasui, Y., Kobayashi, T.: J. Ferment. Technol. **45**, 860 (1967).
14) Nesterova, N. G., Lesnikova, A. V., Khokhlova, Yu. M.: Izv. Acad. Nauk SSSR, Ser. Biol. **1**, 153 (1972).
15) Okazaki, H.: J. Ferment. Technol. **50**, 405 (1972).
16) Okazaki, H.: J. Ferment. Technol. **50**, 580 (1972).
17) Mizusawa, K., Yoshida, F.: J. Biol. Chem. **247**, 6978 (1972).
18) Unpublished data.
19) Mizusawa, K., Yoshida, F.: J. Biol. Chem. **248**, 4417 (1973).
20) Borgia, P., Campbell, L. L.: J. Bacteriol. **120**, 1109 (1974).

PARTIAL CHARACTERIZATION OF A THERMOPHILIC ACTINOMYCETE RENNIN

S. LAXER, A. PINSKY and B. BARTOOV
Department of Life Sciences,
Bar Ilan University,
Ramat Gan, Israel

The thermophilic rennin isolated from a thermophilic actinomycete which was described previously by Laxer et al. (1972) and Pinsky et al. (1973) was further purified and characterized.

The original preparation was an acetone precipitate of a dialyzed ammonium sulphate precipitate (60% saturation). This material separated into at least five bands on electrophoresis in 7% acrylamide gel at pH 8.3 and at pH 6.0. The protein stain was amido black.

Molecular filtration on Sephadex G-100 gave the result shown in Fig.1. The conditions of the adsorption and elution are described in this figure.

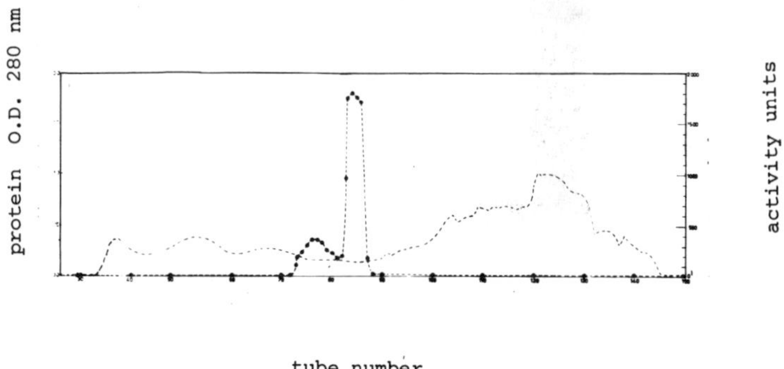

Fig. 1 Molecular filtration of an acetone precipitate of thermophilic rennin on Sephadex G-100. The column was 2.5x80 cm and 1.5 g material in 3 ml 0.01M calcium chloride was applied. The eluent was 0.01M calcium chloride.

Two active fractions may be seen. When Ca^{++} was omitted from the eluent, the eluted enzyme was inactive and added Ca^{++} could not restore its activity. The more active fraction on electrophoresis in acrylamide gel as described above showed one band only to have moved from the origin. Figure 2 shows the electrophoretic picture after acetone precipitation and after molecular filtration.

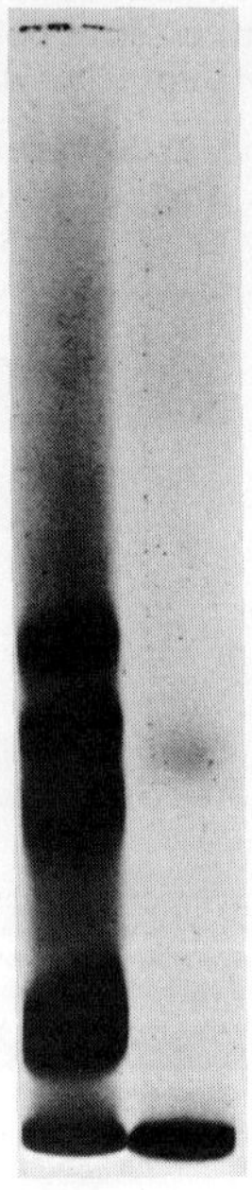

1 2

Fig. 2 Electrophoretic diagram of 1) acetone precipitate and 2) Sephadex G-100 eluate of the thermophilic rennin. The electrophoresis was in 7% acrylamide at pH 8.3. Staining was with amido black.

The yield and enrichment of ten batches of the active enzyme are given in Table I. The final product, after molecular filtration, represents 65% of the total original activity (both fractions). The increase in activity of the acetone precipitate over the previous stage in purification indicates the separation of an acetone soluble inhibitor.

Table I

PREPARATION OF CRUDE THERMOPHILIC RENNIN BY AMMONIUM SULPHATE PRECIPITATION

Stage	Volume Ml	Total Activity Units	Specific Activity Units/mg	Enrichment	Yield %	Protein Mg
Cell Free Culture	10000	8.2×10^4	2.92	1.0	100	2.8×10^4
Ammonium Sulphate Precipitate		8.6×10^4	3.73	1.27	104	2.3×10^4
Dialysis	1000	8.9×10^4	4.68	1.60	108	1.8×10^4
Acetone Precipitate		1.2×10^5	7.50	2.56	146	1.6×10^4
Sephadex G 100						
fraction 1	220	3.8×10^4	9×10^2	300	46	142
fraction 2	270	1.6×10^4	4×10^2	136	19	38

The hydrolyzed purified enzyme gave an amino acid analysis shown in Table II. Tryptophan was determined by the method of Spande and Witkop (1967). The molecule contains 78 amino acid residues with no cysteine. The amino end as determined by the dansyl chloride method as described by Gray (1967), was phenylalanine.

The amino acid analysis gave a calculated molecular weight of 9'700 as compared with 10'500 as determined by analytical ultracentrifugation or by dodecyl sulphate electrophoresis. The S value of 0.8 obtained for this molecule corresponds to this molecular weight.

The effects of various reagents on the activity of the purified enzyme are given in Table III. It may be seen that sulphhydryl reagents were ineffective, thus confirming the amino acid analysis. Specific trypsin and chymotrypsin inhibitors did not inhibit.

Table II

Thermophilic Rennin
Amino Acid Composition

Amino Acid	Molar Content
Lysine	5
Histidine	2
Arginine	3
Serine	3
Proline	9
Glycine	6
Alanine	6
Valine	6
Isoleucine	4
Leucine	5
Tyrosine	4
Phenylalanine	5
Tryptophan	2
Half Cystine	0
Aspartic Acid	6
Glutamic Acid	6
Threonine	4
Total	78

Further work showed that disopropylphosphofluoridate, 0.01M, did not inhibit. The serine residues in the molecule are therefore not in the active site. Dibromosuccinimide at a molecular ratio of 1:1 completely inhibited the clotting activity. Tryptophan is therefore in the active site.

At pH 6.4, the active enzyme caused a slight hydrolysis of a casein substrate as determined by absorption at 280 nm of the trichloracetic acid soluble extract after enzyme action. This hydrolysis was complete within 60 min.

Table III

Effect of various reagents on clotting activity of the thermophilic rennin like enzyme at 70°C when tested on a 10% skim milk preparation at pH 6.4. The reagent was incubated with the enzyme at 70°C for 30 minutes before testing.

Reagent	Concentration	% activity of original	Reagent	Concentration	% activity of original
NaCl	10^{-2}M	100	iodoacetamide	10^{-3}M	20
$CaCl_2$	10^{-2}M	130	iodoacetate	10^{-3}M	20
$MnSO_4$	10^{-2}M	115	EDTA	10^{-3}M	100
$BaCl_2$	10^{-2}M	100	p chloromer- curibenzoate	10^{-3}M	100
$CuSO_4$	10^{-2}M	50	L cysteine HCl	10^{-3}M	100
$ZnSO_4$	10^{-2}M	70	TPCK	10^{-3}M	100
$MgCl_2$	10^{-2}M	100	TLCK	10^{-3}M	100
$AgNO_3$	10^{-2}M	78	ovomucoid	10^{-3}M	100
			normal human serum	20%	0
			human albumin	20%	100
			soy bean trypsin inhibitor	10^{-3}M	100
			urea (0.2 to 8M)		0

TLCK N tosyl L lysine chloromethyl ketone

TPCK N tosylphenylalanine chloromethyl ketone

EDTA Ethylene diamine tetra acetic acid

When compared with calf rennin aside from the differences in molecular weight 10'500 as compared to 40'000 and the temperatures of activity, obligate thermophile with optimum at 75° as compared to body temperature, the thermophilic enzyme did not hydrolyze carbobenzyoxyglutamyl tyrosine which was hydrolyzed by the animal enzyme. In addition, the hydrolysis products of K casein, insulin, α insulin and β insulin are different as shown by paper chromatography. The results are shown in figures 3, 4, and 5.

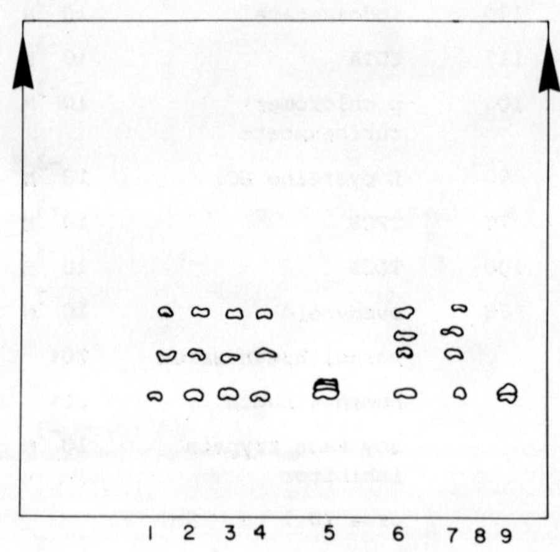

Fig. 3 Paper chromatography of the hydrolysis products of insulin by the thermophilic enzyme and by pepsin and calf rennin after 20 hours at pH 7.6. The paper was Whatman 4 MM and the solvent was n butanol, acetic acid, water at a volume to volume ratio of 40:80:40, respectively. Staining was with 0.2% ninhydrin in acetone. 1,2 insulin control, 3,4 animal rennin at 35°, 5 pepsin control, 6,7 thermophilic rennin, 8,9 pepsin at 35°.

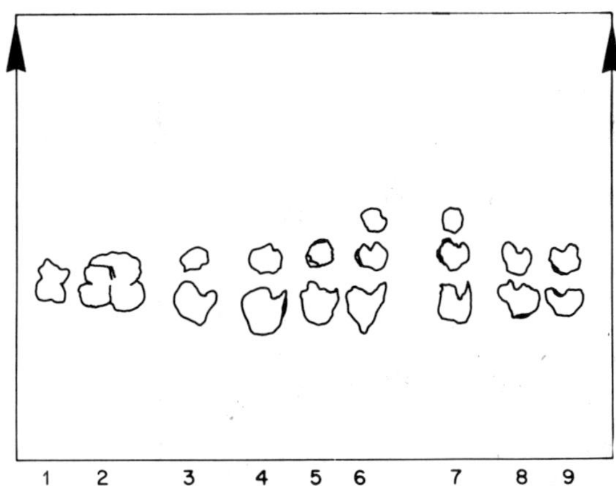

Fig. 4 Thin layer chromatography of the hydrolysis products of whole insulin after treatment with thermophilic rennin or pepsin at pH 3.5. The hydrolysis was for 24 hours at 35° for pepsin and at 50° for the thermophilic enzyme.

The chromatography was on thin layer kieselguhr plates. The solvent was n butanol : acetic acid : water 40:80:40, respectively. Staining was with 0.2% ninhydrin in acetone.

1,2 thermophilic enzyme 6 hours hydrolysis
3,4 thermophilic enzyme 12 hours hydrolysis
5 insulin control
6,7 pepsin 12 hours hydrolysis
8,9 pepsin control 12 hours incubation

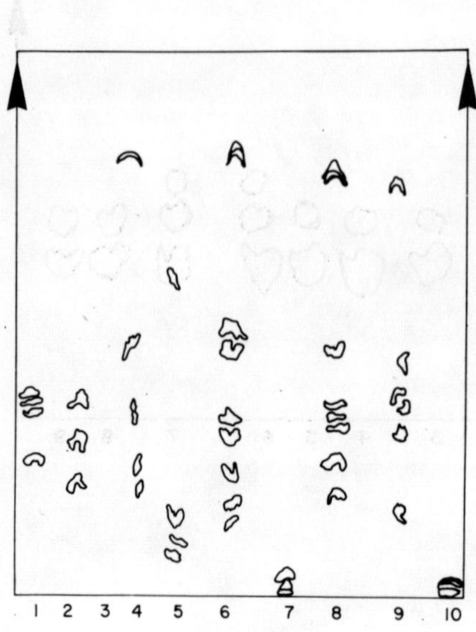

Fig. 5 Paper chromatography of the hydrolysis products of K casein, oxidized α insulin, and oxidized β insulin after treatment with thermophilic rennin or calf rennin.

The hydrolysis was done at pH 3.5 at 35° for the animal rennin and at pH 3.5 and at 50° for the thermophilic enzyme. The chromatography was on Whatman 4 MM paper and the solvent was n butanol, acetic acid and water at a volume ratio of 2:1:4, respectively. Staining was with 0.2% ninhydrin in acetone.

1,2 K casein 5 hours thermophilic, animal rennin, respectively
3,4 oxidized α insulin 10 hours animal, thermophilic rennin, resp.
5,6 oxidized β insulin 24 hours animal, thermophilic rennin, resp.
7 β insulin control 24 hours
8,9 K casein 20 hours thermophilic, animal rennin, respectively
10 K casein control

When 0.5 % skim milk was added to the original tryptic soy broth used as medium for growth of the actinomycete, a much greater quantity of the clotting enzyme was synthesized. This enzyme, however, differed from the original enzyme which could not be detected in the culture. Whereas the new enzyme had considerable proteolytic activity at pH 9.0 and none at all in the acid range, the alkaline protease in the original crude enzyme preparation could be separated from the milk clotting activity by molecular filtration. The original rennin showed hydrolytic activity at pH 3.5 and 2.8. The adaptive enzyme is being further investigated.

REFERENCES

Gray W.R., Methods in Enzymology (1967)
C.H.W. Hirs ed., vol. II, Academic Press p. 139

Laxer S., A. Pinsky and M.D. Tendler (1972)
Abst. 8th Meeting FEBS, Amsterdam 424

Pinsky A., S. Laxer and B. Bartoov (1973)
Abst. 9th International Congress of Biochem., Stockholm

Spande, T.F. and B. Witkop (1967)
Methods in Enzymology, C.H.W. Hirs ed., vol. II,Academic Press p. 496

AMYLASE ACTIVITY AND STABILITY AT HIGH AND LOW TEMPERATURE DEPENDING ON CALCIUM AND OTHER DIVALENT CATIONS

W. HEINEN and A.M. LAUWERS

Department of Exobiology, Faculty of Science, Toernooiveld, University of Nijmegen, Nijmegen, The Netherlands

INTRODUCTION

The extracellular amylase produced by *Bacillus caldolyticus* at 70 C has previously been characterized as a Ca-dependent enzyme (Grootegoed et al., 1973); Heinen and Lauwers, 1975). Furthermore it had been shown that the enzyme consists of subunits with a MW of less than 10.000, which could probably be defined as microenzymes (Schenk and Bjorksten, 1973) because they are enzymically active in the low temperature range. From the ultrafiltration experiments we had also concluded that the subunits associate to a higher MW form when Ca is added to the low MW fraction, and that it is this divalent cation which binds the subunits. All bacterial amylases from mesophilic producers have been found to need Ca for full activity, and many contain also Zn or other metals (Vallee et al., 1959). But in general the role of Ca seems not to be specific, so that it can be substituted by various other cations (Vallee et al., 1959, Smolka et al., 1971, Levitzki and Reuben, 1973). So for us there remained the question whether the effect of Ca on the amylase from *B. caldolyticus* would also be unspecific. For this reason we have studied the effect of various cations at different temperatures. From these experiments we could expect to get some insight into the structural properties of the enzyme.

MATERIALS and METHODS

The organism was grown at 70 C in batch cultures as previously described, and for the preparation of the enzyme we also followed the methods reported earlier (Grootegoed et al., 1973, Heinen and Lauwers, 1975). The same holds for the ultrafiltration method applied. In order to remove as much Ca as possible from the enzyme, it was either subjected to prolonged dialysis, or washed with buffer on a UM-05

Diaflo membrane. According to previous experiments one clear indication for a sufficient removal of Ca was the observation that from a certain point on the enzyme would pass through a UM-10 membrane, which retains molecules that are bigger than 10.000 MW. The low MW fraction which then accumulates on the UM-05 or UM-2 membrane was used throughout the experiments reported here. The methods applied to allow the various cations to react with the subunit fraction are given in the text.

For determination of enzyme activity in absence or presence of Ca or other cations we used again the method of Pfueller and Elliott (1969). Depending on the question under study these assays were carried out in a wide temperature range, from 30 C to 80 C.

RESULTS

1. Amylase activity, and the substitution of Ca at high temperature.

With the Ca-dependence of the amylase at its optimum temperature of 70 C being well established by our previous experiments (Heinen and Lauwers, 1975), the following question was of course whether the function of the Ca ion could be replaced by other divalent cations. Because Sr is known as a substitute for Ca in many biological systems (Heinen, 1974), this element could be expected to be the most likely candidate, and was chosen therefore for the preliminary experiments. In this first approach the amount of Ca added to the enzyme was gradually lowered from 100.0 to 20.0 μg of $CaCl_2$ and substituted by increasing amounts of Sr. These experiments seemed to indicate that Ca could indeed be replaced by Sr to a certain extent. The following series was then set up in such a way that either Ca or Sr would be allowed to react with the enzyme in a 2 minute preincubation period at 70 C, and that the substituting cation was added immediately afterwards. A typical result of this type of experiments is given in Fig. 1: We found that after preincubation with either 55 or 140 μg of $CaCl_2$ and the addition of 140 or 55 μg of $SrCl_2$ the reaction curves were almost identical. In the reversal of this experiment, pre-incubation was done with 55 μg of $SrCl_2$, followed by the addition of 140 μg of $CaCl_2$. This also showed full activity. But when the amount of $SrCl_2$ present during pre-incubation was increased to 140 μg the reaction levelled off after 5 min., in the same way as in absence of Ca ions. Because 55 μg of $CaCl_2$ had been added after the pre-incubation period, and this amount was found to be sufficient for a normal reaction, we had to conclude that the high concentration of $SrCl_2$

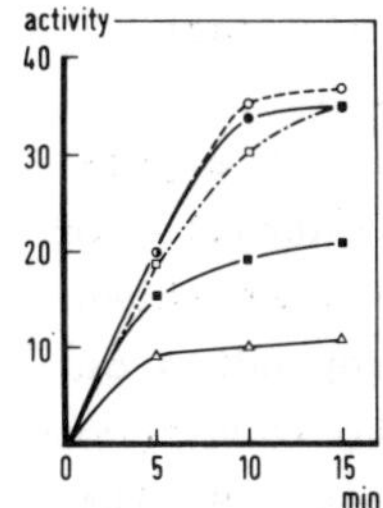

Fig. 1: Amylase activity after pre-incubation with different concentrations of Ca or Sr, and assaying after addition of either Ca or Sr (●—● = 55 µg $CaCl_2$ during pre-incubation, 140 µg $SrCl_2$ in assay; ●---● = reversed situation; o--o = 55 µg of $SrCl_2$, then 140 µg of $CaCl_2$ added before assay; o—o = reserved situation; Δ ... Δ = no addition).

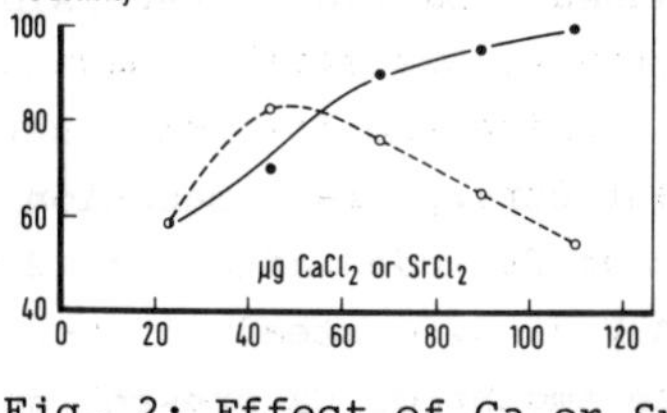

Fig. 2: Effect of Ca or Sr concentration (solid or broken line resp.) during pre-incubation on the activity determined after addition of 80 µg Sr or Ca

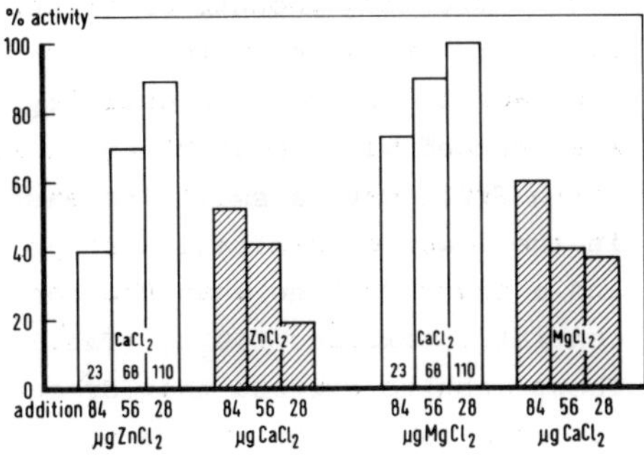

Fig. 3: Amylase activity after pre-incubation with Ca, and determination after addition or Zn or Mg (light bars), or pre-incubation with either Zn or Mg and addition of Ca (shaded bars).

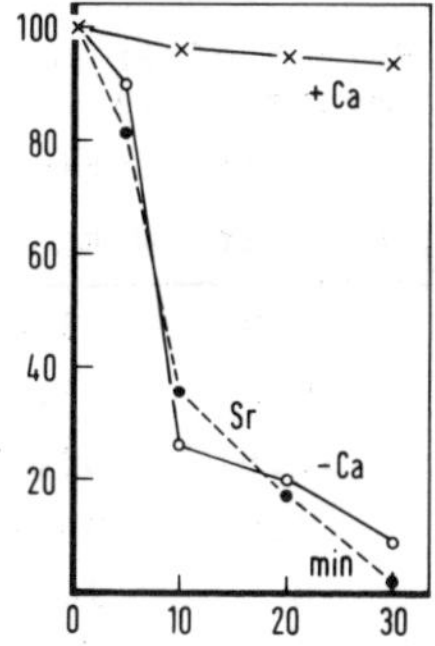

Fig. 4: Amylase stability in absence or presence of Ca or Sr.

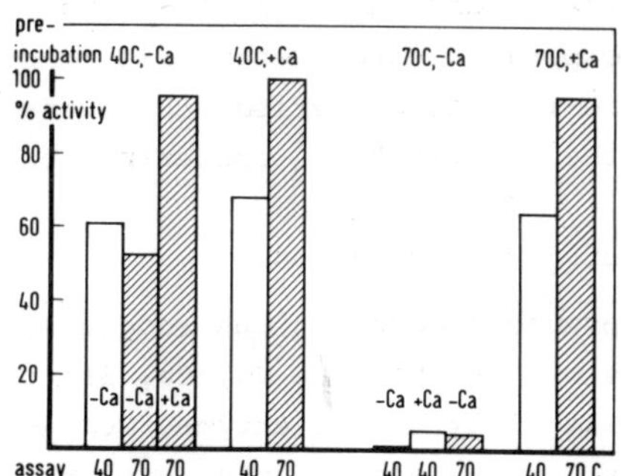

Fig. 5: Enzyme activity after a 30 min. pre-incubation period in absence or presence of Ca at either 40 or 70 C, followed by determination at either 40 C (light bars) or 70 C (shaded bars) again without or with Ca.

caused a partial irreversible loss of activity, though the rest activity was still somewhat higher than in absence of Ca.

During the following series, the amount of either Sr or Ca present during pre-incubation was varied, whereas the concentration of Ca or Sr added after the 2 min. period was kept constant. As shown in Fig. 2, a steady increase of activity was observed depending on the amount of Ca present during pre-incubation. With the lowest concentration of $SrCl_2$ (23 µg) we found almost the same value as with the corresponding $CaCl_2$ concentration. A distinct increase was observed with 45 µg $SrCl_2$, but further increase of the Sr concentration lead to a successive decrease of activity.

Tab. 1: Effect of Sr or Ca on amylase activity

µg $SrCl_2$ or $CaCl_2$	activity with Sr	activity with Ca
27.7	16.0	53.0
55.5	16.2	66.7
83.2	18.6	100.0
111.0	23.9	96.2
139.0	8.0	98.5

When increasing amounts of either Ca or Sr were given without further additions, maximum activity was reached with about 90 µg $CaCl_2$, while $SrCl_2$ gave a small increase in the lower concentration range, and a sharp decline when the concentration exceded 110 µg (Tab.1). These results suggest that Sr can act as a substitute for Ca to a very limited extent only, and that it has an inhibitory effect at higher concentrations.

In analog experiments with Zn and Mg quite similar results were obtained with the latter (Fig. 3), whereas Zn showed a certain inhibitory effect: After pre-incubation with Ca and addition of Zn, the activity stays lower than in the corresponding experiments with Mg or Sr, and if Zn is present during the pre-incubation period, the activity declines to only 14% in case of the highest $ZnCl_2$ concentration applied.

2. Amylase stability and the effect of divalent cations.

The foregoing results as well as previous observations suggested that the competition for binding sites by cations like Mg, Ca, Sr, or Zn might have a severe effect on the stability of the enzyme. Comparative experiments with either Ca, Sr, or Zn, added during pre-incubation periods of 10, 20, and 30 min. at 70 C, and successive determination of the activity at 70 C made it clear that in presence of Sr the enzyme looses its activity almost as fast as in absence of Ca (Fig. 4). When Zn was used instead of Ca or Sr, the same effect was observed. Though these data show that at 70 C an absolute requirement for Ca exists in order to stabilize the enzyme at this high temperature, the question remained whether this would also be the case at lower temperatures.

With regard to this problem we did the following experiment: The enzyme was pre-incubated at 40 C for 30 min., either in absence or presence of Ca, and then assayed at 40 and 70 C. If the pre-incubation had been performed without Ca, the ion was added in the assay procedure. The same treatment was then carried out at 70 C, again without or with Ca, and determination of the activity at 40 and 70 C, with Ca added if necessary. As indicated in Fig. 5, there was no loss of activity after the pre-incubation at 40 C in absence of Ca, as can be seen from the data obtained with the assay at 40 and 70 C. After pre-incubation at 70 C in presence of Ca the enzyme also showed the rates of activity expected at 40 and 70 C. If, however, Ca was omitted during the 70 C pre-incubation, we found a dramatic loss of activity both at high and at low temperature, which even at 40 C could not be restored by addition of normally sufficient concentrations of the ion. This means that once the enzyme has become unstable as a result of the 70 C treatment in absence of Ca, it is irreversibly denatured. Addition of Sr during the 30 min. treatment at 70 C gave the same results as obtained with the omission of calcium.

These results lead us to a series of experiments in which the stability of the enzyme in absence or presence of Ca was determined in relation to temperature. In the first of these experiments we determined the activity under these conditions in the range from 30 to 80 C. As shown in Fig. 6 a, we found that up to 55 C the difference between the activity obtained in absence or presence of Ca is never more than 8 - 12%. From 60 C on, however, the effect of the omission of Ca becomes very evident, with a difference of 50% at 70 C. This effect can be enhanced by a short pre-incubation which loweres the stability of the enzyme (Fig. 6 b): The decrease of the activity beyond 60 C is more drastic than without the short pre-incubation.

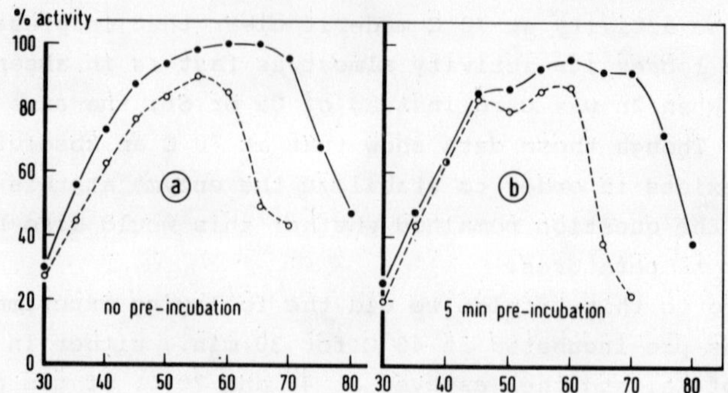

Fig. 6: Enzyme stability at different temperatures, in presence (solid line) or absence (broken line) of Ca. Determinations were either carried out directly (a), or after a 5 min. pre-incubation of the enzyme at the given temperature (b).

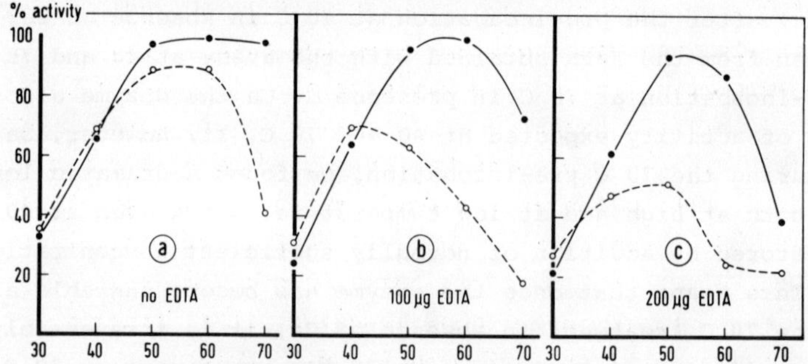

Fig. 7: Effect of two concentrations of EDTA (b and c) on amylase activity at different temperatures, in presence (solid lines) or absence of Ca (broken lines). The controle (a) was done without EDTA.

In a further variation of this experiment we added two concentrations of EDTA (Fig. 7, b and c), and found that up to 40 C there is very little effect on the activity. At higher temperatures however, we observed a fast decline, due to the binding of Ca ions by the chelator, which was also evident for the samples with added Ca, where the concentration of the ion obviously becomes suboptimal. Finally we checked increasing amounts of $CaCl_2$ in presence of either 100 or 200 µg of EDTA, within the temperature range from 40 to 70 C (Tab. 2): With 100 µg of EDTA the decline of the activi-

Tab. 2: Correlation between temperature, Ca-concentration and EDTA-concentrations. Values given in % of maximum activity.

EDTA-concentrations:	100 µg				200 µg			
µg $CaCl_2$	40	50	60	70	40	50	60	70 C
0	49.0	51.1	30.4	14.2	35.2	36.0	22.1	5.8
10	62.7	67.2	51.0	19.0	37.0	47.5	27.7	14.0
20	68.6	82.0	84.7	25.3	46.5	51.5	37.4	12.0
40	68.3	85.5	90.2	35.0	45.8	68.7	47.5	16.8
60	69.2	96.3	100.0	46.8	52.7	74.0	61.0	15.1

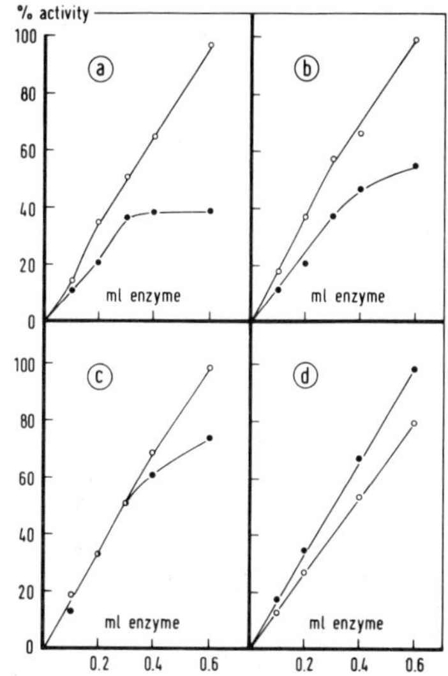

Fig. 8: Relationship between enzyme concentration, Ca supply, and activity at 40 C (open circles) and 70 C (circles). The Ca concentrations in the four plots are a = 15, b = 20, c = 60, and d = 120 µg $CaCl_2$

ty starts at 50 C with either no or 10 µg $CaCl_2$ present, whereas with all higher concentrations the decrease begins at 60 C. With twice the amount of EDTA we found that the decrease starts at 50 C with all the Ca concentrations tested. These results suggest that the lower the temperature the less dependent the enzyme becomes on the availability of Ca ions. Together with the data from Fig. 6 and 7, one can conclude from these values that at least up to 40 C neither the stability nor the activity of the amylase is markedly affected by the omission of Ca ions. Beyond this temperature, however, the stability of the enzyme becomes increasingly dependent on the presence of Ca, which is reflected by an increasing decline of activity.

3. Relationship between temperature, ion concentration and enzyme stability.

To further emphasize the foregoing point, we set up an experiment in which the enzyme concentration was varied, and the supply of Ca kept constant for each determination at 40 and 70 C. This was carried out for Ca concentrations from 5 to 140 µg. Four graphs out of this series are shown in Fig. 8. They indicate that at 40 C very small amounts of Ca are sufficient for full activity, for the entire amount of enzyme protein that is present in the sample. At 70 C the Ca-limitation becomes very evident: The lower Ca concentrations (up to 15 µg, and still quite typical with 20 µg $CaCl_2$) applied in these experiments are sufficient only to stabilize a limited portion of the enzyme concentration given, so that further increase of the protein concentration does not result in higher activity. That this is due to a Ca limitation can be deduced from the data obtained with higher Ca concentrations: The amount of enzyme that can be stabilized clearly increases with the successive increase of available Ca ions. As summarized in Fig. 9 (with 0.6 ml enzyme applied), these data show that 15 µg of $CaCl_2$ are sufficient to obtain full activity at 40 C. At 70 C, however, there is a drastic increase up to 20 µg, followed by a steady increase with the higher Ca concentrations applied. About 120 to 140 µg are necessary at 70 C to reach the level of activity obtained with 15 µg at 40 C. This means that the less Ca is available to support stability, the more protein denatures at 70 C, but that at 40 C the stability of the enzyme is practically indepedent of Ca according to the trace amounts of the ion which are needed to obtain full activity at this tempereture. The next question then was whether Ca could be substituted by other cations at lower temperature.

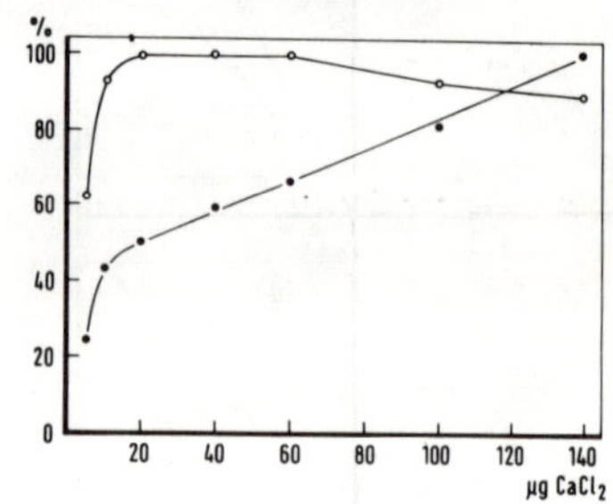

Fig. 9: The effect of Ca concentrations on amylase activity at 40 C (open circles) and at 70 C (circles) with a given enzyme concentration.

To determine the effect of cations other than Ca on the enzyme at lower temperature, we decided to do these experiments at 50 C. At this temperature, the enzyme is still stable even in presence of 100 µg of EDTA, which was added in order to eliminate the small amounts of Ca still present in this enzyme preparation. Increasing concentrations of either Na, Mg, Ca, Sr, or Zn were added to the enzyme before starting the assay, and the determination itself was carried out without the addition of Ca ions, so that solely the activity with the particular cation was measured. From the observed data (Fig. 10) we can conclude, that Ca and Zn have an almost equal effect, while Sr has to be present in higher concentrations in order to reach about the same level of activity. With Mg the activity reached only about 80% of the maximum obtained with either Ca or Zn, and Na obviously has no effect, because the values obtained in absence of any of these cations were identical with these data. These results suggest that at a temperature of 50 C and below, the stability of the enzyme does not depend on the presence of Ca ions. Full activity is reached with very small concentrations (20 µg) of either $CaCl_2$ or $ZnCl_2$, which equals 7.2 µg of Ca or 9.6 µg of Zn. In contrast with the absolute requirement for Ca at higher temperature in order to keep the enzyme stable, the effect of Ca at lower temperature is unspecific. Under these conditions the ion can be substituted by other divalent cations as for instance Zn, albeit that higher concentrations might be necessary, which would apply for Sr, or that the activity is maintained on a lower level, as observed for Mg. But even without either of these cations present, the enzyme still operates at about 60% of its maximum activity at 50 C.

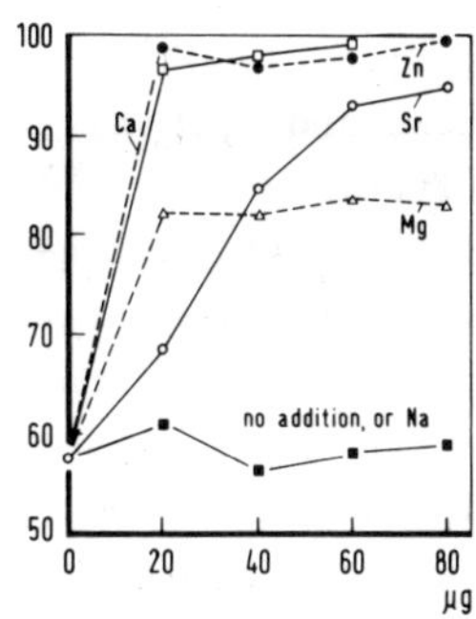

Fig. 10: Effect of increasing concentrations of various cations on the amylase activity at 50 C.

DISCUSSION

Bacterial amylases represent a group of metallo-enzymes which all contain calcium bound to the protein moeity (Vallee et al., 1959; Fischer and Stein, 1960). In many cases Zn has been shown to accompany Ca, and in some amylases the Ca can be partly or totally replaced by other divalent cations. The lanthanides have on the other hand been shown to be unable to replace Ca, with the exception of Gd^{3+} which could be bound by a B. subtilis amylase (Levitzki and Reuben, 1973). It is generally assumed that Ca or it's substitutes act as cofactors for the enzyme. But the tightly bound Ca and Zn, which cannot be removed from the protein by dialysis, or replaced by other cations, are also thought to be to a certain extent responsible for the structural integrity of the enzyme. Our experiments with the amylase from a thermophilic producer strongly suggest that in this case the metal acts in both ways, but that at high temperature it's involvement in maintaining the appropriate structure of the enzyme is much more pronounced than previously observed with amylases from mesophilic producers.

The experiments concerning the substitution of Ca at 70 C and the resulting consequences for the activity of the enzyme show that Ca is indeed essential for full activity of the caldo-active enzyme. If other cations compete for the binding sites normally taken by Ca, the final activity obviously depends on the number of sites occupied by cations other than Ca: If only a few sites are taken by non-Ca cations, and the majority of the sites can be occupied by Ca, this will result in only a minor effect on the enzyme activity. But if Ca is in the minority, and most sites are captured by another cation, the activity declines drastically. This is also observed when competing cations are available in excess: In this case, the sites which are occupied by another cation cannot be recaptured by Ca, even if it is supplied in excess, and the functional breakdown is then irreversible.

Since we knew from previous experience that the enzyme requires Ca for its stability, we could assume that the observed loss of activity was actually due to the instability of the enzyme in absence of Ca ions, and that the high-temperature enzyme thus had an absolute requirement for Ca specifically. The question however, remained, at which temperature instability begins, and whether Ca could be replaced by other ions at lower temperature. From our experiments concerning this question we can draw the following conclusions: First of all the Ca requirement for stability at 70 C is an absolute one, and the ion cannot be replaced by others. But on the other hand the enzyme is stable at 40 C in absence of Ca, and if supplied with Ca will exhibit full activity at 70 C. When the temperature range from 30 to 80 C was determined, we found almost no difference between the Ca-free and Ca-containing samples up to 45 C, very little instability in absence of Ca up to 60 C, but a fast decrease of the stability beyond this temperature. Addition of a chelator in order to bind the trace amounts of Ca in the preparation which cannot be removed by dialysis, showed, that the crucial point where the enzyme becomes unstable is between 40 and 45 C. This means that the amylase is not a thermostable enzyme in the classical sense, which by definition are fully active at their low optimum temperature after being submitted to a heat treatment. The Ca-free amylase is, in the contrary, just as thermo-sensitive as any enzyme from a mesophilic source, but is in its caldoactive state, i.e. in presence of Ca, far more active at its 70 C optimum temperature than at low temperature.

That the supply with Ca is the limiting factor for the formation of the caldoactive state of the enzyme, was proven by the experiments where increasing amounts of enzyme were checked for their activity with a given concentration of Ca. With the stepwise increase of the Ca concentration it is obvious that the amount of protein that can be stabilized increases, because more and more of the subunits can be associated into the thermostable form of the enzyme, while the unassociated excess of protein denatures. The trace amounts of Ca or other divalent cations necessary to obtain the level of activity expected, at 40 C suggest that these cations at this temperature function primarily as cofactors. At the lower temperature range obviously other binding forces are responsible to achieve the correct confirmation for the thermosensitive state of the enzyme. Alternatively, the small subunits might not be assembled to a greater molecule, and be fully active in their low molecular weight form as "microenzymes" (Schenk and Bjorksten, 1973). The assumption that at low temperatures Ca is not involved with the structural arrangement of the enzyme, is supported by the substituting experiments at 50 C, which showed that the effect of Ca at this temperature is unspecific, and that it can be replaced by a variety of other divalent cations.

Two main conclusions can be drawn from our results: 1) That Ca plays a double role for the functioning of the amylase. The first one is to act as a cofactor at temperatures up to at least 50 C, in the same way as generally assumed for amylases from mesophilic producers. In this function it is unspecific, and can be substituted by other divalent cations. At high temperature, however, it takes over its second function, which is to stabilize the enzyme by participating in the configurational arrangement of the protein, leading to the formation of the caldoactive state of the enzyme. In this role, it cannot be replaced by other cations. Most likely it acts simultaneously also as a cofactor at high temperature, and this function can most probably be taken over by other cations, provided that sufficient amounts of Ca are available to ensure the structural integrity of the enzyme. 2) The enzyme can exist in two forms: One is the caldo-active, thermostable, Ca-dependent state, and the other the Ca-independent thermolabile state. One can assume that binding forces that cannot act at 70 C, such as H-bridges, could at low temperature take over the role of Ca as a mediator for the correct configurational arrangement.

ABSTRACT

Since the extracellular amylase from _B. caldolyticus_ had been characterized as a Ca-dependent enzyme, we wanted to determine the function and specificity of this ion, and it's possible significance for the thermophilic properties of the enzyme.

The first question concerned the substitution of Ca with regard to the activity at 70 C. Our data show that Sr or other divalent cations can only to a limited extent replace Ca, and that only in presence of Ca full activity can be achieved. When the majority of the binding sites of the subunits are occupied by cations other than Ca, an almost total loss of activity results.

The reason for this kind of inactivation was found in the responsibility of Ca for the stability of the enzyme at high temperature. The omission of Ca then leads to an irreversible denaturation of the enzyme, so that after this kind of treatment no activity is detectable at either high or low temperature, which means that the amylase is not a thermostable enzyme in the classical sense. The stabilizing effect of Ca could not be substituted by any other of the cations tested.

Experiments concerning the stability of the enzyme at various temperatures in absence or presence of Ca revealed that the enzyme is stable up to 55 C as long as trace amounts of Ca are available. If these are omitted by a chelator, the enzyme becomes unstable between 40 and 45 C. Experiments in which a certain protein concentration range was tested at 40 and 70 C with given concentrations of Ca showed that the stability, and with it the activity, at high temperature is directly related to the amount of Ca available: The more the Ca supply is limited, the less enzyme protein can be kept in the right configuration, and irreversibly denatures as a result. At low temperature, however, the enzyme becomes almost independent of Ca, and the small amounts necessary to obtain full activity can be replaced by other divalent cations.

The main conclusions concerning the role of Ca are that at high temperatures it participates in achieving the correct configuration of the thermostable form of the enzyme, and that in this function it is irreplacable. At low temperature it acts as an unspecific cofactor. The other conclusion is that the amylase can exist either as a caldo-active, thermostable, Ca-dependent enzyme, or in a Ca-independent thermostable state with a lower activity.

LITERATURE

Fischer, E.H., Stein, E.A.; in: P.D. Boyer, H. Lardy and K. Myrbäck, Eds., The Enzymes, Vol. IV, pp. 313-343, New York - London: Academic Press 1960.

Grootegoed, J.A., Lauwers, A.M., Heinen, W.: Arch. Mikrobiol. 90, 223 - 232 (1973).

Heinen, W.: Biosystems 6, 133 - 151 (1974).

Heinen, W., Lauwers, A.M.: Arch. Mikrobiol., in press (1975).

Levitzki, A., Reuben, J.: Biochemistry 12, 41 - 44 (1973).

Pfueller, S.L., Elliott, W.H.: J. Biol. Chem. 244, 48 - 54 (1969).

Schenk, R.U., Bjorksten, J.: Finska Kemists. Medd. 82, 26 - 46 (1973).

Smolka, G.E., Birnbaum, E.R., Darnall, D.W.: Biochemistry 10, 4556 - 4561 (1971).

Vallee, B.L., Stein, E.A., Sumerwell, W.N., Fischer, E.H.: J. Biol. Chem. 234, 2901 - 2905 (1959).

Role of Calcium Ion in the Thermostability of α-Amylase Produced from *Bacillus Stearothermophilus*

Katsuhide Yutani
Institute for Protein Research, Osaka University, Yamadakami, Suita, Osaka, Japan

INTRODUCTION

In 1961, Campbell and Manning provided one clue to elucidate the molecular basis for thermostability of the enzyme from thermophile(1-3). They reported that the α-amylase of *B. stearothermophilus* in the native state exists in a semi-random or random coiled and well hydrated molecule with slight extent of secondary structure formed by disulfide bonds (2). This less-ordered structure was postulated as the reason for thermostability of the enzyme.

Pfueller and Elliot (4), and Ogasahara *et al.* (5-7), however, have reported that the α-amylase produced from *B. stearothermophilus* does not have such a semi-random coiled structure as reported by Campbell and Manning, and the enzyme resembles many other α-amylases which are known to have molecular weights around 50,000 and to need calcium ion for stability. Then, why is the enzyme from *B.stearothermophilus* thermostable?

α-Amylases isolated from various sources have been shown to contain a few firmly bound calcium ions (8) and to be stabilized by the addition of calcium (9). Imanishi has shown that the contribution of calcium binding to the stability of α-amylase is a half to a quarter of the total free energy change for folding of unfolded enzyme and the bound calcium considerably enhances the stability of the native conformation of α-amylase from *B. subtilis* (9). The maintenance of the stability of an enzyme molecule by calcium also has been reported for *B. subtilis* neutral protease (10) and thermostable thermolysin (11) and some other enzymes.

Although it is well known that the calcium ion plays an important role in the thermostability of the α-amylase from *B. stearothermophilus* as shown above (4,6), the mechanism has not yet been elucidated. In the present work, we studied the origin of thermostability of α-amylase from *B. stearothermophilus* (thermophile α-amylase) by examining of the following items. 1) The denaturation of thermophile α-amylase in urea or guanidine hydrochloride (GuHCl) was compared with that of α-amylase from *B. subtilis* (mesophile α-amylase). 2) The stability of the calcium-free thermophile and mesophile α-amylases obtained by dialysis was examined. 3) The numbers of the calcium ions bound to a molecule of α-amylase in various conditions were estimated by equilibrium dialysis technique and the association constants of the calcium ion to the enzyme were determined.

EXPERIMENTAL

Materials --- The preparation of thermophile α-amylase from *B. stearothermophilus* grown at 43°C were described in the previous paper (12). Crystaline *B. subtilis* α-amylase was supplied by Daiwa Kasei Co. Ltd. GuHCl (Guranteed reagent, Nakarai) and urea (Reagent grade) were used after recrystalization from 90 % aqueous methanol and 70 % aqueous ethanol, respectively. All other reagents were the best ones commercially available.

Removal of Calcium --- Removal of calcium from thermophile and mesophile α-amylases was done by dialysis in sodium EDTA. Enzyme solutions (about 0.1 mg/ml) were enclosed in Visking cellulose tubings which were treated for 30 min in boiling water containing 0.01 M EDTA before use. The dialysis bags were immersed in a minimum of 50 volumes of pH 7.0, 0.01 M EDTA (4°C) which was renewed at least once every 24 hrs. Calcium contents of thermophile and mesophile α-amylases were below 0.1 moles calcium per mole of enzyme after dialysis for 5 and 7 days, respectively.

Analysis of Calcium --- The amount of calcium was determined with a Seiko atomic absorption spectra photometer SAS-721 at 422.7 nm. A standard solution of $CaCl_2$ was prepared from CaO by neutralization with HCl. The concentration of the standard calcium solution was in the range of 10^{-6} to 10^{-4} M.

Equilibrium Dialysis --- A teflon cell for dialysis was used. The cell consisted of two compartments which were separated by Visking cellulose membrane; one for protein and the other for calcium ion. The volume of each compartment was 2 ml. Before use, the membrane was treated with boiling water containing 0.01 M EDTA for 30 min. The concentration of enzymes was around 10^{-5} M throughout the dialysis. Enzyme and calcium for the dialysis were dissolved in the pH 8.0, 0.1 M Tris-Cl buffer pretreated with a Dowex chelating resin.
The cell was agitated in a thermostat for 24 hrs to attain equilibrium with respect to the calcium ion concentration.

Circular Dichroism --- Circular dichroism (CD) was measured by a JASCO J-20 automatic recording spectropolarimeter. Constant temperatures under measurements of CD were maintained by circulation of water in a cell holder from a thermostated bath. Specially, the data as shown in Fig. 5 were treated by a JEC-5 spectrum computer lining on JASCO J-20 (13). The measurements were carried out by the following conditions; Time const. 4 sec, CD scale 0.001, Smoothing point 25, Number of accumulating 5. Total experimental time was 2 hrs and the standard deviation of the error was ± 2.1 deg cm^2/dmol.

The concentration of thermophile and mesophile α-amylase were estimated from the absorption at 280 nm assuming $E_{1\ cm}^{1\ \%}$ = 28.7 and 25.6, respectively.

Amylase activity was determined by the measurment of saccharifying power using amylose as a substrate, essentially according to the method of Fuwa (14).

RESULTS and DISCUSSION

Effect of Temperature on Enzymatic Activity --- The enzymatic activities of thermophile and mesophile α-amylases were tested at various temperatures. Fig. 1 shows the temperature profiles of the activities of these enzymes below 25°C. The cross over point, where the activity efficiencies of both enzymes are the same, was observed at 17°C. This shows that the efficiency of activi-

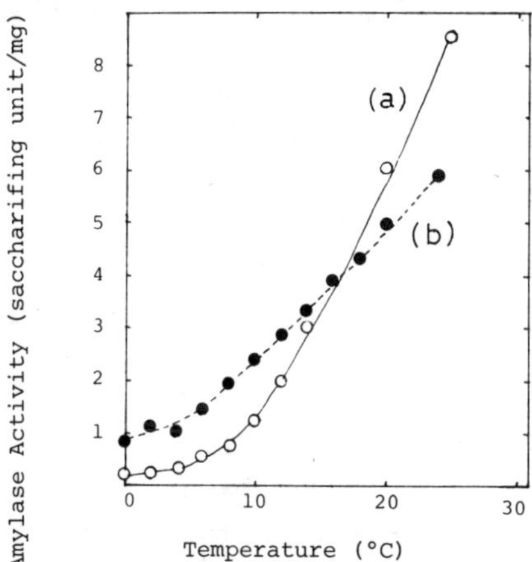

Fig. 1 The effect of temperature on the activities of thermophile (a) and mesophile (b) α-amylases. The enzymatic activity was measured at pH 5.0 for 4 min at each temperature, using 0.17 % amylose as a substrate. Concentrations of two enzymes under assay were 9.7 x 10^{-9} M.

ty for thermophile α-amylase is higher than that of activity for mesophile α-amylase over 17°C, but it is reverse below 17°C. The value of maximal activity for thermophile α-amylase at optimal temperature (65°C) was 14.5 times larger than that at the cross over point (17°C). But, the value of maximal activity for mesophile α-amylase at the optimal temperature (50°C) was 2.6 times larger than that at 17°C. Moreover, the value of enzymatic activity for thermophile and mesophile α-amylases at the respective optimal temperature was 200 and 10 times larger, respectively, than those at 4°C. The enzymatic activity of thermophile α-amylase was extremely depressed at lower temperature around 4°C. This may mean that the structure of thermophile α-amylase is more rigid than that of mesophile α-amylase at lower temperature. In other words, the rigid conformation of thermophile α-amylase may change to the flexible state which is essential to enzymatic function at higher temperatures.

Denaturation by GuHCl at Different Temperatures --- It has been reported that the α-helical conformation of thermophile α-amylase is not destroyed in the presence of 8 M urea at room temperature(6). The variation of the ellipticity at 225 nm with GuHCl concentration is given in Fig. 2 for mesophile and thermophile α-amylases at different temperatures. When the enzyme was incubated in various concentration of GuHCl at 10°C, the value of Θ in the denaturation curves showed itself to be constant at higher concentration of GuHCl, as if the enzyme was in a meta-stable state. The values of Θ at the plateaus for thermophile and mesophile α-amylases reduced with time to be half of the native one after 3.1 and 1.9 days, respectively. After incubation for one and four days in these conditions, apparent transition midpoints, as defined by the concentration of GuHCl at midpoint between the native and denatured (or the plateau) state from the denaturation curves, of mesophile α-amylase were 0.46 and 0.41 M after one and four days, respectively. Those of thermophile α-amylase were 1.42 and 1.12 M after one and four days, respectively. At 40°C, the apparent transition midpoints of thermophile and mesophile α-amylases decreased to be 0.69 and 0.06 M, respectively. These denaturations were irreversible.

From these results, it is evidently found that thermophile α-amylase is more resistant to GuHCl than mesophile α-amylase at low and high temperatures, and both enzymes in lower temperatures are more resistant to GuHCl than they are in higher temperatures.

Effect of Calcium Ion on the Stability --- When thermophile and mesophile α-amylases were incubated in various concentrations of GuHCl in the presence of calcium ion, the transition midpoint of denaturation of both enzymes shifted

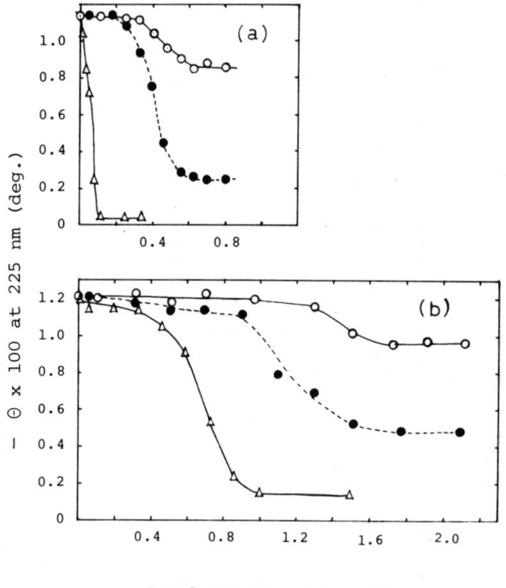

Fig. 2 The dependence of the ellipticity at 225 nm of mesophile (a) and thermophile (b) α-amylases on the concentration of GuHCl. The enzyme concentrations were 3.1×10^{-7} (a) and 3.5×10^{-7} (b) M, respectively. Open and closed circles represent the data for incubation for one and four days, respectively, at 10°C and pH 8.0. Triangles represent for one day at 40°C.

to higher concentration of GuHCl as shown in Fig. 3. In mesophile α-amylase, the apparent transition midpoint of denaturation was 0.24 M in the absence of calcium ions, but this value was increased to be 0.38 and 0.54 M in the presence of 10^{-4} and 10^{-3} M $CaCl_2$, respectively. Morepver, while mesophile α-amylase was completly denatured after incubation in 0.4 M GuHCl for 20 hrs, the enzyme was not completely denatured in the presence of calcium ions. In thermophile α-amylase, the effect of calcium ion was larger than in mesophile α-amylase (Fig. 3). Thermophile α-amylase was little denatured in 3 M GuHCl containing 10^{-3} M calcium ion for 20 hrs at 25°C. This means that calcium ions play a significant role on the stabilization of the conformation of both α-amylases.

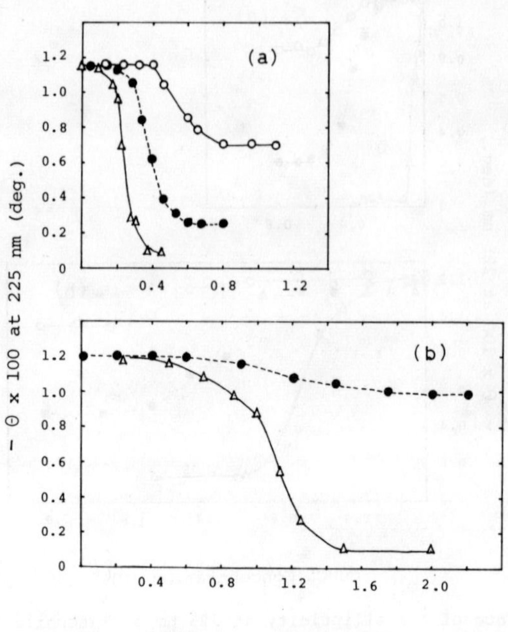

Fig. 3 The dependence of the ellipticity at 225 nm of mesophile (a) and thermophile (b) α-amylases on the concentration of GuHCl. Enzyme concentrations were 3.2 x 10^{-7} (a) and 3.5 x 10^{-7} M (b), respectively. Open and closed circles and triangles represent the data for incubation with the denaturant containing 10^{-3}, 10^{-4} and 0 M $CaCl_2$, respectively, for 20 hrs at pH 8.0 and 25°C.

Stability of Calcium-depleted Enzymes --- Imanishi has reported that the calcium binding considerably enhances the stability of the native conformation of mesophile α-amylase (9). Calcium ions play a significant role for thermophile α-amylase to maintain native conformation as shown in Fig. 3. The CD spectra in the 210-250 nm region for calcium-free thermophile and mesophile α-amylases were not different from those of the respective native enzyme at 10°C. This means that the secondary structure of both α-amylases in a calcium-free state is not destroyed at 10°C. Then, we examined the difference in stability between calcium-free thermophile and mesophile α-amylases. The variation of the ellipticity at 225 nm with urea concentration is given in Fig. 4 for cal-

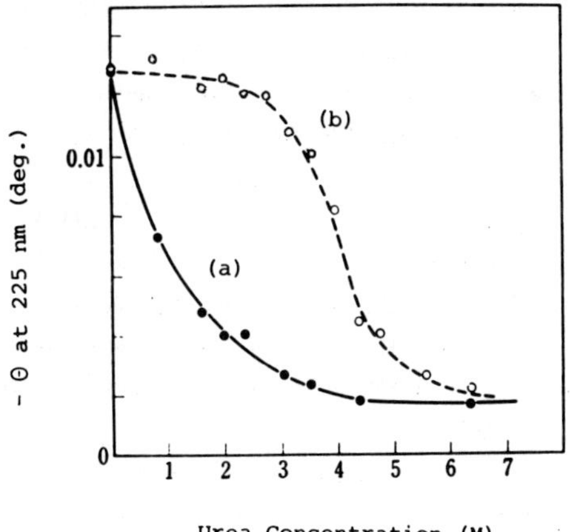

Fig. 4 Comparison of the denaturation of calcium-free mesophile (a) and thermophile (b) α-amylases. Each calcium-free enzyme was incubated with the respective concentrations urea in pH 8.0,0.1 M Tris-Cl buffer containing 10^{-3} M EDTA at 10°C. The ellipticity was measured at 225 nm after the incubation of 24 hrs. The concentrations of calcium-free mesophile and thermophile α-amylases were 4.1×10^{-7} and 4.0×10^{-7} M, respectively.

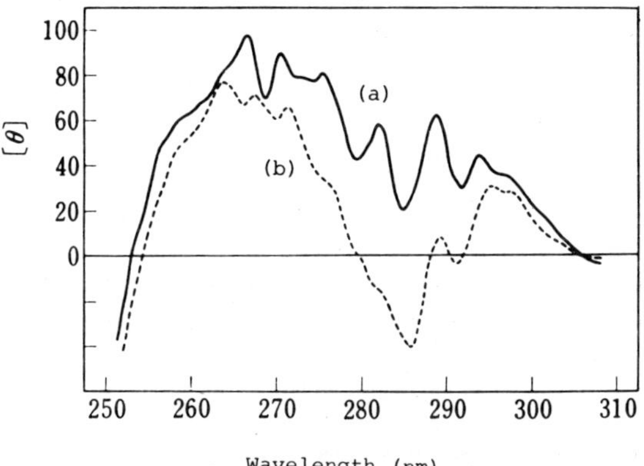

Fig. 5 The CD spectra of calcium-free (a) and native (b) thermophile α-amylases in the 250-310 nm region. The concentrations of enzymes were 0.046 (a) and 0.047 (b) %, respectively, in pH 8.0, 0.1 M Tris-Cl buffer at 10 °C. This CD spectra were smoothed by the computer.

cium-free thermophile and mesophile α-amylases at 10°C. The apparent transition midpoints of mesophile and thermophile α-amylases were 0.9 and 4.0 M, respectively, from the denaturation curves. About 70 % of these denatured enzymes was recovered at the ellipticity when the solution was diluted 10 times with pH 8.0, 0.1 M Tris-Cl buffer containing 0.001 M EDTA. These results show that the conformation for the both of calcium-free enzymes is sensitive to denaturants, and the conformation of thermophile α-amylase is more stable than that of mesophile α-amylase when calcium ions are removed from the enzyme. Therefore, it can be presumed that the difference in the stabilities of thermophile and mesophile α-amylases is caused by the difference in the respective amino acid sequences in the protein molecules.

When calcium ions are removed from thermophile α-amylase, the secondary structure seems not to be destroyed judging from the CD spectra in the 210-250 nm region at 10°C. Fig. 5 shows the CD spectra smoothed by the computer in the 250-310 nm region of calcium-free and native thermophile α-amylases at pH 8.0 and 10°C. Although the CD spectra of both enzymes were very complexity, it should be noted that the spectra of two enzymes are different from each other. This means that the local conformations in these enzyme molecules are different from each other.

Calcium Binding by Calcium-depleted Enzyme --- In order to elucidate the correlation between the calcium content and the conformation stability, the number of the calcium ions bound to thermophile and mesophile α-amylases was estimated by the methods of equilibrium dialysis at different temperatures. A series of reversible binding of calcium ions to the enzyme is represented in a whole by the following relation;

$$PCa_{i-1} + Ca^{++} \rightleftharpoons PCa_i \quad (1)$$

where PCa_i represents the enzyme-calcium complexes containing i moles of calcium ions per mole of enzyme. The average number of bound calcium per molecule of enzyme, r, is obtained by the following (15);

$$r = \frac{\text{moles of bound Ca}}{\text{total moles (p) of enzyme}} \quad (2)$$

$$r = \frac{K_1C + 2K_1K_2C^2 + \cdots + nK_1K_2\cdots K_nC^n}{1 + K_1C + K_1K_2C^2 + \cdots + K_1K_2\cdots K_nC^n} \quad (3)$$

where the K_i represent the binding constants for the successive reactions in the series of equilibria, and n the maximum number of bound calcium ions, and

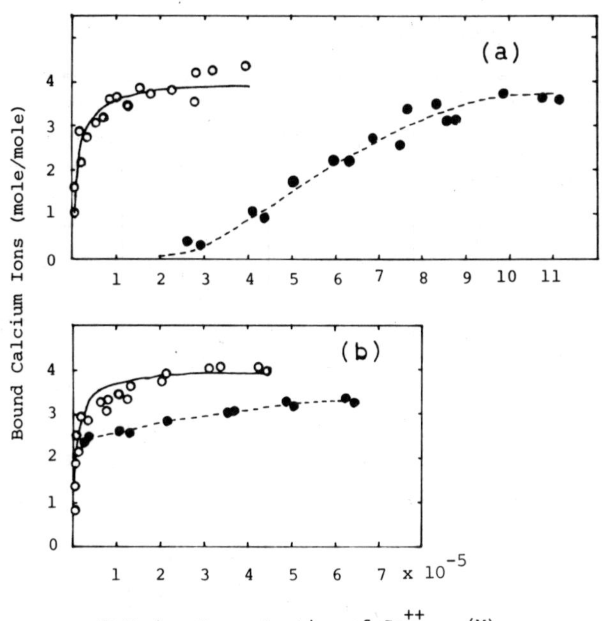

Fig. 6 Binding isotherms of calcium ions to mesophile (a) and thermophile (b) α-amylases in pH 8.0, 0.1 M Tris-Cl buffer at 10°C (open circles) and 40°C (closed circles). Concentration of both enzymes under equilibrium dialysis was about 0.5 mg per ml. The solid lines in (a) and (b) are theoretical curves calculated from the equation (3) using the values of K_i in Table I. The dotted line in (a) is the same theoretical curve as shown in Fig. 7. The dotted line in (b) is discretionary.

C equilibrium concentration of calcium ions. Fig. 6 shows the binding isotherms of calcium ions to mesophile and thermophile α-amylases at different temperatures.

At 10°C, the binding isotherm of calcium ions to mesophile α-amylase was similar to that of thermophile α-amylase. The number of binding sites of both enzymes was the same, namely 4 per molecule. The maximum number of calcium binding sites for mesophile α-amylase was different from the value, about 3 moles per mole of the enzyme, reported by Imanishi (9). The difference can not be elucidated.

Table I Binding Constants and Thermodynamic Data for Binding of Calcium Ions by Mesophile and Thermophile α-Amylases at 10°C

	Mesophile α-amylase		Thermophile α-amylase	
i	$K_i\ (10^6/M)$	ΔF(Kcal/mole)	$K_i\ (10^6/M)$	ΔF(Kcal/mole)
1	3.8	-8.5	5.2	-8.7
2	1.4	-8.0	2.0	-8.1
3	0.6	-7.5	0.9	-7.7
4	0.2	-6.9	0.3	-7.1
(sum)		-30.9		-31.6

Binding constants of calcium ions to the enzymes were calculated from Scatchard plots (16) on the assumption that the four binding sites of these enzymes are identical and have no effect upon each other. Table I shows the binding constants and the free energy changes of the formation of each enzyme-calcium complex obtained usual by usual thermodynamic procedures. The free energy changes for the binding of four calcium ions by thermophile and mesophile α-amylases were -31.6 and -30.9 Kcal per mole of enzyme, respectively. These data show that thermophile α-amylase is stabilized by the binding of calcium ions to a similar extent to that of mesophile α-amylase at 10°C. Tanford has reported that typical values of the conformational free energy of unfolding lie near 10 Kcal per mole (17). From these, it is clear that the calcium binding to the enzymes considerably enhances the stability of the native conformation of both enzymes.

The solid lines in Fig. 6 (a) and (b) are theoretical binding isotherms which were obtained from the equation (3) using the data of binding constants in Table I. The theoretical binding isotherm of mesophile α-amylase fits the experimental data well. That of thermophile α-amylase also fits the experimental data well excluding that the number of bound calcium ions per moles enzymes is between 3 and 4 moles. This unfitness means that the mode of calcium binding by thermophile α-amylase differs for each step of calcium binding.

The binding isotherm of calcium ions to mesophile α-amylase at 40°C was very different from that of thermophile α-amylase as shown in Fig. 6. At the equilibrium calcium concentration of 3×10^{-5} M, thermophile α-amylase was able to bind with about 3 moles calcium ions per mole of enzyme, but mesophile α-amylase was not able to bind with calcium ions. The binding isotherm of calcium ions to mesophile α-amylase exhibited a sigmoidal curve with respect the equilibrium concentration of calcium ions. These difference may suggest the reason why thermophile α-amylase is thermostable. The binding isotherm of thermophile α-amylase at 40°C shows that the binding constants for the fourth calcium binding by this enzyme are smaller than that of other three, which is similar to the phenomenon at 10°C.

The mesophile α-amylase may be denatured in the absence of calcium ions at 40°C. Then, mesophile α-amylase may be denatured through the following steps;

$$PCa_4 \overset{K_4}{\longleftrightarrow} PCa_3+Ca \overset{K_3}{\longleftrightarrow} PCa_2+Ca \overset{K_2}{\longleftrightarrow} PCa_1+Ca \overset{K_1}{\longleftrightarrow} P+Ca \overset{K_d}{\longleftrightarrow} D \quad (4)$$

where D represents denatured enzyme. And the denaturation constant, K_d, may be described by the following;

$$K_d = \frac{(D)}{(P)} \quad (5)$$

where parentheses represent concentrations, and when equation (4) is substituted in equation (2), we get

$$r = \frac{K_1C + 2K_1K_2C^2 + 3K_1K_2K_3C^3 + 4K_1K_2K_3K_4C^4}{1 + K_1C + K_1K_2C^2 + K_1K_2K_3C^3 + K_1K_2K_3K_4C^4 + K_d} \quad (6)$$

Equation (6) may be applicable to the analysis of the data obtained from the equilibrium dialysis of mesophile α-amylase at 40°C (closed circles in Fig. 6 (a)). The denaturation constant, K_d, was calculated by the method of least squares from equilibrium dialysis data to be 7.2×10^6, if the values of K_i were the same as the values calculated at 10°C (Table I). The curve C in Fig. 7 was calculated by using above values for K_i and K_d, and fits the experimental data well. This means that mesophile α-amylase is denatured at 40°C when bound calcium ions are depleted. If the value of K_d is changed, the binding isotherms of calcium ions to the enzyme will change as shown in Fig. 7. When the value of K_d comes to be near zero, the binding isotherm will coincide with the curve at 10°C (Fig. 6).

The Mechanism of thermostability of thermophile α-Amylase --- Thermophile and mesophile α-amylases may be denatured through the following steps;

$$\underset{\text{(Native state)}}{N} \overset{\text{Step 1}}{\longleftrightarrow} \underset{\text{(Meta-stable state)}}{N' + Ca^{++}} \overset{\text{Step 2}}{\longleftrightarrow} \underset{\text{(Denatured state)}}{D} \quad (7)$$

where N represents native enzyme containing bound calcium, N' calcium-free en-

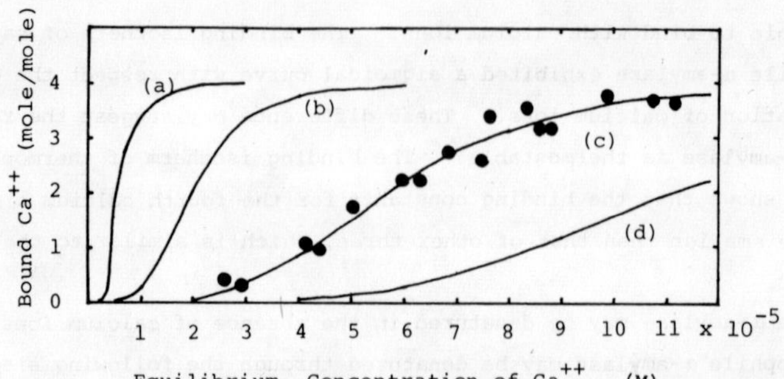

Fig. 7 Theoretical binding isotherms of calcium ions to mesophile α-amylase. The solid lines in (a), (b), (c), and (d) are theoretical curves calculated using the values of 10^3, 10^5, 7.2×10^6, and 10^8, respectively, for K_d and the curves of Table I for K_i from equation (6). Closed circles represent the binding isotherme of calcium ions to mesophile α-amylase at 40°C.

zyme which has the similar conformation to native enzyme, and D denatured enzyme.

1) The equilibrium system for step 1 will be affected by the concentration of calcium ions and circumstances such as heating or being in denaturant, where the calcium-free enzyme is denatured.

2) The binding constants of calcium ions to thermophile α-amylase was similar to that of mesophile α-amylase on the conditions where calcium-free enzymes were not denatured. It means that the contribution of calcium binding to the stability of thermophile α-amylase is not different from that of mesophile α-amylase and both α-amylases are considerably stabilized by the binding calcium ions. That is, the stabilization by calcium ion is not characteristic of thermophile α-amylase (step 1).

3) Calcium-free thermophile α-amylase was easily denatured, but was more resistant to heat and denaturant than calcium-free mesophile α-amylase. The denaturation constant of calcium-free thermophile α-amylase in step 2 is smaller than that of calcium-free mesophile α-amylase on heating or the addition of denaturant. The difference in the denaturation constant may be caused by the difference in the amino acid sequence in the enzyme molecules.

4) The difference in the denaturation constants of calcium-free thermophile and mesophile α-amylases in step 2 will affect the equilibrium for step 1. If the calcium-free enzyme is easily denatured, the equilibrium at step 1 should go to the right. When the enzyme is bound with calcium ion, it is considerably

stabilized. Calcium ion may play a role in amplifying the small difference in the stability of the both calcium-free enzymes.

5) It may be given as a conclusion that calcium ions play a significant role for thermostable α-amylase in maintaining the native conformation, and the difference in thermostability of the α-amylases from *B. stearothermophilus* and *B. subtilis* is caused by the difference in the affinity to the calcium ion at higher temperatures.

The author wishes to express his gratitude to Prof. Y. Kyogoku, Drs. K. Ogasahara, I. Urabe, and T. Takagi for valuable discussions.

Refference

1. Manning, G. B. and Campbell, L. L., *J. Biol. Chem.*, 236, 2952 (1961)
2. Manning, G. B., and Campbell, L. L., and Foster, R. J., *J. Biol. Chem.*, 236, 2958 (1961)
3. Campbell, L. L. and Manning, G. B., *J. Biol. Chem.*, 236, 2962 (1961)
4. Pfueller, S. L. and Elliott, W. H., *J. Biol. Chem.*, 244, 48 (1969)
5. Ogasahara, K., Imanishi, A., and Isemura, T., *J. Biochem.*, 67, 65 (1970)
6. Ogasahara, K., Imanishi, A., and Isemura, T., *J. Biochem.*, 67, 77 (1970)
7. Ogasahara, K., Yutani, K., Imanishi, A., and Isemura, T., *J. Biochem.*, 67, 83 (1970)
8. Vallee, B. L., Stein, E. A., Summerwell, W. N., and Fischer, E. H, *J. Biol. Chem.*, 234, 2901 (1959)
9. Imanishi, A., *J. Biochem.*, 60, 381 (1966)
10. McConn, J. D., Turu, D., and Yasunobu, K. T., *J. Biol. Chem.*, 239, 3706 (1964)
11. Feder, J., Garrett, L. R., and Wildi, B. S., *Biochemistry*, 10, 4552 (1971)
12. Yutani, K., Sasaki, I., and Ogasahara, K., *J. Biochem.*, 74, 573 (1973)
13. unpublished
14. Fuwa, H., *J. Biochem.*, 41, 583 (1954)
15. Tanford, C., *Phys. Chem. Macromolecules*, 526 (1961)
16. Scatchard, G., *Ann. New York Acad. Sci.*, 51, 660 (1949)
17. Tanford, C., *Advan. Protein Chem.*, 24, 1 (1970)

THERMOPHILIC DEHYDROGENASES

THERMOPHILIC GLYCERALDEHYDE-3-P DEHYDROGENASE

Remi E. Amelunxen and Rivers Singleton, Jr.
Dept. of Microbiology, Univ. Kansas Med. Ctr., Kansas City, Kansas
Dept. Biological Sciences, Univ. Delaware, Newark, Delaware

INTRODUCTION

D-Glyceraldehyde-3-P:NAD oxidoreductase (phosphorylating) (E.C. 1.2.1.12) (abbr. GPDH) is a ubiquitous enzyme, found in most living systems, where it functions in the glycolytic pathway. There were several reasons for studying the enzyme in thermophilic bacteria. First, at the time this enzyme was characterized in the obligate thermophile _Bacillus stearothermophilus_ (1503), no intracellular enzymes from thermophilic bacteria had been isolated in highly purified form. Second, the enzyme had been well characterized in many other systems, thereby providing a sound basis for comparison with the thermophilic enzyme. Third, GPDH is almost always present within cells at very high concentrations, thereby providing a large quantity of potential material necessary for extensive physicochemical analyses. Finally, was the demonstration by Amelunxen and Lins (1), that the enzyme in crude extracts from _B. stearothermophilus_ was indeed remarkably thermostable, as seen in Table I.

TABLE I

SPECIFIC ACTIVITY OF GPDH FROM _B. stearothermophilus_ AND _B. cereus_ AFTER HEATING AT VARIOUS TEMPERATURES

	TEMPERATURE OF HEATING FOR 10 MIN.				
ENZYME SOURCE	NONE	60°	70°	80°	90°
B. stearothermophilus	1.4	--	4.1	13.4	28.9
B. cereus	1.2	0.2	0	0	0

In this experiment, extracts were prepared from the two organisms and heated at the same protein concentration for ten minutes at various temperatures. The extracts were then cooled and assayed for remaining enzymatic activity and protein concentration. As seen, the mesophilic enzyme is rapidly inactivated, whereas a marked purification of the thermophilic enzyme occurs.

GPDH FROM _Bacillus stearothermophilus_

The increase in specific activity that occurred upon heating was taken advantage of in the purification scheme for the enzyme, which is summarized in Table II (2). As seen, the purification is straightforward and involves a heat step (in which an apparent increase in units occurs, due to removal of the NADH oxidase), fractionation with ammonium sulfate between 72 and 90% saturation, followed by crystallization from ammonium sulfate. The crystallization step is rather specific for pH, protein concentration and amount of ammonium sulfate. Trace impurities can be removed by recrystallization. The resulting needle-like crystals, seen in Fig. 1, exhibit a constant specific activity upon recrystallization and the preparation is homogeneous by most criteria, such as disc electrophoresis and sedimentation velocity and sedimentation equilibrium centrifugation.

TABLE II

SUMMARY OF PURIFICATION PROCEDURE

STEP	SPEC. ACT.	TOTAL PROTEIN	TOTAL UNITS	RECOV.	PURIF.
1. Sonic extract	3.4u/mg	4,200mg	14,280	---	1.0
2. Supn. after heat step	11.4	1,300	15,100	100%	3.4
3. Supn. at 72% $(NH_4)_2SO_4$	43.5	248	10,800	71	13.0
4. Dissol. ppt. of 90% $(NH_4)_2SO_4$	61.0	144	8,800	58	17.9
5. First crystals	108.0	47	5,000	33	31.8
6. Second crystals	113.0	43	4,900	32	33.2
7. Third crystals	113.0	39	4,400	29	33.2

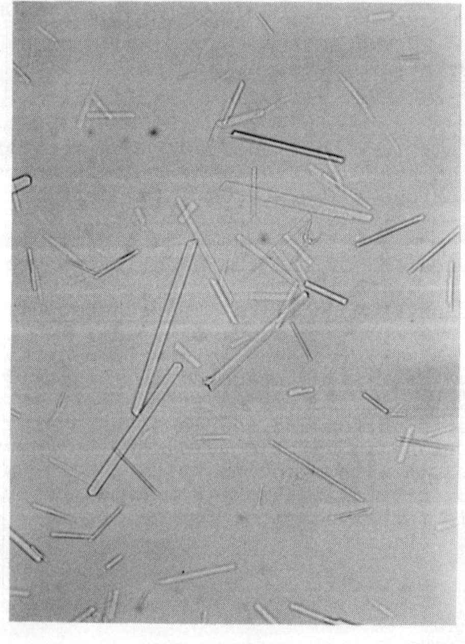

Figure 1. Crystals of GPDH from B. stearothermophilus, mag. 250X. From (3), used with permission.

In evaluating the thermostability of the thermophilic GPDH (3), crystals were dissolved in deionized water to a protein concentration of 10-11 mg/ml; little or no inactivation could be demonstrated after heat treatment at 80° for 10 minutes. After 10 minutes at 90°, there was only around 10% inactivation, clearly demonstrating the remarkable thermostability of the enzyme. By contrast, non-thermophilic GPDHs under similar conditions generally show extensive inactivation after 10 minutes at 70°. Further evidence for the thermostability of the purified enzyme is reflected by the effect of temperature on enzymatic activity, as seen in Figure 2 (4).

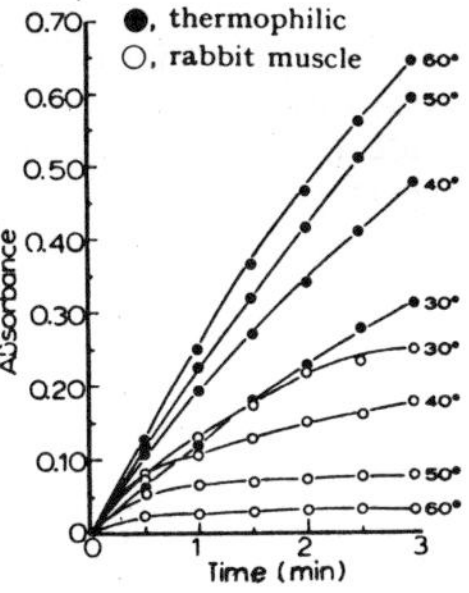

Figure 2. The effect of temperature on the enzymatic activity of thermophilic (o) and rabbit muscle (o) GPDHs. From (4), used with permission.

These data represent the effect of temperature on the activites of the rabbit muscle (open circles) and thermophilic (closed circles) enzymes. The marked difference in thermostability of these two enzymes is immediately apparent from these data. Unfortunately, at temperatures above 60°, the assay system becomes quite unstable and it is difficult to obtain reliable data at higher temperatures. In this case, the concept of optimal temperature is without meaning.

There is a slight difference in the pH profiles for the muscle and thermophilic enzymes, as seen in Figure 3 (4). The thermophilic

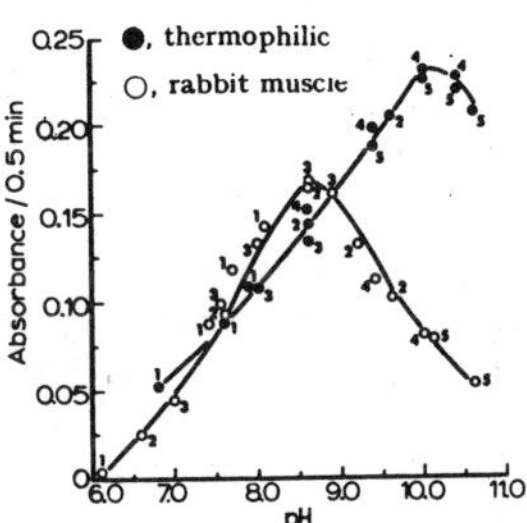

Figure 3. The effect of pH on the enzymatic activity of thermophilic (o) and rabbit muscle (o) GPDHs. From (4), used with permission.

enzyme has a pH optimum at pH 10, whereas the optimum for the muscle enzyme is maximal around pH 8.6. The pH stability of the two enzymes also differs slightly, with the thermophilic enzyme exhibiting a broader range of pH stability from pH 5 to 10.

The crystalline enzyme sediments at several protein concentrations, as a single boundary in the ultracentrifuge, as seen in Figure 4 (5). These data yield a sedimentation constant of 7.17 S,

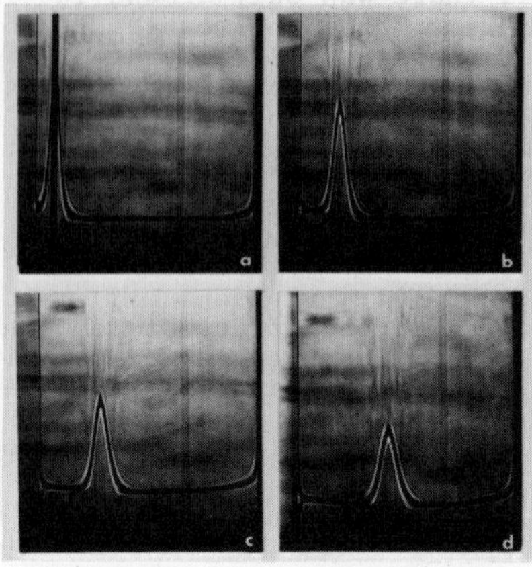

Figure 4. Sedimentation velocity pattern for thermophilic GPDH. From (5), used with permission.

corrected to zero protein concentration in water. The plot of s_{obs} versus protein concentration was linear with a slight decrease in slope, typical of globular proteins. The diffusion constant was determined by means of the synthetic boundary method in the ultracentrifuge and a diffusion constant of 3.95 F was obtained. Using the Svedberg equation, the sedimentation and diffusion constants yielded a molecular weight of 178,900, a value higher than that obtained by sedimentation equilibrium centrifugation. The partial specific volume was determined by density gradient techniques and a value of 0.754 ml/g was found, which is in close agreement with that obtained from the amino acid analysis. High speed meniscus depletion sedimentation equilibrium centrifugation in dilute KCl was employed to determine the molecular weight and a value of 148,500 Daltons was obtained, which is completely consistent with that found for other GPDH's. In phosphate buffer, a higher value was obtained, a phenomenon also observed with other GPDHs.

In the presence of 8 M urea or 6 M guanidine-HCl, the enzyme dissociated into subunits of molecular weight 36,000 Daltons, as determined by sedimentation equilibrium centrifugation (6). The tetrameric nature of the protein is supported by the observation of four critical sulfhydryls and four moles of bound NAD^+ per native molecular weight. That the subunits might be identical was suggested by the finding of a single C-terminal amino acid, leucine.

A summary of the physical properties of the thermophilic enzyme and a comparison with those of the yeast and rabbit muscle enzymes is shown in Table III. The data for the thermophilic enzyme are characteristic of a typical, tetrameric, globular protein.

Furthermore, as these data suggest, it bears a remarkable resemblance to its counterparts from rabbit muscle and yeast, enzymes which are markedly thermolabile by comparison.

TABLE III

SUMMARY OF PHYSICAL PROPERTIES

ENZYME SOURCE	$s^{o}_{20,w}$(S)	$D^{o}_{20,w}$(S)	$M_w x10^4$(D)	$\bar{v}$(ml/g)	f/fo	SUB-UNITS
Thermophile	7.71	3.95	14.9	0.754	1.38	4
Rabbit muscle	7.71	4.97	13.7	0.725	1.26	4
Yeast	6.80	5.19	12.2	0.739	1.25	4

The amino acid composition of the enzyme is not greatly different from other GPDHs and is shown in Table IV. The data here are given in residue percent (residues per 100 residues) to facilitate

TABLE IV

SUMMARY OF AMINO ACID COMPOSITION

RESIDUE	THERMOPHILIC ENZYME	NON-THERMOPHILIC ENZYMES[1]	DIFFERENCE
ASP	11.56	11.85(0.70)	0.28
THR	5.78	6.43(0.98)	0.65
SER	5.20	5.95(1.08)	0.75
GLU	8.09	6.16(0.87)	-1.93
PRO	3.18	3.81(0.52)	0.63
GLY	7.51	9.44(0.57)	1.93
ALA	11.85	10.45(0.49)	-1.40
½CYS	0.87	1.60(2.25)	0.74
VAL	11.85	8.96(2.55)	-2.89
MET	1.45	1.99(0.63)	0.55
ILE	5.78	6.27(0.93)	0.49
LEU	8.09	5.90(0.44)	-2.19
TRY	2.31	2.88(0.46)	0.57
PHE	1.45	3.93(0.48)	2.48
LYS	6.94	7.96(0.40)	1.03
HIS	2.60	2.36(0.73)	0.24
ARG	4.62	3.19(0.32)	-1.43
TRP	0.87	0.89(0.50)	0.02

[1]Mean for GPDH from 13 procaryotic and eucaryotic organisms

comparisons of compositions from proteins of differing molecular weight. The non-thermophilic composition was calculated as the mean composition for thirteen different GPDHs from various sources. The values in parenthesis are the standard deviation from the mean. If one applies the rather crude statistical test of assuming differences greater than three units of standard deviation are significant, then we must consider LEU and ARG to be significantly elevated, whereas PHE and GLY are significantly decreased in the thermophilic enzyme, relative to this "average" non-thermophilic enzyme. Although the level of GLU is not significantly elevated (by this test), it does appear to be high, and most likely exists as the acid, rather than the amide. This is based on the observation of an acidic isoelectric point (pI=4.6) for the enzyme.

Table V (5) shows a comparison of the amino acid composition of the subunits of thermophilic, lobster muscle, and pig muscle GPDHs. The latter two are taken from sequence data and are more accurate than that of the thermophilic analysis, which was calculated using a molecular weight which has been subsequently shown to

TABLE V

AMINO ACID COMPOSITION OF GPDH SUBUNITS

AMINO ACID	THERMOPHILIC	LOBSTER MUSCLE	PIG MUSCLE
LYS	24	28	26
HIS	9	5	11
ARG	16	9	10
ASP	40	32[1]	38[1]
THR	20	20	22
SER	18	25	19
GLU	28	24[1]	18[1]
PRO	11	12	12
GLY	26	30	32
ALA	41	32	32
½CYS	3	5	4
VAL	41	38	34
MET	5	10	9
ILE	20	18	21
LEU	28	18	19
TYR	8	9	9
PHE	5	15	14
TRP	3	3	3
TOTAL RESIDUES	346	333	332

[1]Combined values of free acids and amides

be elevated. A more realistic value for the total residue number is probably 333, similar to that observed for the muscle enzymes. Again, no striking differences in composition are seen, although a true comparison can only be made using sequence data.

Several structural parameters calculated for the thermophilic enzyme from its amino acid composition are summarized in Table VI. These are again compared with the parameter calculated for the mean composition determined for thirteen non-thermophilic GPDHs. The three measures of hydrophobicity are average hydrophobicity ($H\phi_{ave}$) (7), sum of non-polar side chains (NPS) (8), and the polarity index (p) (9). All of these parameters are based on the same assumption, i.e., that all of the hydrophobic side chains are on the interior of the molecule, and all of the polar side chains are on the exterior of the protein. However, current evidence suggests that

TABLE VI

SUMMARY OF STRUCTURAL PARAMETERS

PARAMETER	THERMOPHILIC GPDH	NON-THERMOPHILIC GPDHs[1]	ALL THERMOPHILIC[2]	ALL NON-THERMOPHILIC[3]
$H\Phi_{ave}$	1.06	1.10(0.04)	--	--
NPS	0.34	0.34(0.01)	--	--
P	0.97	0.92(0.07)	--	--
% Helix	36.23	30.90(2.82)	31.79(9.72)	30.83(6.47)
% Beta	27.04	27.85(4.89)	16.20(11.15)	23.64(1.16)
% Turn	37.03	36.26(3.85)	40.45(5.31)	37.75(6.12)

[1]Mean for GPDH from 13 procaryotic and eucaryotic organisms (lsd indicated in parenthesis)
[2]Mean for 13 thermophilic enzymes
[3]Mean for 56 comparable non-thermophilic enzymes

this model of protein structure may not be entirely correct (10). The predicted helical, beta, and turn contents were calculated from the equations of Krigbaum and Knutton (11).

These data demonstrate, as have numerous other studies (12), that no significant difference exists in the hydrophobic character of the thermophilic enzyme, at least no difference that is measurable within the assumptions of these parameters.

There is a slight suggestion in these calculations of an increased helical content for the thermophilic GPDH. This, however, may not be a generalized prediction for all thermophilic proteins. On the right side of this table are calculations for the mean amino acid composition for 13 different thermophilic proteins which are compared to those of 56 corresponding non-thermophilic proteins. These data suggest that there is no difference in the helical content of thermophilic proteins as a group, but do suggest that these proteins may have a decreased beta content. Perhaps of greater interest is the observation that the standard deviations for both of these parameters are quite large, suggesting that the tertiary and quaternary structures of these proteins will tolerate a large amount of variance in secondary structure.

In addition to having similar amino acid compositions, the thermophilic enzyme, like the muscle enzyme, has four moles of firmly bound NAD^+ (6). The NAD^+ can be removed by charcoal treatment, and the apoenzyme can be crystallized, crystals of which are seen in Figure 5 (13). Note the difference in crystalline structure of the apoenzyme, compared with the needle-like crystals of the holoenzyme. This suggests that some changes in structure occur upon NAD^+ removal. However, these changes are not major nor

Figure 5. Crystals of NAD^+-free thermophilic GPDH. From (11), used with permission.

irreversible, as the NAD^+ can be replaced with full restoration of activity. Furthermore, NAD^+ removal does not effect the thermostability of the enzyme.

The sequence around the critical sulfhydryl in the active site has been determined and is shown in Figure 6 (14). The S-carboxymethylated peptide was prepared by reaction with ^{14}C-iodacetic acid, followed by tryptic digestion. After purification, the peptide sequence was determined by conventional methods.

Figure 6. Active site peptides from several GPDHs.

1 = RABBIT & PIG 17 AA
2 = LOBSTER 20 AA
3 = THERMOPHILE 20 AA

1- • Ser • Leu • Lys • Ile • Val • Ser • Asn • Ala • Ser • Cys(SH) • Thr • Thr • Asn • Cys(SH) • Leu • Ala • Pro • Leu • Ala • Lys •

2- • Asp • Met • Thr • Val • Val • Ser • Asn • Ala • Ser • Cys(SH) • Thr • Thr • Asn • Cys(SH) • Leu • Ala • Pro • Val • Ala • Lys •

3- • Ala • His • His • Ile • Val • Ser • Asn • Ala • Ser • Cys(SH) • Thr • Thr • Asn • Cys(SH) • Leu • Ala • Pro • Phe • Ala • Lys •

The sequence shown here is co-aligned with the corresponding sequence of lobster, rabbit and pig muscle GPDHs. In most GPDHs, the active-site tryptic peptides consist of 17 residues (1 in Fig. 6) and the sequence of this segment is highly conserved. However, the lobster muscle enzyme has a 20 residue peptide (2 in Fig. 6) where LYS is replaced by THR (position 3 of lobster muscle sequence). The thermophilic tryptic peptide also consists of 20 residues (3 in Fig. 6), and differs from the prototype sequence only in position 18, where LEU is replaced by PHE. In the thermophilic peptide HIS (position 3) replaces LYS of the prototype sequence. The adjacent HIS residue (position 2 of the thermophilic sequence) occurs in a highly variable portion of the protein chain. The two adjacent HIS residues represent a novel feature of the thermophilic active-site sequence. Although the amino acid substitutions observed in the thermophilic sequence are probably major from the genetic viewpoint, it remains to be established if these structural alterations are involved in conferring thermostability to the enzyme. Evidence to be presented is suggestive that the active-site area does play a role in stabilizing the enzyme.

Dr. J. I. Harris is currently working on the sequence and three dimensional structure of the enzyme. His group has reported that approximately 50% homology exists between the thermophilic and lobster muscle enzymes for 230 of the 332 residues sequenced thus far. Of interest is the observation that approximately 60% homology is found between the enzymes from B. stearothermophilus and the extreme thermophile, Thermus aquaticus.

As previously noted, the enzyme dissociates into subunits in either 8 M urea or 5 M guanidine-HCl. The dissociation in guanidine-HCl is rapid and irreversible at room temperature. However, the dissociation in urea has the characteristics of a first order decay process, as seen in Figure 7 (6). These data were obtained by heating

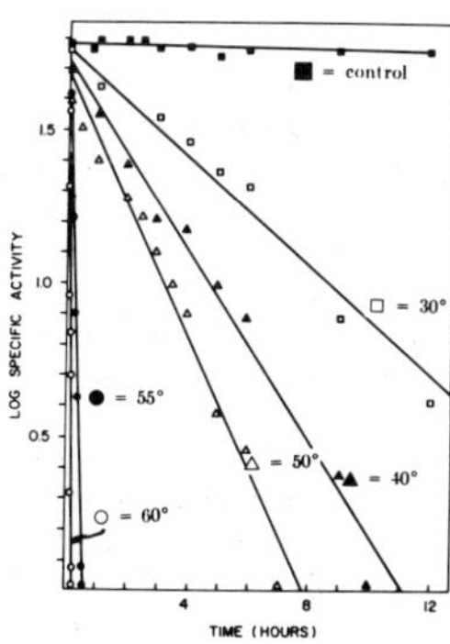

Figure 7. Inactivation of thermophilic GPDH in 8 M urea at various temperatures. From (6), used with permission.

the enzyme in 8 M urea at the indicated temperature. After a period of time, the enzyme was removed, diluted and assayed for remaining enzymatic activity. The activity remaining was taken as a measure of the degree of dissociation. Note that at temperatures of 55° and higher, the dissociation becomes very rapid. An Arrhenius plot of these data is linear up to around 55°, at which point the data become non-linear. It was also noted, that up to the transition temperature of around 55°, the dissociation seemed to be at least partially reversible; however, at temperatures beyond this the dissociation was irreversible.

This conformational transition at 55° is also indicated in the optical rotatory dispersion (ORD) data summarized in Table VII (6). Upon heating from 30° to 55°, no changes occur in either ORD parameter. However, at 60°, pronounced changes occur in both a_o and

TABLE VII
SUMMARY OF ORD DATA

EXPERIMENT	TEMP/SOLVENT	$-a_o$	$-b_o$
1	30° (water)	129(14)	134(11)
2	55° "	119(6)	135(8)
3	60° "	176(10)	206(14)
4	30° " [1]	128(10)	211(15)
5	30° (urea)	526(19)	66(11)
6	60° "	462(17)	68(8)
7	30° (guHCl)	589(18)	64(8)
8	60° "	540(20)	66(9)

[1]Enzyme first heated at 60° for 16 hours, then cooled at 30°.

b_o. These changes are partially reversible, as suggested in experiment 4, where the value of a_o returned to that of the unheated enzyme. The values for the enzyme in 5 M guanidine-HCl most likely reflect that of the completely unfolded enzyme. If this is true, then two points are apparent. First, urea does not cause the complete unfolding of the protein. Second, temperature also does not cause the complete unfolding of the protein structure. Thus, at 60°, although a large bulk of the protein is unfolded, a part of it appears not to unfold.

We feel these data demonstrate two primary points about the enzyme structure. First, at temperatures below 55° little change occurs in the enzyme structure, with changes in temperature. Second, as the enzyme is heated past this transition region, unfolding begins to occur as evidenced both in the susceptibility to urea denaturation and in changes in the optical rotatory parameters. However, this unfolding is not complete, as evidenced by the difference in a_o or b_o in urea or guanidine HCl and at 60°. Indeed, the increase in b_o (experiment 3) could be interpreted as an indication of increased helix formation at 60°. We have concluded that the enzyme maintains a core of organized structure around the active site which provides a basis for refolding of the protein upon denaturation. This conclusion is further supported by the unpublished observation that approximately 50% of the protein can be digested with trypsin with little loss in activity. Furthermore, the observation of protein unfolding at temperatures where it is enzymatically active, suggests that perhaps the entirely folded structure is not necessary for activity; that is, the protein will tolerate a certain amount of unfolding without losing enzymatic activity. It is possible this unfolding may be required for thermal stabilization of the enzyme.

GPDH FROM *Bacillus coagulans*

Work now in progress at the University of Kansas Medical Center involves a study of the GPDH from *B. coagulans* KU as a model for evaluating the mechanism of thermophily in facultative thermophilic bacteria. This organism grows well at both 37° and at the lower thermophilic temperature of 55°; the growth rate shows surprisingly little variation over this temperature range. Novitsky et al. (15) in studies with the cell walls of *B. coagulans* KU reported that glycolytic enzymes from crude extracts of cells grown at either 37° or 55° were as thermolabile as those from mesophilic organisms suggesting that enzymes from facultative thermophiles do not possess intrinsic thermostability.

In the course of experiments to purify the GPDH from this organism grown at both 37° and 55°, a marked lability of the enzyme was noted in buffers of low ionic strength. In addition, the enzyme was heat inactivated well below the thermophilic growth temperature in extracts of cells grown at either temperature. As shown in Table VIII, when the ionic strength of extracts of cells grown at 37° and 55° is increased by a factor of 1.7 to 1.8, the GPDH is thermostabilized at temperatures significantly beyond the thermophilic

TABLE VIII

INDUCED THERMOSTABILITY OF GPDH IN CRUDE EXTRACTS OF *B. coagulans* GROWN AT 37° and 55°

SAMPLE	% ACTIVITY REMAINING AFTER 10 MINUTES AT					
	37°	50°	55°	60°	65°	70°
1) 37° Extract						
No adn.	70	29	0	0	0	0
$(NH_4)_2SO_4$, 0.61 M	93	87	95	97	52	0
NaCl, 1.7 M	93	98	95	89	31	0
2) 55° Extract						
No adn.	97	19	0	0	0	0
$(NH_4)_2SO_4$, 0.61 M	90	92	98	97	52	0
NaCl, 1.7 M	102	99	95	99	31	0

growth temperature, whereas extensive inactivation occurs below this temperature in the absence of added salts (16). This observation was recently extended to the crystalline enzyme from cells grown at 55° as shown in Table IX. In crude extracts which contain the total soluble macromolecules of the cell, there is little or no difference in the protective effects exerted by $(NH_4)_2SO_4$ or NaCl suggesting that increased charge is the sole requirement for stabilization. However, in the case of the crystalline enzyme, the known stabilizing effect of $(NH_4)_2SO_4$ on a single molecular species is more pronounced

TABLE IX

INDUCED THERMOSTABILITY OF CRYSTALLINE GPDH FROM B. coagulans GROWN AT 55° GPDH UPON ADDITION OF SALT

ADDITION	% ACTIVITY REMAINING AFTER 10 MINUTES AT 50°	55°	60°	65°
None	21	0	0	0
$(NH_4)_2SO_4$, 0.61 M	--	100	--	96
NaCl, 1.7 M	--	95	83	30

than NaCl, even though this latter neutral salt affords a considerable degree of stability. Furthermore, this does not appear to be a general salt effect as the GPDH from mesophiles, such as B. cereus and B. subtilis are not significantly stabilized by the addition of salt. Although it is speculative to assume that cells grown at 37° and 55° would contain salt concentrations at the level employed in these studies, Damadian (17) has determined that the combined internal molality of fixed charge in Escherichia coli is 1.0 molal, a value approximated by the salt concentrations used above.

Based on these experiments, we propose that facultative thermophiles survive at elevated temperatures because their proteins are stabilized by a highly charged internal environment that can induce thermostability to proteins that otherwise would be heat inactivated.

CONCLUSIONS

What can these data obtained from thermophilic GPDHs tell us about how these organisms survive in such an extreme environment? For the obligate thermophiles, we believe that a general answer is now apparent (12). The organism survives by synthesizing intrinsically heat stable proteins, i.e, proteins which are stable without the influence of exogeneous factors (however, this should not be taken to exclude such factors as tightly bound metals or cofactors). In this respect, GPDH is a typical enzyme from these organisms. (We must note that the ultimate mechanism, i.e., what stabilizes the protein is still not resolved.) However, the GPDH is also typical of other proteins from obligate thermophiles, in that its physical and chemical properties are quite similar to its non-thermophilic counterparts. What this says is that thermostability is a more subtle property than we have previously believed; it is not necessary for a protein to be a random coil to be thermostable. That changes of a single amino acid can cause dramatic changes in thermostability was nicely demonstrated by Langridge (18) with the E. coli β-galactosidase. He induced a series of amber mutations in the enzyme, which were then suppressed by insertion of SER into the protein chain at the point of the nonsense mutation. This resulted in 52 mutationally altered forms of the enzyme. The most significant result of these mutations was a loss of thermostability, rather than major losses of enzymatic activity.

This theme was further supported recently by Perutz and Raidt (19), who superimposed the sequence of thermophilic ferrodoxin on the three-dimensional model for the mesophilic protein. Alignment of the thermophilic sequence with that of the mesophilic protein required the shifting of only a few residues, which increased the number of potential salt bridges. The energy yield from these minor changes seemed to be sufficient to stabilize the protein with respect to heat.

On the other hand, the picture does not appear as clear cut for the survival of the facultative thermophiles. Proteins from these organisms often appear to be quite heat labile, indeed often more heat labile than their mesophilic counterparts. In the absence of added salt, the GPDH from _B. coagulans_ is less stable than the enzyme from _B. cereus_ or _B. subtilis_. It seems possible that this stabilization by salt may be a reflection of the fact that within the cell, the enzyme is stabilized by a highly charged intracellular environment. The validity of this hypothesis is presently being tested by further experimentation.

REFERENCES

1. R. E. Amelunxen and M. Lins, Arch. Biochem. Biophys. _125_, 765 (1968).

2. R. E. Amelunxen, in _Methods in Enzymology_, Vol. _XLI(B)_ (Ed. W. A. Wood, Academic Press, New York, 1975), p. 268.

3. R. E. Amelunxen, Biochim. Biophys. Acta _122_, 175 (1966).

4. R. E. Amelunxen, Biochim. Biophys. Acta _139_, 24 (1967).

5. R. Singleton, Jr., J. R. Kimmel and R. E. Amelunxen, J. Biol. Chem. _244_, 1623 (1969).

6. R. E. Amelunxen, M. Noelken and R. Singleton, Jr., Arch Biochem. Biophys. _141_, 447 (1970).

7. C. C. Bigelow, J. Theoret. Biol. _16_, 187 (1967).

8. D. F. Waugh, Adv. Prot. Chem. _9_, 326 (1954).

9. H. F. Fisher, Proc. Nat. Acad. Sci. USA _51_, 1285 (1964).

10. I. M. Klotz, Arch. Biochem. Biophys. _138_, 704 (1970).

11. W. R. Krigbaum and S. P. Knutton, Proc. Nat. Acad. Sci. USA _70_, 2809 (1973).

12. R. Singleton, Jr., and R. E. Amelunxen, Bacteriol. Revs. _37_, 320 (1973).

13. R. E. Amelunxen and J. Clark, Biochim. Biophys. Acta _221_, 650 (1970).

14. J. Bridgen, J. I. Harris, P. W. McDonald, R. E. Amelunxen and J. R. Kimmel, J. Bacteriol. 111, 797 (1972).

15. T. J. Novitsky, M. Chan, R. H. Himes and J. M. Akagi, J. Bacteriol. 117, 858 (1974).

16. J. W. Crabb, A. L. Murdock, and R. E. Amelunxen, Biochem. Biophys. Res. Comm. 62, 627 (1975).

17. R. Damadian, Ann. N. Y. Acad. Sci. 204, 211 (1973).

18. J. Langridge, J. Bacteriol. 96, 1711 (1968).

19. M. F. Perutz and H. Raidt, Nature 255, 259 (1975).

GLYCERALDEHYDE 3-PHOSPHATE DEHYDROGENASE FROM AN EXTREME THERMOPHILE, THERMUS AQUATICUS

J.D. HOCKING and J. IEUAN HARRIS

MRC Laboratory of Molecular Biology
Hills Road, Cambridge, U.K.

Most of the concepts concerning the thermal stability, or instability, of biologically active systems are based upon studies carried out with organisms that live exclusively at moderate (< 40°C) temperatures. Temperature is an important environmental factor in controlling the evolution and activities of organisms and although the existence of living cells is presumably limited to temperatures at which water exists as a liquid it is nevertheless well established that microbial life does exist at temperatures that approach the boiling point of water (1). As chemists and biochemists we are therefore left to ponder the mechanisms that allow biologically active molecules to be synthesised and to survive in a biologically active form at temperatures that are lethal to their counterparts in mesophiles.

Of the more extreme thermophiles that have been grown successfully in large scale culture *Thermus aquaticus*, first isolated (2) from a thermal spring in Yellowstone National Park, a gram-negative, non-sporulating, obligate aerobe that grows optimally at 70-75°C is among the more extensively studied (2,3).

Previous work on thermophile enzymes in this laboratory has been mainly concerned with glycolytic enzymes, principally glyceraldehyde 3-phosphate dehydrogenase (GPDH) from the more moderate thermophile *B. stearothermophilus* e.g. (4,5,6,7). GPDH from mesophiles has been studied in considerable detail (for review see (8)). Its enzymatic properties are well documented and complete amino acid sequences have been determined for enzyme obtained from pig and lobster muscle and from yeast (cf. (9)). Moreover, the three-dimensional structure to 3 Å of the lobster muscle enzyme is also known (10) while a similar study of the amino acid sequence and X-ray crystallographic structure of GPDH from *B. stearothermophilus* is already in progress in the Laboratory of Molecular Biology at Cambridge.

Thermophile GPDHs were originally chosen for study in the hope that their increased stability could be utilised to advantage in studies that were in progress on the structure and mechanism of this key glycolytic pathway enzyme. Although the three-dimensional structure of a mesophile holoenzyme is known (10) attempts to determine the structure of NAD-free (apo) enzyme have been frustrated by its instability. A structural analysis of apoenzyme is necessary so as to identify structural changes at the active site that occur when NAD^+ binds to the individual subunits in the tetrameric enzyme and in this connection the successful crystallisation of a stable apoenzyme from *B. stearothermophilus* (4) has now enabled structural work to proceed (A. Wonacott and colleagues Unpublished Results). In the course of this work unusual properties, such as the stability of both halo and apo forms to heat, as well as to a number of denaturing solvents such as 8 M urea and SDS were recognised (4) (cf. (11)). At the same time preliminary chemical studies (12) showed the *B. stearothermophilus* enzyme to be structurally homologous to mesophile GPDHs. A detailed structural study of thermophile GPDHs is therefore being undertaken with the aim of comparing the tertiary and quaternary structures of the *same* enzyme from a mesophile and a thermophile, and of identifying structural changes that confer thermal stability necessary for biological adaptation to high temperature.

This paper summarises our results on the isolation, properties and structure of GPDH from *T. aquaticus*. Full experimental details are given in references (13) and (14).

PURIFICATION OF GPDH BY AFFINITY CHROMATOGRAPHY

NAD^+ was attached to a Sepharose matrix and the resulting immobilised NAD^+ derivative of benzamidohexyl-Sepharose was used as the final step in the purification of GPDH from cell extracts of both *B. stearothermophilus* and *T. aquaticus* as described previously (13,14). The results of a typical purification scheme for *T. aquaticus* GPDH are summarised in Table 1. The content of GPDH varied with different batches of cells but yields of pure enzyme were usually in the range of 100 to 200 mg per 500 g of washed frozen cells. From *B. stearothermophilus* cells higher yields ranging between 1 to 2 g per Kg cells were obtained.

The material eluted from NAD^+-Sepharose gave a single band when examined by gel electrophoresis both in the presence and absence of 0.1% SDS, and the pattern of protein bands present in SDS-gels of samples examined before and after NAD-Sepharose chromatography are shown in Fig. 1. The purified enzyme preparations gave single N-terminal residues (alanine for *B. stearothermophilus* GPDH, methionine for the *T. aquaticus enzyme*) and unique N-terminal sequences when submitted to automatic sequencer analysis (cf. 6). Based on these criteria the respective proteins were adjudged to be pure.

PROPERTIES OF *T. AQUATICUS* GPDH

Molecular weight. The subunit molecular weight of *T. aquaticus* GPDH was estimated by SDS-gel electrophoresis to be 37,000 ± 2,000. The molecular weight of the native enzyme was found to be 150,000 by gel-filtration on Sephadex G-200 and it may therefore be concluded that *T. aquaticus* GPDH, like its mesophile counterparts, is a tetrameric enzyme comprising similar and probably identical subunits. The results of amino acid analysis (14) showed that the protein subunit comprises 330 amino acid residues corresponding to a molecular weight of about 36,000.

The pure microcrystalline enzyme was found to contain 3.5 to 4.0 moles of bound NAD^+ and its isoelectric point was estimated by gel electrophoresis to be 4.8 (as compared with 4.3 for the *B. stearothermophilus* enzyme (14)).

Kinetic parameters. *T. aquaticus* GPDH is less active at 20° than its counterparts from mammalian muscle (18-20 units/mg compared with approximately 50 units/mg for the rabbit muscle enzyme under comparable assay conditions). The variation in reaction rate with substrate concentration (G-3P) and temperature was investigated and the results of Lineweaver-Burk plots (Fig. 2) showed that whereas V_{max} increases from 25 units/mg at 25°C to 60 units/mg at 35°C the K_m remains constant at 7.5×10^{-4} M. The specific activity increases with temperature and an Arrhenius plot (Fig. 3) is linear to at least 60°C showing that there is probably no change in enzyme conformation or mechanism over this range of temperature. The activation energy was calculated to be 15 Kcal/mole which although relatively high for an enzyme reaction is not altogether unexpected in view of the high (70-75°C) *in vivo* temperature at which the enzyme normally operates.

Thermal stability. *T. aquaticus* GPDH is stable at 90°C and the half-life at 98°C is greater than 30 min. A comparison of the thermal stabilities of *T. aquaticus*, *B. stearothermophilus* and rabbit muscle GPDHs is shown in Fig. 4.

TABLE 1: Purification scheme for *T. aquaticus* GPDH

Fraction	Volume	Protein Conc. (mg/ml)	Specific Activity GPDH	Total Units GPDH	Percent Recovery GPDH
DNase treated crude extract	1 L	42	0.061	2600	100%
DEAE-cellulose 400 mM NaCl washes	3 L	4.05	0.21	2550	98%
Hydroxyapatite 200 mM $NaPO_4$ wash	320 ml	12.1	0.77	3000	115%
Ammonium sulphate supn.	340 ml	9.8	1.0	3400	131%
Affinity column eluant	157 ml	≈0.50	≈ 20	1600	62%
Concentrated GPDH	10.5 ml	7.8	16.5	1350	52%

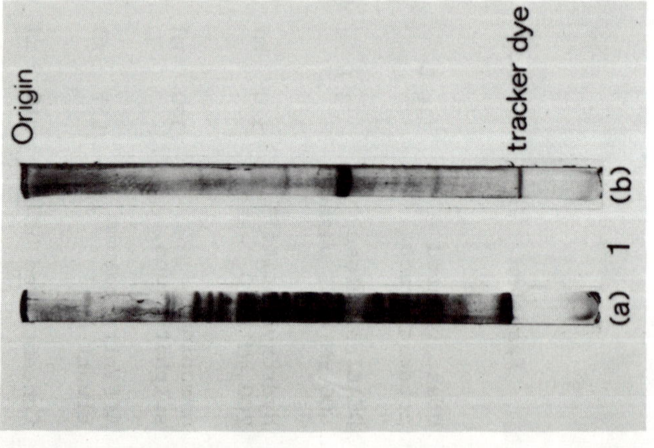

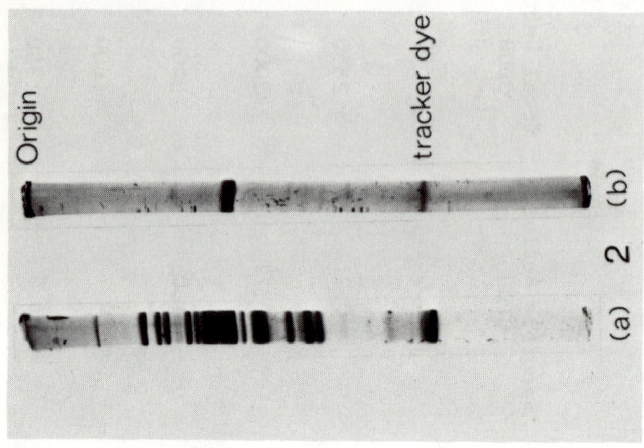

Fig. 1. Purification of T. aquaticus (1) and B. stearothermophilus (2) GPDH: SDS-gel electrophoresis (a) before and (b) after NAD-Sepharose.

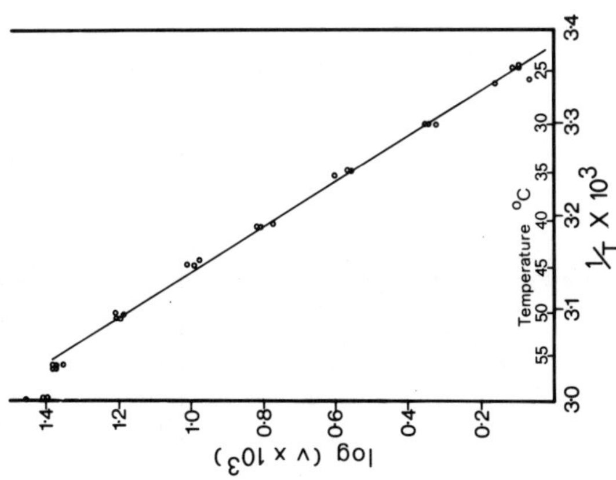

Fig. 3. Effect of temperature on *T. aquaticus* GPDH activity: Arrhenius plot.

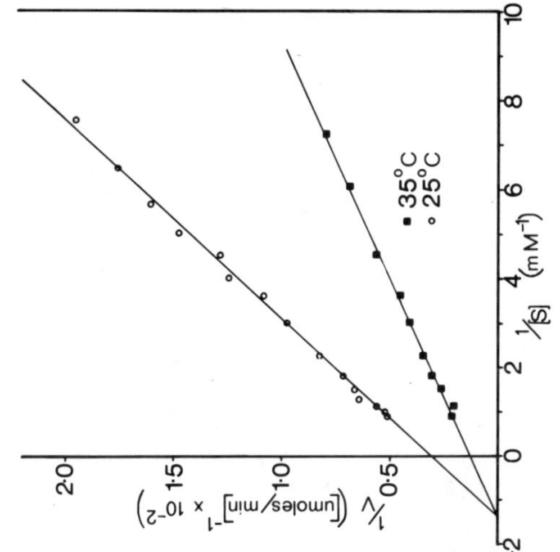

Fig. 2. Variation of *T. aquaticus* GPDH activity with concentration of G-3P.

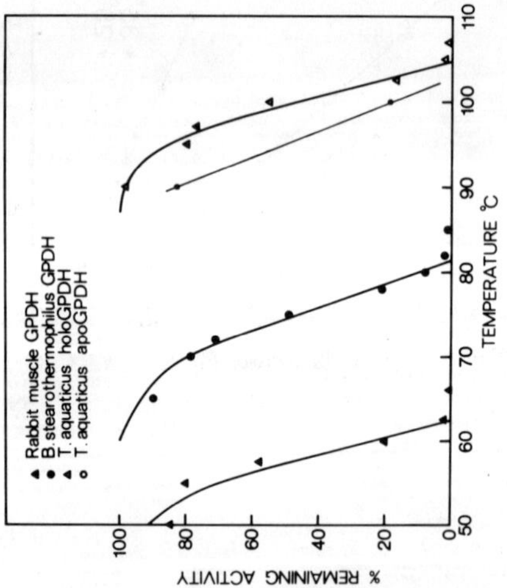

Fig. 4. Thermal inactivation of GPDHs.

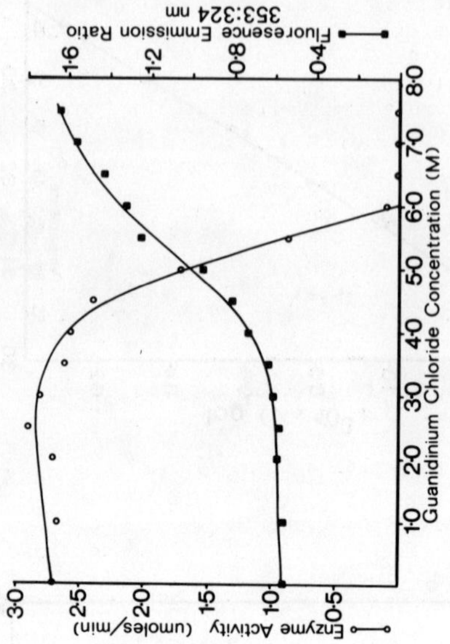

Fig. 5. Effect of guanidine chloride on T. aquaticus GPDH

Samples of the respective enzymes were heated at the indicated temperatures for 20 min, cooled in ice, and subsequently assayed at 20°C. The "melting temperatures" (i.e. the temperature at which 50% activity was lost in 20 min) were 55°C for the rabbit holo-enzyme, 75°C for the B. stearothermophilus enzyme and 100°C for the T. aquaticus enzyme. It should also be noted that removal of the coenzyme (NAD^+) by treatment with charcoal, or following chromatography on phosphocellulose (14), had no appreciable effect on the thermal stability of the enzyme.

Stability in denaturing solvents. T. aquaticus GPDH is also remarkably stable in denaturing solvents such as urea, guanidine-HCl and SDS. Thus it retains its activity fully in 8 M urea and in 0.1% SDS at pHs >7.0. The enzyme is also stable in up to 4 M Gu.HCl but inactivates at higher concentrations of the denaturant. As shown in Fig. 5 loss of activity is correlated with exposure to the solvent of "buried" tryptophan residues indicating that the enzyme is denatured in 4 to 6 M Gu.HCl. The inactivation is however reversible, and dilution of a totally inactive sample in 6 M Gu.HCl into the standard assay mixture leads to an appreciable recovery of activity following a lag period of approximately 45 sec (for details see (14)).

PRIMARY STRUCTURE OF T. AQUATICUS GPDH

T. aquaticus GPDH was carboxymethylated with [2-^{14}C]iodoacetic acid in 6 M Gu.HCl and the S-carboxymethylated derivative was submitted to amino acid sequence analysis using standard procedures. Full details of the work are given in reference (14) and are summarised only briefly here.

A provisional sequence of 35 residues from the N-terminus, as determined by automated analysis in a Beckman 890B Sequencer, was as follows:

Met-Lys-Val-Gly-Ile-Asn-Gly-Phe-Gly-Arg-Ile-Gly-Arg-Glx-Val-Phe-Arg-

Ile-Leu-His-Ser-Arg-Gly-Val-Glu-Val-Ala-Leu-Ile-Asx-Asp-Leu-Thr-Asx-Asx-

The remainder of the sequence was deduced from sequences of peptide fragments obtained by tryptic (before and after citraconylation of lysines), peptic, and CNBr cleavage of the S-carboxymethylated protein. The sequence information obtained from fragments that were isolated from these digests was sufficient to establish the sequence of 330 of the estimated 332 residues in the protein subunit. This provisional sequence is given in Fig. 6. Comparison of the sequences of the T. aquaticus and lobster muscle enzymes (Fig. 7) indicates a sequence homology of 50% between the mesophile (eucaryotic) and thermophile (procaryotic) enzymes. A similar comparison of the amino acid sequences of T. aquaticus and B. stearothermophilus (unpublished results of J. Walker, J. Bridgen and J.I. Harris) GPDH shows a sequence homology of 59% (over the 230 residues that have so far been compared). A further comparison of the T. aquaticus, lobster and pig, and yeast sequences reveals that 125 residues (38%) are identical in the four enzymes.

DISCUSSION AND GENERAL CONCLUSIONS

A. GPDH structure and function. The overall structure of enzyme from T. aquaticus is very similar to that of enzyme from mesophiles. Thus the number (4) and molecular size (36,000) of subunits is the same. Moreover the sequence homology between T. aquaticus GPDH and each of the three mesophile enzymes so far studied is in the range of 48 to 50% which is in general agreement with the figure of

10 20
Met-Lys-Val-Gly-Ile-Asn-Gly-Phe-Gly-Arg-Ile-Gly-Arg-Gln-Val-Phe-Arg-Ile-Leu-His-

30 40
Ser-Arg-Gly-Val-Glu-Val-Ala-Leu-Ile(Asn,Asp)Leu-Thr(Asn,Asp)Lys-Thr-Leu-Ala-His-

50 60
Leu-Leu-Lys-Tyr-Asp-Ser-Ile-Tyr-His-Arg-Phe-Pro-Gly-Glu-Val-Ala-Tyr-Asp-Asp-Gln-

70 80
Tyr-Leu-Tyr-Val-Asp-Gly-Lys-Ala-Ile-Arg-Ala-Thr-Ala-Val-Lys-Asp-Pro-Lys-Glu-Ile-

90 100
Pro-Trp-Ala-Glu-Ala-Gly-Val-Gly-Val-Val-Ile-Glu-Ser-Thr-Gly-Val-Phe-Thr-Asp-Ala-

110 120
Asp-Lys-Ala-Lys-Ala-His-Leu-Glu-Gly-Gly-Ala-Lys-Lys-Val-Ile-Ile-Thr-Ala-Pro-Ala-

130 140
Lys-Gly-Glu-Asp-Ile-Thr-Leu-Val-Met-Gly-Val-Asn-His-Glu-Ala-Tyr-Asp-Pro-Ser-Arg-

150 160
His-His-Ile-Ile-Ser-Asn-Ala-Ser-Cys-Thr-Thr-Asn-Ser-Leu-Ala-Pro-Val-Met-Lys-Val-

170 180
Leu-Glu-Glu-Ala-Phe-Gly-Val-Glu-Lys-Ala-Leu-Met-Thr-Thr-Val-His-Ser-Tyr-Thr-Asx-

190 200
Asx-Glx-Arg-Leu-Leu-Asp-Leu-Pro-His-Lys-Asp-Leu-Arg-Arg-Ala-Arg-Ala-Ala-Ala-Ile-

210 220
Asn-Ile-Ile-Pro-Thr-Thr-Thr-Gly-Ala-Ala-Lys-Ala-Thr-Ala-Leu-Val-Leu-Pro-Ser-Leu-

230 240
Lys-Gly-Arg-Phe-Asp-Gly-Met-Ala-Leu-Arg-Val-Pro-Thr-Ala-Thr-Gly-Ser-Ile-Ser-Asp-

250 260
Ile-Thr-Ala-Leu-Leu-Lys-Arg-Glu-Val-Thr-Ala-Glu-Glu-Val-Asn-Ala-Ala-Leu-Lys-Ala-

270 280
Ala-Ala-Glu-Gly-Pro-Leu-Lys-Gly-Ile-Leu-Ala-Tyr-Thr-Glu-Asp-Glu-Ile-Val-Leu-Glx-

290 300
Asx-Ile-Val-Met-Asp-Pro-His-Ser-Ser-Ile-Val-Asp-Ala-Lys-Leu-Thr-Lys-Ala-Leu-Gly-

310 320
Asn-Met(? ? Lys)Val-Phe-Ala-Trp-Tyr-Asp-Asn-Glu-Trp-Gly-Tyr-Ala-Asn-Arg-Val-

330
Ala-Asp-Leu-Val-Glu-Leu-Val-Leu-Arg-Lys-Gly-Val

Fig. 6. Amino acid sequence of T. aquaticus GPDH.

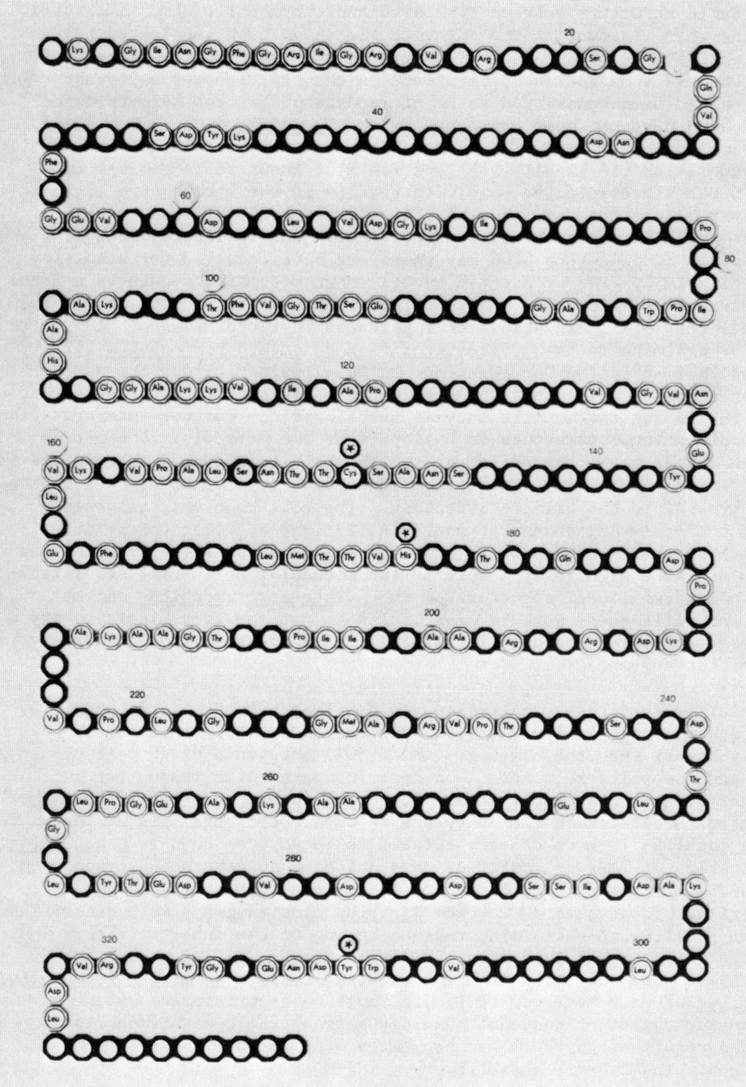

Fig. 7. Residues common to lobster muscle and T. aquaticus GPDH (50% sequence homology).

two point mutations per 100 residues per 10^8 years calculated for GPDH (14). The similarity in structure between mesophile and thermophile GPDH is particularly striking when residues in the sequence that have been implicated in forming the active site of the lobster muscle enzyme (10) are considered (cf. (8)). Nevertheless some residues, invariant in the pig, lobster and yeast enzymes, which had been considered to be of importance for the structure and mechanism of GPDH have not been conserved in the _T. aquaticus_ enzyme. Among these are Cys-153 and Lys-183. The former had been conserved in all the GPDHs hitherto investigated (14 in all (15)) and was also known to be capable of forming an intrachain disulphide bond with Cys-149 in the active site of these enzymes. Nevertheless the three-dimensional structure of the lobster enzyme (10) does not indicate any special role for Cys-153 and its replacement by a serine residue in _T. aquaticus_ GPDH may therefore have occurred for stability reasons. Similarly Lys-183, conserved in all known mesophile sequences and the site of a specific acyl-transfer reaction involving Cys-149 in both muscle and fungal enzymes (cf. 8,9), had also been implicated in the binding of NAD^+. Its replacement by arginine in the _T. aquaticus_ (and _B. stearothermophilus_ (12)) enzymes shows that acyl-transfer reactions to this lysine residue are not important for enzyme function but the significance (if any) of its replacement by arginine in the two thermophile enzymes must await the further interpretation of the three-dimensional structure in this part of the molecule. Another sequence change that has occurred in the two thermophile enzymes has resulted in the introduction of a His-His sequence (residue numbers 141-142, Fig. 6) in the vicinity of Cys-149 in the primary structure. The fact that this sequence change (like the Lys$\rightarrow$Arg change at position 183) had occurred (presumably independently) in the two thermophile enzymes raised the possibility (15) that changes such as these might play a role in the mechanism of thermal stabilisation. The three-dimensional structure of the _B. stearothermophilus_ enzyme (A Wonacott and colleagues, work in progress) does not, however, indicate any role for these particular residues in maintaining the tertiary and quaternary structures of the enzyme.

It may be recalled that the overall sequence homology between the _T. aquaticus_ and _B. stearothermophilus_ enzymes (59% of the compared (230) residues) is significantly higher than the homology (48 to 50%) between a given pair of thermophile and mesophile sequences. There is no particular reason for supposing that _B. stearothermophilus_ and _T. aquaticus_ are necessarily more closely related to each other than they are to eucaryotic organisms or that _B. stearothermophilus_ is more closely related to eucaryotes than is _T. aquaticus_. The classification of _Thermus aquaticus_ does not seem to have been agreed upon, but as a gram negative non-spore forming rod, it could belong to any one of several orders of procaryotes (2) whereas _Bacillus_ species are a well classified group of gram positive spore forming rods belonging to the Eubacteriales. Thus _a priori_ there is no way of predicting from evolutionary relationships the degree of sequence homology that might be expected between _B. stearothermophilus_ and _T. aquaticus_ GPDHs. Moreover it is difficult to ascertain the extent to which the greater degree of sequence homology between the two thermophile enzymes is the result of phylogenetic proximity - or an index of a common approach to achieving thermal stability.

The ability of living organisms, and of procaryotes in particular, to adapt to extreme environmental conditions suggests that insofar as it involves the synthesis of thermostable proteins the mechanism of thermal adaptation among thermophiles has ample precedent in the world of mesophiles. Adaptation of organisms to a thermophilic environment is likely to have occurred on many separate occasions in evolutionary history. Thus there is no reason to consider organisms to be necessarily closely related phylogenetically simply because they are adapted to a thermal environment. In particular, the taxonomic position of a thermophilic *Bacillus* is amongst other predominantly mesophilic *Bacilli* and not amongst thermophiles of other genera, so that *B. stearothermophilus* and *T. aquaticus* may have acquired thermal stability independently and from different starting points. For this reason a detailed comparison of the three-dimensional structures of GPDHs from the two thermophiles is of considerable interest.

There have been numerous attempts to correlate differences in the amino acid compositions of proteins with their thermal stability. In the case of GPDHs such correlations have not been particularly fruitful (14). Thus, there is no obvious correlation between thermal stability and content of hydrophobic amino acids or of amino acids with side chains capable of forming hydrogen bonds. On the other hand there appears to be a positive correlation between thermal stability and acidic isoelectric points (4.3 and 4.8 for *B. stearothermophilus* and *T. aquaticus* GPDHs compared to values close to 7.0 for mesophile enzymes) and a decreased content of cysteine residues (cf. 6). In the GPDHs there are five cysteines in the lobster enzyme, four in pig, two in yeast, *E. coli* and *B. stearothermophilus* but only one in the *T. aquaticus* enzyme. Other enzymes from extreme thermophiles have also been found to be entirely devoid of cysteine residues. These include enolases from *Thermus* X-1 and *T. aquaticus* YT-1 (16) and phosphofructokinase from *T. aquaticus* (17). This finding that non-essential cysteines are frequently replaced in enzymes from extreme thermophiles (in GPDH one cysteine, Cys-149, is essential since it participates directly in the catalytic reaction) is perhaps not surprising since thiol groups are susceptible to oxidation and to disulphide bond formation and are therefore potential sites for inactivation of enzymes that are required to function *in vivo* aerobically and at temperatures in excess of 70°C. Another positive correlation has been made between thermal stability and content of leucine and alanine. Thus in GPDHs there are 18 leucine residues in the pig and lobster enzymes, 22 in yeast, 28 in *B. stearothermophilus* and 31 in the *T. aquaticus* enzyme. Similarly there are 32 alanine residues in the pig and lobster enzymes, 33-34 in yeast, 39 in *B. stearothermophilus* and 42 in the *T. aquaticus* enzyme. The interest in these residues, and particularly in the leucine, stems from work by Chou and Fasman (18) who concluded on the basis of statistical data gathered from the tertiary structures of 15 proteins, that certain amino acids occur predominantly in characteristic types of secondary structure. Thus, for example, an increased content of leucine would be expected to result generally in a more stable secondary structure while alanine gives rise to more stable helices. On the other hand comparison of the amino acid compositions of mesophile and thermophile enolases (16) and phosphofructokinases (17) has not revealed a similar trend with respect to leucine and alanine and if the increased content of these residues is connected with the stabilisation of thermophile GPDHs it is obviously not a general factor in the heat stabilisation of all thermophile proteins. Complete three-dimensional structure analysis will be required in order to ascertain whether the structure forming amino acids (cf. 18) are located in regions of appropriate secondary structure.

B. General considerations. In considering the phenomenon of thermophily it is important to recognise that thermophile proteins must possess thermodynamic as well as kinetic stability, since newly synthesised polypeptide chains are required to fold correctly so as to form stable tertiary (and in the case of subunit enzymes, quaternary) structures *in vivo* at the elevated growth temperatures of thermophiles. On the other hand, to put the problem into perspective it is also important to recognise that, on the *Absolute* scale, temperatures of 80°C (353°A) are only of the order of 15% higher than temperatures of normal mesophile growth (i.e. 35° to 40°C, 308° to 313°A). Moreover since ΔG^* (free energy of activation), which may be taken as a measure of the degree of thermal denaturation of a protein, is logarithmically proportional to the rate constant for inactivation, k, relatively modest differences in ΔG^* can account for large differences in the thermal stabilities of proteins.

For example, comparison of the thermal stabilities of rabbit muscle and *B. stearothermophilus* triose phosphate isomerase (TIM) has shown that a seemingly large (30-fold) difference in stability at 60°C results from a difference in ΔG^* of only 2.2 kcal/mole (19, and unpublished results of R.C. Fahey and J.I. Harris). Similarly the total additional free energy of stabilisation provided by extra salt bridges and hydrogen bonds in the three-dimensional structure of thermophile ferredoxins amounts to no more than 4.5 to 8.5 kcals/mole (20). It follows from considerations such as these that the three-dimensional structure of a thermophile protein could be expected to be closely similar to that of its mesophile counterparts. Moreover it becomes clear that thermal stabilisation of protein structures to temperatures close to the boiling point of water (373°A, i.e. barely 20% higher than the growth temperatures of mesophiles) does not require the formation of highly specialised or novel bonds. Bond energies of up to 10 kcals/mole may be gained in a variety of different ways and the types of bonds that are commonly involved in stabilising proteins at mesophilic temperatures (such as salt bridges, hydrogen bonds and hydrophobic bonds) appear to be perfectly adequate for stabilising proteins in thermophiles. Implicit in this conclusion is that different proteins, even in the same thermophilic organism, will acquire thermal stability by a variety of different combinations of bonding features. These in turn will reflect different sequence changes in the primary structure such that the sought after "common structural mechanisms of thermophily" are unlikely to be discovered.

Comparison of the three-dimensional structures of GPDH from mesophiles and thermophiles may ultimately result in the identification of the structural elements that have evolved in order to stabilise the tertiary and quaternary structure of *this particular* enzyme. More important, however, is the knowledge already gained that the structure of an intracellular multi-subunit enzyme, such as GPDH, from an extreme thermophile such as *T. aquaticus*, can be expected to be closely similar to the structure of the same enzyme from a mesophile. We suggest that it is no longer necessary to consider a thermophile enzyme to be a "special case", but rather as a legitimate prototype that may be studied in the knowledge that its structure will be closely similar to and therefore representative of the structure of that enzyme throughout Nature.

Thermophile enzymes will be seen to possess often decisive practical advantages for the study of the structure and function of enzymes. As proteins they greatly extend the range of conditions that may be used for purification and crystallisation, as well as for physico-chemical and organo-chemical studies of proteins of biological interest. The existence in living organisms of proteins that are both thermodynamically and kinetically stable at elevated temperatures poses several interesting questions. Did life originate in a thermophilic environment and are present day thermophiles direct descendants of more primitive ancestral forms? Why are mesophile proteins heat labile? Attitudes towards the structural stability of proteins appear to have been unduly influenced by the properties of mesophile proteins that not only fail to fully exploit the potential stabilisation energy that is inherently available in a folded polypeptide structure but which may also have been deliberately destabilised to ensure adequate turnover of cellular constituents at normal physiological temperatures.

"Plus ca change, plus c'est la même chose."

REFERENCES

1. Brock, T.D. (1967). Science 158, 1012.

2. Brock, T.D. and Freeze, H. (1969). J. Bact. 98, 289.

3. Freeze, H. and Brock, T.D. (1970). J. Bact. 101, 541.

4. Suzuki, K. and Harris, J.I. (1971). FEBS Letters 13, 217.

5. Kolb, E. and Harris, J.I. (1971). Biochem. J. 124, 76P.

6. Bridgen, J., Kolb, E. and Harris, J.I. (1973). FEBS Letters 33, 1.

7. Pearse, B.M.F. and Harris, J.I. (1973). FEBS Letters 38, 49.

8. Harris, J.I. and Waters, M. (1975). In The Enzymes Vol. XIII, ed. P.D. Boyer (Academic Press, New York), in Press.

9. Harris, J.I. (1972). In Structure and Function of Oxidation-Reduction Enzymes, eds. A. Akeson and A. Ehrenberg, p.639 (Pergamon Press, Oxford).

10. Buehner, M., Ford, G.C., Moras, D., Olsen, K.W. and Rossmann, M.G. (1975). J. Mol. Biol. 90, 25.

11. Singleton, R., Kimmel, J.R. and Amelunxen, R.E. (1969). J. Biol. Chem. 244, 1623.

12. Bridgen, J. and Harris, J.I. (1973). Abs. 2el, 9th Int. Cong. Biochem., Stockholm, p.59.

13. Hocking, J.D. and Harris, J.I. (1973). FEBS Letters 34, 280.

14. Hocking, J.D. (1974). Ph.D. Dissertation, University of Cambridge.

15. Bridgen, J., McDonald, P., Harris, J.I., Amelunxen, R.E. and Kimmel, J. (1972). J. Bact. 111, 797.

16. Barnes, L.D. and Stellwagen, E. (1973). Biochemistry 12, 1559.

17. Hengartner, H. and Harris, J.I. (1975). FEBS Letters 34, 280.

18. Chou, P.Y. and Fasman, G.D. (1974). Biochemistry 13, 211.

19. Fahey, R.C., Kolb, E. and Harris, J.I. (1971). Biochem. J. 124, 77P.

20. Perutz, M.F. and Raidt, H. (1975). Nature 255, 256.

THERMAL PROPERTIES OF GLYCERALDEHYDE 3-PHOSPHATE DEHYDROGENASE FROM ESCHERICHIA COLI

A. Fontana, C. Grandi, E. Boccu, and F.M.Veronese

Institute of Organic Chemistry and of Pharmaceutical Chemistry, University of Padova, 35100 Padova, Italy

Summary: The molecular properties of glyceraldehyde 3-phosphate dehydrogenase from *E. coli* have been evaluated by circular dichroism and fluorescence emission spectroscopy measurements, with the purpose of studying the structural properties which are relevant for a comparison with the enzyme from the obligate thermophile *Bacillus stearothermophilus*.

The enzyme is moderately resistant to heat treatment, being pratically stable when heated for 10 min at 50°C and completely inactivated when heating was performed at 60°C. The secondary structure of the *E. coli* GPDH appears to be predominantly β-structure as judged by circular dichroism, showing a negative band centered at about 219 nm. The emission fluorescence of the enzyme shows a maximum at 333 nm upon excitation at 295 nm. In the native *E. coli* enzyme the tryptophan residues seem to be buried in a hydrophobic region rather than exposed to a polar environment.

The structure of the enzyme did not change up to about 50°C, at which temperature thermal inactivation takes place . Upon denaturation the circular dichroic signal at 219 nm gradually decreases, and a red shift of the emission maximum from 333 nm to *ca.* 345 nm upon heating is indicative that the native structure of the enzyme is unfolded, the tryptophan being exposed to the solvent medium.

Since it has been found that the *E. coli* GPDH closely resembles in many of its properties the *B. stearothermophilus* enzyme, this bacterial enzyme seems to be useful for comparison with the thermophilic enzyme in studies of its thermostability.

1. Introduction

Studies on the thermostable glyceraldehyde 3-phosphate dehydrogenase (GPDH)* from the obligate thermophile *B. stearothermophilus* are under way in several laboratories with the aim of clarifying the mechanism of thermophily of enzymes. Sequence analysis of the subunit protein chain and determination of the three-dimensional structure of the crystalline enzyme by X-ray methods are being carried out by Harris and coworkers [1]. Recently, an extensive physico-chemical investigation of this thermophilic GPDH using circular dichroism and emission fluorescence measurements has been reported by Suzuki and Imahori [2].

It seemed to us worthwhile to carry out a physico-chemical study on the structural properties, including thermostability, of GPDH from *E. coli*, because

*Abbreviations: GPDH, D-glyceraldehyde 3-phosphate dehydrogenase (EC 1.2.1.12)

of widespread interest on the structural, catalytic and allosteric properties of GPDH in general and also because of the importance of having available structural data on GPDH from bacterial mesophilic source.

The basic properties of GPDH from E. coli have been studied [3-5]. It has been shown that the enzyme closely resembles its counterparts from other sources in many of its properties including molecular weight, tetrameric quaternary structure, amino acid composition and behaviour toward sulfhydryl reagents. Earlier studies by Allison [3] have shown that the E. coli enzyme has the same active center peptide sequence as 13 other dehydrogenases from diverse sources. However data are not so far available on the secondary structure of this mesophilic GPDH and on its thermal properties.

From the above considerations it appears that the E. coli GPDH could be usefully employed in comparative studies of its properties which are relevant for a comparison with the enzyme from B. stearothermophilus. In the work presented here, the effect of temperature on the stability of the E. coli GPDH has been examined. The results on the stability of activity of the enzyme have been related to the stability of its structure, as determined by circular dichroism (CD) and fluorescence emission measurements.

The aim of these studies is to provide useful information on the structural stability of the mesophilic bacterial enzyme as compared to the thermophilic B. stearothermophilus GPDH, in an effort to contribute to the elucidation of the problem of thermostability of proteins.

2. Materials and methods

2.1. Reagents. The barium salt of diethyl acetal glyceraldehyde-3-phosphate was obtained from Sigma and hydrolyzed in the manner suggested by the manufacturer. NAD, β-mercaptoethanol, tris(hydroxymethylamino)-methane(Tris), EDTA, charcoal(Norit A) were all obtained from Serva. The charcoal was extensively washed with acid, alkali and ethanol in turn and activated by heating. All other chemicals were reagent grade and used without further purification.

2.2 Enzyme preparation. GPDH was prepared from E. coli (MRE 600) by a multi-enzyme purification procedure leading to the simultaneous isolation of glutamate dehydrogenase, 6-phosphogluconate dehydrogenase, glyceraldehyde-3-phosphate dehydrogenase and enolase.

Preparations of the E. coli GPDH used in this study have been demonstrated to be homogeneous by disc gel electrophoresis with and without sodium dodecylsulfate. In addition, a single coincident peak of protein and activity was obtained from a Bio-Gel A-0.5 m column at the last step of purification.

The enzyme was stored as an $(NH_4)_2SO_4$ suspension at 4°C. Under these conditions the enzyme is stable for months.
Solutions of the enzyme were prepared by centrifugation of an aliquot of the (NH) SO suspension, removal of the supernatant solution and addition of a proper amount of buffer. The solution was then dialyzed against the same buffer containing 3 mM β-mercaptoethanol.

The GPDH from E. coli used in this study exhibited an $A_{280/260\,nm}$ ratio of 1.5. The apoenzyme was prepared by charcoal treatment; the preparation gave an $A_{280/260\,nm}$ ratio of 1.8.

The enzyme containing the full complement of NAD (4 moles per molecule of enzyme) was prepared by adding 10 equivalents of NAD to a solution of the enzyme in the Tris/EDTA/β-mercaptoethanol buffer. The solution was then passed through a Sephadex G-25 (1 x 30 cm) column equilibrated with the same buffer. The enzyme fractions were collected and the spectrum of the NAD-loaded enzyme recorded. The ratio $A_{280/260nm}$ was found to be 1.08. This value is in good agreement with the 1.15 figure reported earlier [5], and corresponds to the enzyme with 4 moles of NAD bound to the protein molecule.

2.3. Enzyme assay. Enzyme activity was determined following the method used by D'Alessio and Josse [5] for the assay of the *E. coli* GPDH. Cuvettes containing 0.8 ml of the assay mixture were placed in the cell holder of the spectrophotometer equilibrated at 25°C. The assay was initiated by addition of enzyme solution (*ca.* 0.05-0.3 μg/5 μl). The increase in absorbance at 340 nm was determined during the first 30 sec of the reaction in a Varian Tecktron recording spectrophotometer. Activity was calculated from the initial velocity of the reaction. Protein concentrations were determined spectrophotometrically at 280 nm using an extinction coefficient $E^{0.1\%}_{280\ nm} = 1.0$ [5].

2.4. Heat inactivation. Enzyme samples (0.1 mg/ml) in 5 mM Tris-HCl buffer, pH 7.5, 1 mM EDTA and 3 mM β-mercaptoethanol were incubated in a thermostatically controlled bath. Aliquots of the solution were withdrawn on a time schedule and immediately assayed for enzyme activity.

2.5. Circular dichroism. CD spectra were recorded on a Cary 61 spectropolarimeter. The cell compartment was flushed with dry nitrogen. Spectra were routinely corrected for base-line shifts by running samples of solvent buffer. The temperature of the cell compartment was controlled by circulating an ethylene glycol-water mixture from a Haake water bath. The temperature inside the cell was measured with a thermistor connected through the cap of the cell. A CD scale setting of 0.02 and 0.05 was used in all experiments. When measurements were performed at higher temperature, the cell was preincubated in a water bath for 3 min at each temperature before measurements. The CD scans were generally run in duplicate.

Results are expressed as mean residue ellipticity, [θ], in degrees cm^2/decimole, calculated using the equation where θ is the observed ellipticity, l is the path length in centimeters, c is the protein concentration in grams per ml

$$\theta = \frac{\theta \times 109}{10 \times l \times c}$$

and 109 is the mean residue weight of the *E. coli* GPDH calculated from its amino acid analysis.

2.6 Fluorescence. The fluorescence measurements were performed with a Hitachi-Perkin Elmer MPF 2A spectrofluorimeter, equipped with a thermostated cuvette holder and connected to a Hitachi QPD_{33} recorder. The temperature inside the cell was controlled and measured as indicated in the CD measurements. The excitation wavelength was 295 nm.

3. Results

The enzyme preparation used in this study was highly homogeneous as judged by various criteria of purity. The ratio of $A_{280/260nm}$ was 1.5, a figure that is in excellent agreement with that (1.6) reported previously [5]. This figure indicates that the enzyme has been isolated with a low NAD content [6].

Since the GPDH from B. stearothermophilus is isolated containing the full complement of NAD (4 moles/mole of enzyme) [1,7], it was of interest to prepare the holoenzyme from E. coli, in order to be able to compare the mesophilic and thermophilic enzyme in similar molecular organization. This was easily achieved by incubating the GPDH from E. coli with 10 equivalents of NAD and then separating the excess of coenzyme by gel filtration. Holoenzyme prepared by this procedure gave an $A_{280/260nm}$ of 1.08, which is indicative of 4 moles of enzyme bound NAD [8,9].

It has been found in this study that the native GPDH from E.coli and the enzyme saturated with NAD (holoenzyme) have the same enzymological and conformational properties, including thermostability. Therefore, the results here reported are referred to the native enzyme.

On the other hand, the apoenzyme prepared from the native enzyme by charcoal treatment [8], showed some differences in respect to the native one, and in particular less stability against heat and protein denaturants.

The heat stability of the mesophilic enzyme was investigated and the results are shown in Fig. 1 and 2. Heating at 48°C for 30 minutes did not affect

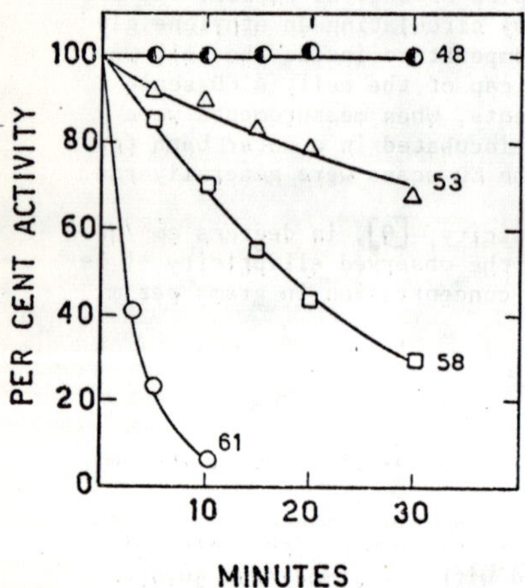

FIG. 1. Heat inactivation of E.coli GPDH. (A) The enzyme (0.1 mg) was dissolved in 2 ml of 5 mM Tris-HCl buffer, pH 7.5, with 1 mM EDTA and 3 mM β-mercaptoethanol and the solution heated in a closed vial at the indicated temperature. Samples were withdrawn on a time schedule and assayed for enzyme activity.

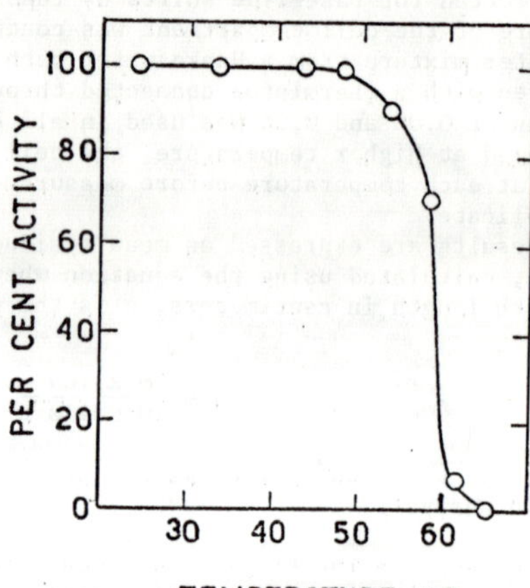

FIG. 2. The enzyme was heated for 10 min at the indicated temperature under the same conditions as reported under Fig. 1 and then the remaining activity was determined.

the enzyme activity, whereas at temperatures above 60°C the enzyme lost activity within a few minutes. The percentage activity remaining after heating for 10 minutes at various temperatures is shown in Fig. 2. The enzyme is stable only up to about 50°C, whereas the B. stearothermophilus enzyme retains its activity after heating at 80°C for 5 minutes [1]. The thermal stability of the E.coli enzyme is of the same order to that found with the B. cereus enzyme recently studied by Suzuki and Imahori [10].

The secondary structure of GPDH from E. coli was examined by measuring CD spectra (Fig. 3).

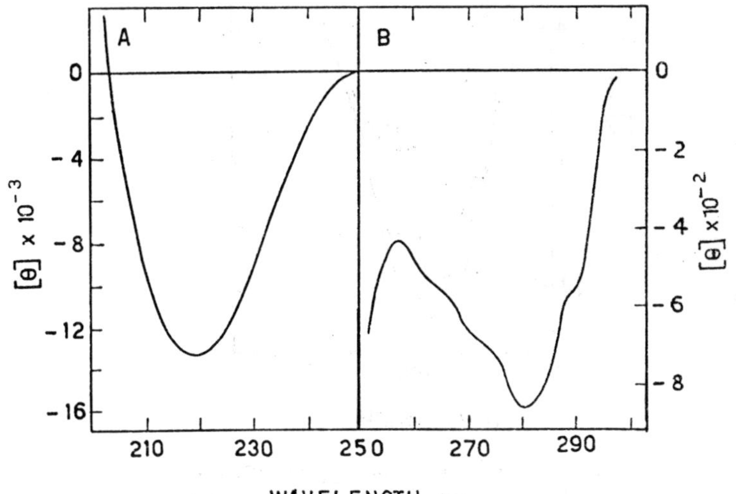

FIG. 3. CD spectra of the E. coli GPDH in 50 mM Tris-HCl buffer, pH 7.5, 1 mM EDTA and 3 mM β-mercaptoethanol. The spectra were measured in 0.1 mm and 100 mm quartz cells at protein concentrations of 0.03 mg/ml and 0.05 mg/ml in the 250-195 nm and 320-250 nm region respectively. Spectra were run at room temperature.

The trough at 219 nm is typical for a polypeptide chain containing β-structure (11). The calculated molar ellipticity value at 219 nm was found 12,800∓300, a figure which is remarkably similar to that found already with rabbit muscle [12] as well as with other bacterial GPDH'S [2,10,13].

The essential features of the spectrum in the 280 nm region agree well with those reported for the B. stearothermophilus [2] as well as for the rabbit muscle enzyme [12]. A trough appears at 280 nm and shoulders at 270 and 290 nm.

The far uv CD spectrum of the native, apo-and holoenzyme were similar, this fact being indicative that the secondary structure of the enzyme is independent of the binding of the coenzyme NAD. Similar observ.ation has been already reported for the rabbit muscle enzyme [12].

The effect of temperature on the structure of GPDH from E.coli was next examined by CD measurements. As shown in Fig. 4, no appreciable changes in shape

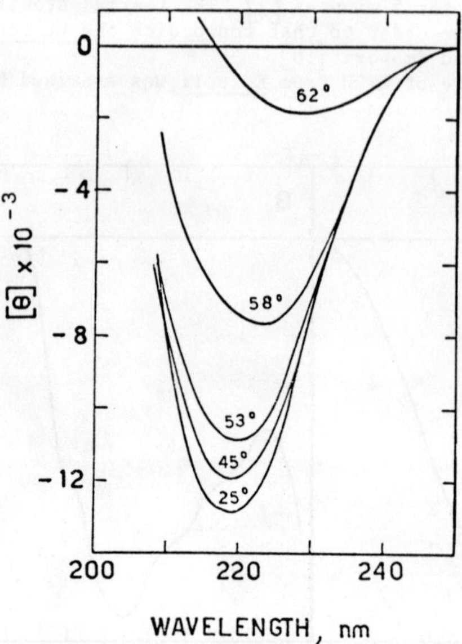

FIG. 4. CD spectra of the E. coli GPDH at various temperatures. Conditions are as described in Fig. 2 and in the Experimental Section. Figures near the curves indicate the temperature at which spectra were recorded.

of the spectrum and ellipticity value at 219 nm have been observed up to about 50°C, but above this temperature significant changes occurred.

The thermostability of the secondary structure of the E. coli enzyme was also studied using the melting profile method recently proposed by Fujita and Imahori [13]. Essentially, the method consists in continuous recording of the CD signal at a fixed wadelength, while the sample is heated at a constant rate. From a plot of ellipticity value and temperature versus time, the direct relationship between ellipticity and temperature can be derived. Fig. 5 shows the results of a melting experiment with the E. coli GPDH, where the ratio of $[\theta]/[\theta]_\circ$ ($[\theta]$, ellipticity at room temperature) is plotted against temperature. The first derivative curve ($d([\theta]/[\theta]_\circ / dT)$, as obtained by evaluating the slope of the line connecting successive data points, is also shown in Fig. 5. The maximum value of the derivative curve (58°C) can be taken as the measure of the melting temperature (Tm). This graphical method of obtaining Tm is at some variance of that proposed by Fujita and Imahori [13].

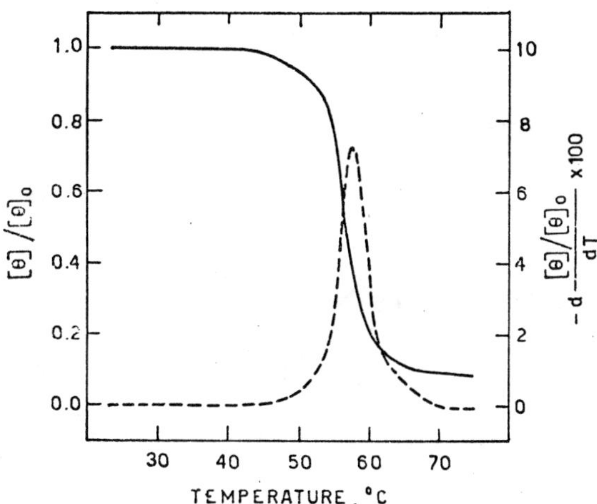

FIG. 5. The melting profile of the E. coli GPDH as determined by CD measurements. The solid line represents the ratio $[\theta]/[\theta]_o$ at 219 nm with $[\theta]_o$ being the ellipticity value at room temperature; the dashed line shows the derivative curve as a function of temperature. The protein concentration was 0.05 mg/ml in 50 mM Tris-HCl buffer, pH 7.5, 1 mM EDTA and 3mM β-mercaptoethanol.

When the melting profile experiment was carried out with an equimolar mixture of both the E. coli and B. stearothermophilus enzyme, the profile of ellipticity versus temperature was clearly biphasic (Fig. 6). The first derivative curve showed two distinct maxima at about 58 and 72°C. Since this last value has been obtained also when the melting experiment was performed on the thermophilic enzyme alone, it can be concluded that the two proteins in a mixture fall apart independently on raising the temperature.

The structural properties, and the thermostability of the E. coli GPDH has been evaluated also by emission fluorescence properties of the 4 tryptophan residues per subunit of the protein [5]. The fluorescence spectrum of the enzyme showed a maximum at 333nm upon excitation at 295 nm (Fig.7). In these conditions the measured fluorescence is pratically due only to the tryptophan residues.

In Fig. 7 the effect of temperature on the fluorescence properties of GPDH from E. coli is also shown. By increasing the temperature the intensity of fluorescence is gradually reduced, this fact being related to thermal quenching [15]. Up to 55°C the protein does not show shifts in the wavelength of maximum emission, but above this temperature a red shift is observed, this fact being indicative of unfolding of the protein molecule [14].

At 68°C the maximum of emission is shifted to ca 345 nm, a figure close to that found with simple tryptophyl compounds [14]. On cooling the protein

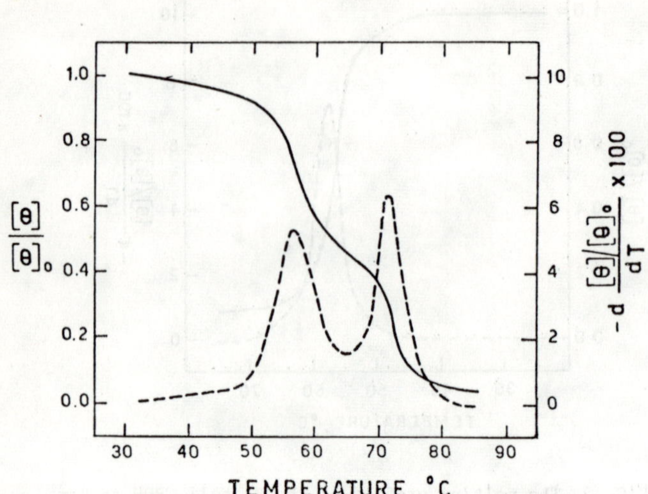

FIG. 6. Melting profile of a mixture of E. coli and B. stearothermophilus GPDH. The experiment was carried out in the same conditions as described under Fig. 5.

sample heated at 68°C the maximum of emission remains constant at 345 nm, signifying an irreversible unfolding of the protein molecule.

A melting profile experiment similar to that above reported using the CD technique has been also carried out by fluorescence measurements. The thermal quenching of free tryptophan has been shown to produce a monotonic and almost linear variation of intensity of fluorescence versus temperature, although the shape of the emission spectra is unaffected by heating [15]. On the other hand, the effect of temperature on the fluorescence properties of various proteins has been found different from that of free tryptophan or simple tryptophyl peptides. In fact, if conformational transitions are occurring in a protein on heating, the chemical environment of tryptophan residues are being changed and consequently the quantum yield of these residues is affected.

Fig. 8 shows the temperature effect on the intensity of fluorescence of GPDH from E. coli. A transition is clearly evident in the 50-60°C region. The corresponding derivative curve more clearly shows the transition occurring. If the maximum of the curve will be taken as a measure of the melting temperature, the same temperature of 58°C is obtained using the fluorescence technique as above reported using CD measurements.

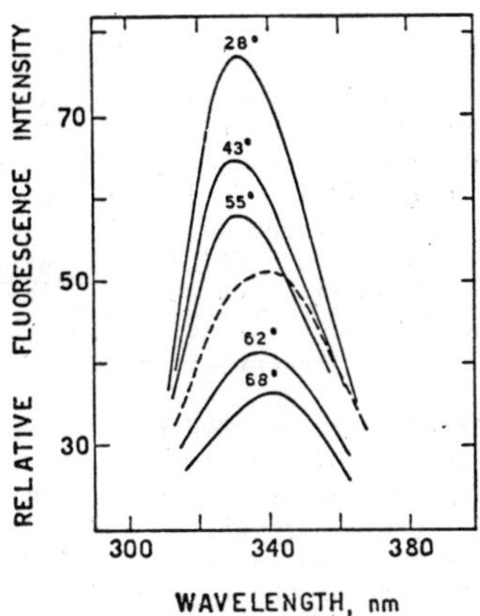

FIG. 7. Fluorescence emission spectra of E. coli GPDH in 50 mM Tris-HCl buffer, pH 7.5, 1 mM EDTA and 3 mM β-mercaptoethanol (0.02 mg/ml) at various temperatures. Figures near curves indicate temperatures at which spectra were recorded. The dashed line shows the fluorescence spectrum of the enzyme sample heated at 68°C and then cooled at room temperature.

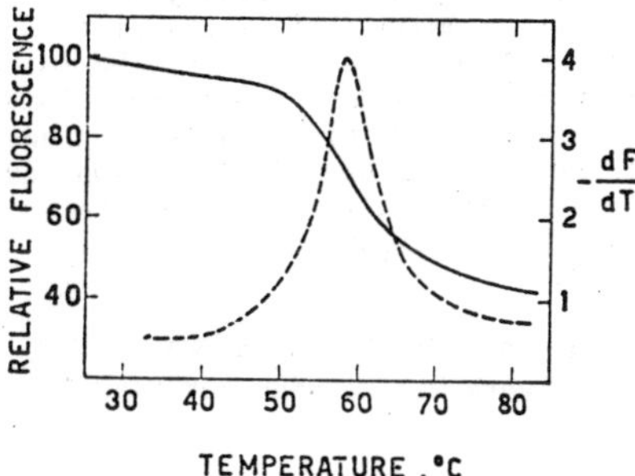

FIG. 8. Corresponding integral (—) and derivative (---) curve of intensity of fluorescence of GPDH from E. coli. Denaturation was produced by continuous heating (2°C/min) a protein sample at a concentration of 0.02 mg/ml in 5 mM Tris-HCl buffer, pH 7.5, 1 mM EDTA and 3mM β-mercaptoethanol.

4. Discussion

The secondary structure of the *E. coli* GPDH, as determined by circular dichroism measurements, resembles that of the enzyme from other sources. The circular dichroic spectrum shows an extremum at 219 nm and the molar ellipticity is *ca.* -12,800 $\pm$300 deg. cm /decimole. It is worth of note that essentially the same far uv circular dichroic spectrum has been reported for GPDH from rabbit muscle [12] and from bacteria, including *B. stearothermophilus* [2], *Thermus thermophilus* [13] and *Bacillus cereus* [10]. The reported figures for the molar ellipticity ranged from -12,000 to -13,500. These results suggest that the characteristically high content of β-structure is common to GPDH from a number of sources.

The fluorescence emission peak of GPDH from *E. coli* at 333 nm is in accord with that observed for many native proteins and is considerably blue-shifted with respect to simple tryptophyl peptides in aqueous solution. The emission maximum of the *B. stearothermophilus* enzyme is at about 327 nm [2]. Since the wavelength of the emission maximum of proteins is usually taken as an indication of the environment of the tryptophan residues, it can be deduced that the tryptophan residues in the *E. coli* enzyme are somewhat more exposed to the solvent than those in the *B. stearothermophilus* enzyme.

Despite the similarities in structure there are subtle and at the present time unexplained differences that do indeed exist when the mesophilic and thermophilic GPDH are compared in their thermal properties. The mesophilic enzyme in comparison to the thermophilic one is less stable towards heat of about 15°C.

In conclusion, the structural data on the *E. coli* GPDH reported here show that the gross structure of the mesophilic enzyme at room temperature resembles that of the enzyme from other sources, including the *B. stearothermophilus* one. This implies that thermostability of the thermophilic enzyme has to be related to a rather small part of the protein molecule, and not to a special structure of the molecule as a whole. The folded structure of a thermophilic enzyme should be made more stable by a combination of factors including hydrophobic interactions, hydrogen bonding and ionic stabilization, but the relative contributions are usually not known and we are obliged to advocate all, emphasizing a particular one depending on the case under study or upon the experimental data available.

Acknowledgment

The authors wish to thank Mr. Zambonin and Mr. F. Miozzo for skilful technical assistance. The work was carried out with the financial support of the Italian C.N.R.

References

[1] K. Suzuky and J.I. Harris, FEBS Letters 13, 217 (1971)

[2] K. Suzuky and K. Imahori, J. Biochem. (Tokyo) 74, 955 (1973).

[3] W.S. Allison, Ann. N.Y. Acad. Sci 148, 180 (1968)

[4] W.S. Allison and N.O. Kaplan, J. Biol. Chem. 239, 2140 (1964).

[5] G. D'Alessio and J. Josse, J. Biol. Chem. 246, 4326(1971).

[6] J.B. Fox, Jr. and W.B. Dandliker, J. Biol. Chem. 221, 1005 (1956)

[7] R. Singleton, Jr., J.R. Kimmel and R.E. Amelunxen, J. Biol. Chem. 244, 1623 (1969).

[8] A.L. Murdock and O.J.Koeppe, J. Biol. Chem. 239, 1983 (1964).

[9] I. Krimski and E. Racker, Biochemistry 2, 512 (1963).

[10] K. Suzuky and K. Imahori, J. Biochem. (Tokyo) 73, 97 (1973).

[11] N. Greenfield and G.D. Fasman, Biochemistry 8, 4108 (1969).

[12] D.W. Darnall and T.B. Barela, Biochim. Biophys. Acta 236, 593 (1969).

[13] S.C. Fujita and K. Imahori, in Peptides, Polypeptides and Proteins, Proc. Rehovot Symp. 1974 (E.R. Blout, F.A. Bovey, M. Goodman and N. Lotan, Eds.), J. Wiley, New York, pag. 217.

[14] L. Brand and B. Witholt, Methods in Enzymology 11, 776 (1967).

[15] J.A. Gally and G.M. Edelman, Biochim. Biophys. Acta 60, 499 (1962)

COMPARATIVE CONFORMATIONAL PROPERTIES OF THERMOPHILIC AND MESOPHILIC 6-PHOSPHOGLUCONATE DEHYDROGENASE

F.M. Veronese, C. Grandi, E. Boccù and A. Fontana

Institutes of Pharmaceutical Chemistry and of Organic Chemistry, University of Padova, Padova, Italy

Summary : The structural properties of 6-phosphogluconate dehydrogenase from the mesophilic bacterium *E.coli* and the thermophilic *B. stearothermophilus* are compared using circular dichroism and fluorescence emission spectroscopy. The enzymes appear to possess a similar structure which does not change on heating up to the respective temperature of stability of the enzyme. The thermostability of the two 6-phosphogluconate dehydrogenases as determined by activity measurements parallels that determined by CD with the melting profile method, indicating that the loss of biological activity in the enzymes is directly related to the unfolding of the protein molecule. The pattern of unfolding of the proteins by the action of 8 M urea suggests that a core of enhanced conformational stability exists in the *B.stearothermophilus* enzyme.

1. INTRODUCTION

Although much is currently being reported on thermophilic enzymes, relatively little is known or published on their conformational properties. Enzyme stability toward heat treatment is usually assessed by determining the residual activity. It seems also important to relate stability of activity of an enzyme with that of its secondary structure.

We have recently isolated and characterized 6-phosphogluconate dehydrogenase (6-PGDH)[*] from the thermophilic bacterium *B.stearothermophilus* (1); furthermore since only 6-PG dehydrogenases from eukaryotes have so far been described, we have also isolated and characterized, for comparative studies, the enzyme from the mesophilic bacterium *E. coli* (2). It is reasonable to think that an ideal enzyme for comparative purposes would be one which presents similar physical, chemical and structural properties but differs only in thermostability. In this respect 6-PG dehydrogenases from

(*) Abbreviations : 6-PGDH, 6-phosphogluconate dehydrogenase (EC 1.1.1.44); CD, circular dichroism.

B.stearothermophilus and E.coli were found to share many characteristics in common such as mol wt, kinetic constants, active site properties and amino acid composition. On the other hand, yeast enzyme, the most intensively studied 6-PGDH, has many general properties in common with the B.stearothermophilus enzyme but recent studies in our laboratory have revealed the former to be distinct in its secondary structure.

Here we report comparative physico-chemical investigations on the stability to temperature and urea of the secondary structure of 6-PGDH from the thermophilic bacterium B.stearothermophilus and mesophilic E.coli.

2. MATERIALS AND METHODS

2.1. Enzymes. 6-phosphogluconate dehydrogenase from B.stearothermophilus (NCA-1503) and E.coli (MRE-600) have been isolated and purified as previously described (1,2).

2.2. Circular dichroism. CD measurements were performed an a Cary 61 automatic recording spectropolarimeter. The temperature of the cell compartment was controlled by circulating ethylene glycol through the cell-holder jacket with the use of a Haake bath. The temperature of the sample was determined using a thermistor attached through the cap of the cell. Measurements were made with a 1-mm path length quartz cell. Spectra were corrected for base-line shifts by running samples of solvent buffer. The mean residue weight of amino acids was taken as 110 in the calculation of the molar ellipticity, $[\theta]$, which is expressed as mean residue ellipticity in degrees . cm^2/dmole.

2.3. Melting profile. For monitoring the thermal transition of proteins, the melting profile method recently proposed by Fujita and Imahori (3) was used. The method consists of continuous recording of the circular dichroic (CD) signal at a fixed wavelength, while the temperature of the sample in the cuvette is raised at a constant rate. The loss of secondary structure of the protein following denaturation is accompanied by a contemporaneous decrease in the intensity of the CD signal. A comparative measure of the thermostability of proteins is the transition temperature midpoint. To estimate transition midpoints, the first derivative plots of ellipticity vs. temperature curves were obtained by evaluating the slope of the line connecting successive data points. Maximum values of the slope in the plots were taken as the $T_{1/2}$. This graphical method of obtaining $T_{1/2}$ is a variation of the original one proposed by Fujita and Imahori (3), which defined $T_{1/2}$ as the temperature at which half change in $[\theta]/[\theta]_o$ has occurred.

2.4. Fluorescence. Fluorescence spectra were measured with a Perkin-Elmer Model MPF-2 spectrofluorimeter connected to a Hitachi, Model QPD_{33} recorder. Protein fluorescence intensities were measured with excitation at 295 nm. Temperature control was accomplished by circulating water from a Haake thermostat bath through a hollow copper block which surrounded the sample cuvette.

In denaturation experiments the 8 M urea potassium phosphate buffer, pH 6.8, 1 mM EDTA and 2-mercaptoethanol were placed in a square quartz fluorescence cell (3 ml) and equilibrated at 35°C for a least 5 minutes. An eliquot (50-100 µl) of an enzyme solution (0.5 mg/ml) was added and the fluorescence intensity at 334 nm was continuously recorded. Excitation was at 295 nm.

3. RESULTS

The thermostability of the two enzymes, determined by the evaluation of the residual activity after 15 minutes heating at various temperatures is reported in Fig. 1. The *B.stearothermophilus*

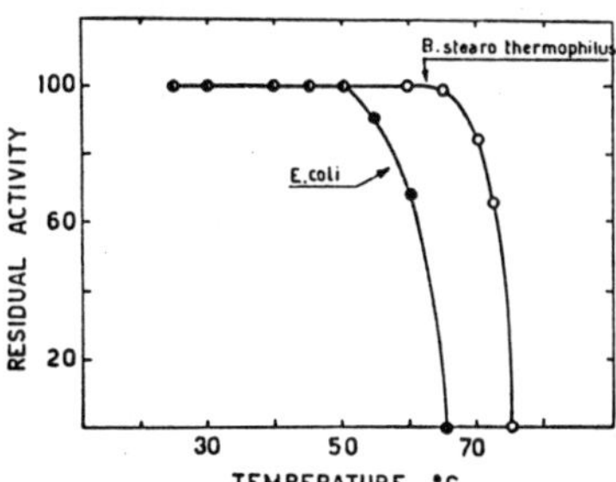

Fig. 1. Heat stability of the *B.stearothermophilus* (○—○) and *E. coli* (●—●) 6-PGDH. The enzyme samples (0.1 mg/ml) were incubated at various temperatures for 15 min in 0.1 M phosphate buffer, pH 7.2, 1 mM EDTA and 1 mM dithiothreitol. The remaining activity after heating was examined using standard assay conditions (1).

6-PGDH appears more thermostable, being inactivated at about 70°C whereas the E.coli enzyme is inactivated at about 55°C.

The CD spectra of 6-PGDH from B.stearothermophilus and E.coli between 200 and 250 nm at 25°C are characterized by a negative band centered at about 219 nm (Fig. 2). This spectrum exhibits

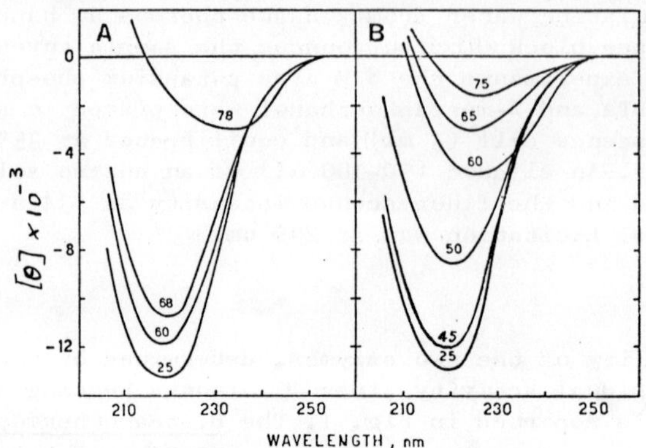

Fig. 2. CD spectra of 6-phosphogluconate dehydrogenase from B.stearothermophilus (A) and E.coli (B) at various temperatures in 0.1 M potassium phosphate buffer, pH 6.8, 1 mM EDTA and 2-mercaptoethanol. The protein concentration was 0.05 mg/ml for the thermophilic enzyme and 0.04 mg/ml for the mesophilic one. Figures near the curves indicate the temperature at which spectra were recorded.

features which are typical for polypeptide chains having predominantly β-structure (4-7). Shoulders at 208 or 222 nm where the α-helix would be expected were not found in the two enzymes. Quantitative estimates of structure following published methods could not be made owing to excessive noise below 205 nm. The molecular ellipticity was calculated ca. -13,300 and -13,000 deg. . cm^2/dmole for the B.stearothermophilus and E.coli enzyme respectively. The overall features of both spectra are remarkably similar at room temperature indicating that, as far as the CD data are concerned, the structure of thermophilic 6-PGDH is indistinguishable from that of the mesophilic enzyme.

Fig. 2 shows that no relevant variations in the general feature of the CD spectra are evident in the temperature range of stability of the enzymes. This would indicate that, even at high temperatures, the structure of neither enzyme undergoes appreciable modification. At temperatures above the stability of the enzymes loss of structure is evident.

A further study of thermostability of the two enzymes was obtained by the melting profile method proposed by Fujita and Imahori (3). This method offers a direct correlation between temperature and stability of the secondary structure of the protein. The results reported in Fig. 3 give $T_{1/2}$ values of 67° and 51°C respectively for the B. stearothermophilus and the E.coli enzyme.

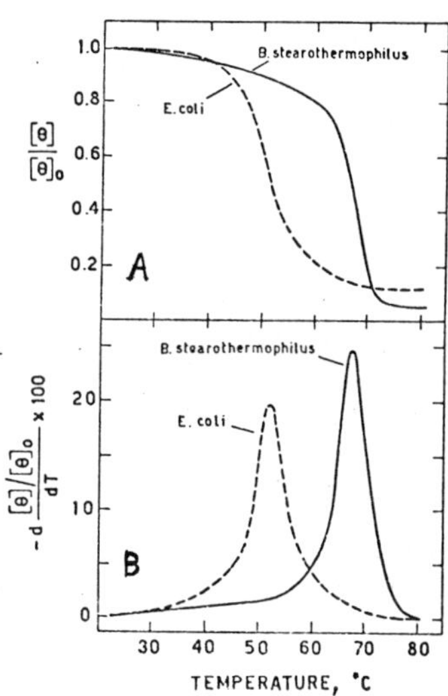

Fig. 3. The melting profile of B.stearothermophilus (——) and E.coli (- - -) as determined by CD measurements. In A is reported the ratio $[\theta]/[\theta]_o$ at 219 nm with $[\theta]_o$ being the ellipticity value at room temperature. In B the derivative curve as a function of temperature is reported. The protein concentration was 0.03 mg/ml in 0.1 M potassium phosphate buffer pH 6.8, 1 mM EDTA and 2-mercaptoethanol.

The fluorescence emission spectra of both the thermophilic and mesophilic enzymes were measured after excitation at 295 nm. At this wavelength the fluorescence spectra of proteins is due to the contribution of tryptophan residues only (8-10). The emission intensity and maxima of tryptophan fluorescence are very dependent on the chemical environment of tryptophan residues, so that measurement of their fluorescence should give information regarding aspects of protein structure (11,12). Fig. 4 A and B show the fluorescence emission spectrum of the thermophilic and mesophilic enzymes, upon excitation at 295 nm at various temperatures.

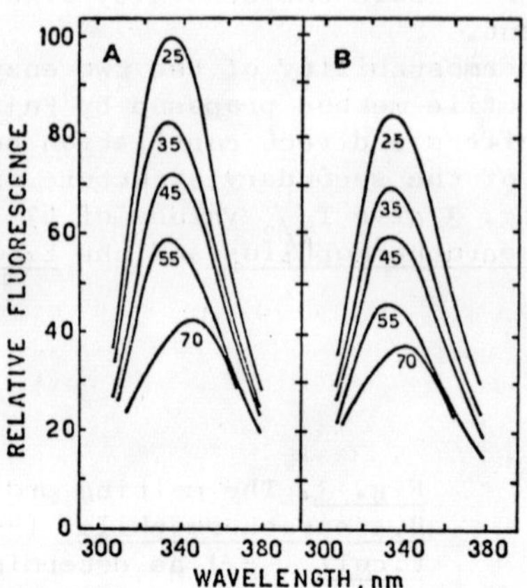

Fig. 4. Fluorescence emission spectra of B.stearothermophilus (A) and E.coli (B) 6-phosphogluconate dehydrogenase at various temperature in 0.1 M potassium phosphate buffer, pH 6.8, 1 mM EDTA and 2-mercaptoethanol. Spectra were measured at a protein concentration of 0.04 mg/ml after 10 min incubation at each temperature. Figures near the curves indicate the temperature inside the cell measured with a thermistor.

On heating, the fluorescence intensity gradually decreases in a virtually linear way due to thermal quenching up to 65°C for the thermophilic enzyme and up to 50°C for the mesophilic one. Up to these temperatures, the maximum wavelength of emission for both enzymes remained constant or only slightly shifted to shorter wavelengths by 1-2 nm.

The pattern of unfolding of a protein molecule by the action of denaturing agents can give some insight into the forces stabilizing the protein itself, and could be useful in detecting even small differences in conformational stability (13-17). The denaturation reaction of the thermophilic and mesophilic 6-PGDH by 8 mM urea solution has been compared.

The apparent fluorescence intensity of the B.stearothermophilus enzyme in 8 mM urea as revealed by spectra at different times gradually decreased to reach a value (after about 5 hours) of nearly 60 % of that the native enzyme, and the emission maximum was shifted to longer wavelength, i.e., 350 nm. The emission maximum did not change any further, even after the enzyme was heated at 100°C for 5 minutes in 8 M urea. This value corresponds with the emission maximum of tryptophan in water, and this fact indicates that by the action of 8 M urea the tryptophan residues of the enzyme are transferred from a moderately hydrophobic environment inside the protein molecule to an exposed polar environment (11). Similar results have been obtained with the E.coli enzyme, with the significant difference that the denaturation process was much faster (by a factor of ten).

The urea-induced unfolding of 6-PGDH gives a linear plot (i.e. log $(F_t-F_\infty)/F_o-F_\infty)\times 100$ vs time) for the E.coli enzyme with a first order rate constant (K obs) of 0.23 min^{-1}, whereas the B.stearothermophilus enzyme shows a biphasic first order kinetics with rate constants respectively of 0.032 and 0.021 min^{-1} (Fig.5).

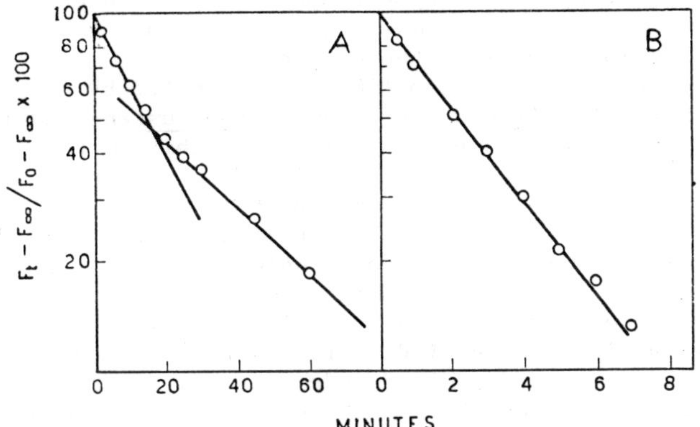

Fig. 5. The kinetics of the unfolding of 6-PGDH from B.stearothermophilus (A) and E.coli (B) in 8 M urea as followed by emission fluorescence. F_o is the initial value of the intensity of relative fluorescence and F_t the intensity of time t and F_∞ the value obtained several hours after mixing. The protein concentration was 0.05 mg/ml in 8 M urea in 0.1 M potassium phosphate buffer, pH 6.8, 1 mM EDTA and 2-mercaptoethanol.

4. DISCUSSION

The experiments here reported clearly indicate that the secondary structure of the thermophilic and mesophilic enzymes,as verified by circular dichroism and fluorescence, are much the same. In addition,their structures do not appear to undergo modifications on heating to the respective temperatures of stability of the enzymes.

The CD spectra do not show distinct differences in shape and [θ] value up to 60°C for the B.stearothermophilus enzyme and about 45°C for that of E.coli. Above these temperatures the negative band at 219 nm is gradually reduced, since denaturation occurs at higher temperatures. These results are in agreement with the fluorescence experiments, which revealed that both enzymes show the same maximum of emission at 334 nm and similar general shape of the spectrum indicating that the tryptophan environment in both proteins is the same. The wavelength of emission suggests that the tryptophan residues are in a moderately hydrophobic environment of about the polarity of dioxan. The fact that the same monotonic decrease in fluorescence, by heating up to 65° for the mesophilic and 50°C for the thermophilic enzyme respectively, with only a slight shift to shorter wavelengths shows that by heating no gross change in conformation occurs in the enzymes.

The differences in thermostability as determined by enzymatic analysis between the E.coli and B.stearothermophilus enzymes were found to correspond to their melting behaviour. This correlation indicates that the loss of biological activity in the enzymes is directly related to the unfolding of the protein molecule. The thermostability of activity of the thermophilic 6-PGDH appears therefore the result of its heat stable conformation,in contrast to that of the enzyme from the mesophilic source.

The pattern of unfolding of the protein molecule by the action of 8 M urea gave an interesting indication on differences in conformational stability of the two 6-PG dehydrogenases. Kinetic examination of the unfolding process revealed that it is monotonic in the case of the E.coli enzyme and biphasic in that of B.stearothermophilus. The biphasic nature of the denaturation process could indicate a core of enhanced stability in the thermophilic protein, as already suggested by Suzuki and Imahori (18) for glyceraldehyde-3-phosphate dehydrogenase from B.stearothermophilus.

These results indicate that, although the thermophilic and mesophilic enzymes are conformationally similar, they show different conformational stability, which can be related to specific side-chain interactions in the protein.

Acknowledgments. The authors wish to thank Mrs. R. Largajolli and Mr. M. Zambonin for skilful technical assistance. The work was carried out with the financial support of the Italian C.N.R.

REFERENCES

1. Veronese, F.M., Boccù, E., Fontana, A., Benassi, C.A. and Scoffone, E. (1974) Biochim.Biophys.Acta 334, 31-44.
2. Boccù, E., Veronese, F.M. and Fontana, A. (1974) FEBS Meeting, Budapest, 25-30 August, Abstr. Comm.F10b10.
3. Fujita, S.C. and Imahori, K. (1975) in Peptides, Polypeptides and Proteins, Proc. Rehovot Symp. 1974 (Blout, E.R., Bovey, A.F., Goodman, M. and Lotan, N.,Eds.) J.Wiley & Sons, New York, 217-228.
4. Townend, R., Kumonishi, T.F., Timasheff, S.N., Fasman, G.D. and Davidson, B. (1966) Biochem.Biophys.Res.Commun. 23, 163-169.
5. Sarkar, P.K. and Doty, P. (1966) Proc.Natl.Acad.Sci.U.S. 55, 981-989.
6. Greenfield, N. and Fasman, G.D. (1969) Biochemistry 8, 4108-4116.
7. Darnall, D.W. and Barela, Th.D. (1971) Biochim.Biophys.Acta 236, 593-598.
8. Edelhoch, H. and Steiner, R.F. (1964) in Electronic Aspects of Biochemistry (Pullman, B., Ed.) Academic Press, New York, 7-55.
9. Weber, G. and Teale, F.W.J. (1965) in The Proteins (Neurath, H., Ed.), Academic Press, New York 3, 445-521.
10. Brand, L. and Witholt, B. (1967) Methods in Enzymology 11, 776-856.
11. Steiner, R.F., Lippoldt, R.E., Edelhoch, H. and Frattali, V. (1964) Biopolymers Symp. 1, 355-366.
12. Cowgill, R.W. (1966) Biochim.Biophys.Acta 112, 550-558.
13. Tanford, C. (1970) in Advances in Protein Chemistry (Anfinsen, C.B., Jr., Edsall, J.T., Anson, M.L. and Richards, F.M., Eds.), Academic Press, New York 24, 1-95.
14. Aune, K.C. and Tanford, C. (1969) Biochemistry 8, 4579-4585.
15. Aune, K.C. and Tanford, C. (1969) Biochemistry 8, 4586-4590.
16. Puett, D. (1973) J.Biol.Chem. 248, 4623-4634.
17. Puett, D., Friebele, E. and Hammonds, R.G., Jr. (1973) Biochim.Biophys.Acta 328, 261-277.
18. Suzuki, K. and Imahori, K. (1973) J.Biochem. 74, 955-970.

PURIFICATION AND PROPERTIES OF MALATE DEHYDROGENASE FROM THERMUS AQUATICUS

J.H.F. BIFFEN and R.A.D. WILLIAMS

Department of Biochemistry, The London Hospital Medical College, Turner Street, London E1 2AD. England

INTRODUCTION

The extreme thermophile *Thermus aquaticus*[1] grows optimally at 70° to a high cell density with a mean generation time of about one hour. The type strain YT_1, and several strains of similar genera and species have been used for the purification of enzymes for comparison with those of mesophiles[2]. We have purified and investigated malate dehydrogenase (E.C.1.1.1.27). from *Thermus aquaticus* strain YT_1 (T.aq.MDH).

MATERIALS AND METHODS

Cells were grown at 70° by the Microbial Products Section, Microbiological Research Establishment, Porton Down. Frozen cell paste (1 Kg.) was dispersed in three volumes of 10mM "tris", 0.5mM EDTA, 0.5mM mercaptoethanol at pH 7.5, and disrupted by four passes through a Manton-Gaulin (15M-8BA) disintegrator. The crude homogenate was mixed with DEAE cellulose, previously equilibrated with the same buffer, at a rate of 1g. dry ion exchanger per g. protein. The suspension was stirred at room temperature for one hour, allowed to settle, and the supernatant rejected. Much of the yellow pigment was removed from the DEAE cellulose by washing thrice with two volumes of the above buffer, and a further twice with the same buffer containing 0.1M sodium chloride. T.aq.MDH was eluted in three washes of one volume of buffer containing 0.4M sodium chloride. Fractions of eluate corresponding to 3000 units of enzyme were absorbed onto columns of hydroxyapatite (10cm. x 4.4cm.) which were washed with 2.5mM phosphate buffer pH 6.8, and the enzyme eluted with the same buffer at 100mM concentration[3]. Absorption onto DEAE cellulose columns (140cm. x 2.2cm.) was effected in 20mM phosphate buffer pH 7.5 containing 2mM sodium chloride, and elution was achieved with a chloride concentration gradient up to 600mM. Gel filtration on Sephadex G 100 in 20mM phosphate buffer 7.5 gave an enzyme preparation virtually free of pigment. Second DEAE cellulose columns of analytical grade ion exchanged, loaded with only 200 mg of protein per column and using the same elution programme gave a further 4.6 times purification. AMP-Sepharose[4] (Ligand concentration 1.1 μmoles AMP per g. of Sepharose 4B) was used with several buffer systems. Binding of T.aq.MDH was not successful with 10mM potassium phosphate buffer pH 7.5, which has been used with yeast MDH[4]. The thermophile enzyme did bind to the column at pH 6.8, but leaching occurred on washing with the same buffer. Using pH 6.0 phosphate improved the binding capacity (6 units per ml of column), but elution in a pH gradient occurred over a wide pH range. The use of organic buffers[5] in place of phosphate greatly improved the separation on such columns. Enzyme was applied in 5mM PIPES pH 6.3, and washed with 8 volumes of the same buffer. Elution with a PIPES-HEPES[5] gradient to pH 9.1 was used successfully. Enzyme purified to this stage was used for most experiments described here and for kinetic studies[6]. For analytical purposes however three purification stages were repeated with an overall purification factor of 1.53. These stages were: a third DEAE cellulose chromatography, gel filtration on Sephadex G 200 and a repeat AMP-Sepharose chromatography in the above pH gradient.

T.aq.MDH was assayed in 10mM CAPS[5] buffer pH 11.0 containing 1 μmole NAD and 30 μmoles sodium malate in a volume of 3 ml. Reduction of NAD was followed at 340nm using a Perkin-Elmer 402 Spectrophotometer with a Smiths Servoscribe multirange recorder. Cuvette temperature was maintained at 45° with a Haake water circulator. Pig heart soluble MDH (pMDH) was assayed in 150mM "tris" 15mM EDTA pH 9.0 at 27°. The pH of all reagents was measured and adjusted at the temperature at which they were to be used.

Ultracentrifugation was carried out in the An-D rotor of a Beckman analytical ultracentrifuge at 52,640 r.p.m., and photographs taken every 16 min. using the Schleiren optical system. Gel electrophoresis was by the method of Ornstein[7] and Davis[8], and Sephadex gel filtration as described by Andrews[9]. Isoelectric focusing in a 110 ml LKB column was used with 1.6% of pH 3-10 Ampholine to minimise denaturation.

Pure T.aq.MDH was hydrolysed in sealed tubes at 100° with 6N HCl for 24, 48 and 72 hours, and the amino acid composition determined with a Locarte single column amino acid analyser. Serine and threonine were estimated by extrapolation to zero hydrolysis time. Tryptophan was measured in the unhydrolysed protein by titration with N-bromosuccinimide[10].

For studies of inactivation by reagents the incubation mixture contained such a quantity of enzyme that dilution for enzyme assay reduced the reagent concentration to the degree that it had no effect on the enzyme assay.

RESULTS AND DISCUSSION

For most purposes T.aq.MDH purified 449-fold, to the stage of the first AMP-Sepharose column, was used (Table 1), but quantities of further purified enzyme were prepared for analytical purposes. For the sake of comparison the data for the last three

	Total protein (mg.)	Total enzyme (u.)	Specific activity (u./mg.)	Purification (Xfold)	Recovery (%)
Crude extract	176,000	27,200	0.154	-	100
Batch DEAE	30,700	22,900	0.75	4.9	84
Hydroxyapatite	7,800	16,670	2.14	13.9	61
DEAE column1	2,200	12,580	5.7	37.1	46
Sephadex G100	1,060	10,600	10	65	39
DEAE column 2	172	7,880	45.8	297.5	29
AMP Sepharose	57	3,940	69.1	449	14.5
DEAE column 3	35.7	2,758	77.3	502	10.1
Sephadex G200	16.1	1,287	80	520	4.7
AMP Sepharose	7.9	840	106	690	3.1

1 enzyme unit (u.) oxidises 1μ mole of malate per minute at 45°C

Table 1 Purification of T.aq.MDH

stages have been expressed in terms of the whole batch of enzyme, although only small quantities were actually passed through these steps. The final stage was an AMP-Sepharose chromatography (Fig.1). The enzymes purified to the 449-fold stage showed three protein bands on acrylamide gel electrophoresis, but the 690-fold purified sample had only one band, and was also homogenous by ultracentrifugation (Fig.2).

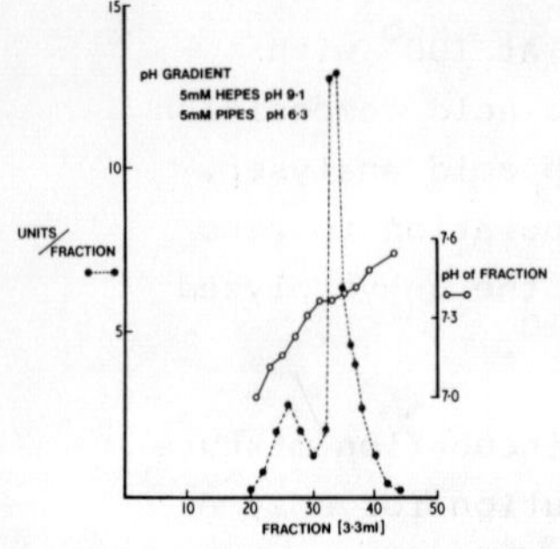

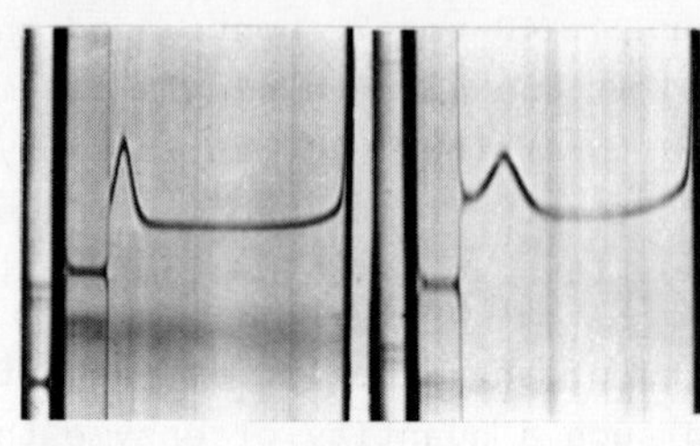

Fig.1 Elution of T.aq.MDH from AMP-Sepharose 4B

Fig.2 Analytical ultracentrifugation of T.aq.MDH

The molecular weight of T.aq.MDH was 73,500 by gel filtration on Sephadex G 100 (Fig.3) and G 200 using human haemoglobin, rabbit muscle lactate dehydrogenase and cytochrome C as markers. T.aq.MDH clearly belongs to the small molecular weight subgroup of MDH[11], commonly about 60,000 daltons, and is thus of a similar size to animal and several bacterial MDH molecules, but much smaller than the bacillus (120,000 daltons) type. Isoelectric focusing (Fig.4) indicated an isoelectric point of 4.40, at which

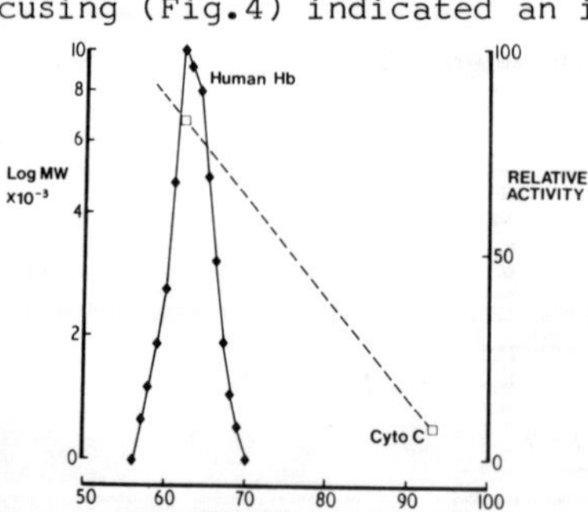

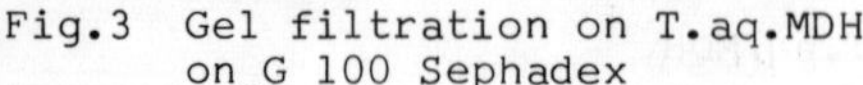

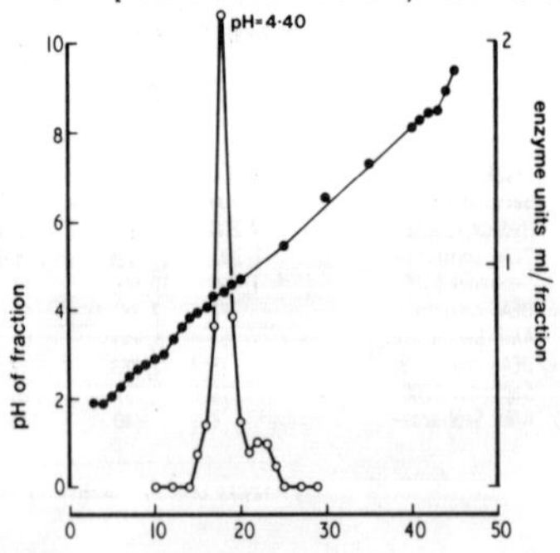

Fig.3 Gel filtration on T.aq.MDH on G 100 Sephadex

Fig.4 Isoelectric focusing of T.aq.MDH

pH the enzyme was unstable and lost 80% of its activity during the 65 hours focusing period. This rendered preparative isoelectric focusing impractical as a purification step.

The isoelectric point of 4.40 is similar to that of 4.78 for pMDH, and to the range of 4.40 - 4.80 for a series of thermophilic bacteria. The amino acid composition of T.aq.MDH (Table 2) seems quite unrèmarkable when the content of amino acids, or of groups of similar amino acids (Table 3) are compared with date for other specimens of MDH.

Amino acid	Residues per mole	Amino acid	Residues per mole	Amino acid	Residues per mole
asp	58.0	lys	35.3	his	12.7
glu	82.2	arg	44.6	pro	32.7
thr	27.9	ala	89.0	met	2.2
ser	30.7	val	58.2	cys	6.8
tyr	0	ile	36.8	gly	68.6
phe	24.9	leu	68.8	try	3.0

Table 2 Amino acid composition of T.aq.MDH (mean of three determinations) residues per 73,500 daltons

Amino acid	T.aq.	B.subtilis[12]	E.coli[12]	Pig[13] mito.	Chick[14] mito.	Pig[15] sol.	Chick[14] sol.
asp, glu	20.4	20.0	18.5	14.3	16.6	20.0	17.2
thr, ser	8.6	14.0	10.8	12.5	13.1	9.4	10.8
tyr, phe	3.7	6.6	5.5	4.8	6.3	5.5	6.2
lys, arg	11.7	10.0	9.2	10.6	10.5	12.8	12.1
ala, val, ile, leu	36.9	32.1	37.0	35.4	32.1	34.3	32.9
his, pro	6.7	4.8	5.3	9.6	7.9	5.4	6.2
met, cys	1.6	2.5	2.4	3.7	4.1	4.4	3.4
gly	10.0	10.0	10.7	9.1	9.3	6.8	9.2
try	0.4	0	0.6	0	0	1.4	1.9

Table 3 Amino acid composition of MDH from various sources

The T.aq.MDH fell within the range of values of other MDH samples in any of four parameters calculated from amino acid composition. These values are the frequency of acidic plus basic amino acid relative to all residues, the frequency of non-polar side chain residues;

$$NSP = \frac{try+ile+tyr+phe+pro+leu+val}{\text{total of amino acid residues}}$$

the polarity ratio;

$$p = \frac{\text{volume of the total polar side chains}}{\text{volume of non-polar side chains}}$$

and the average hydrophobicity ($H\emptyset_{ave}$) calculated from the amino acid composition and data for the free energy of transfer of amino acid side chains from an organic to an aqueous environment. There is no evidence for any correlation between the thermal stability of T.aq.MDH and the mean composition of its protein (Table 4). This is in accord with other results which indicate that the molecular basis of thermal stability must be very subtle and is not necessarily the same for all thermostable proteins.

Enzyme	Acidic and basic residues	Nonpolar[16] side chains	Polarity[17] ratio	Average[18] hydrophobicity (K cal/mole)
T.aquaticus	0.32	0.328	0.828	1.029
B.subtilis	0.30	0.346	1.045	1.021
E.coli	0.28	0.370	0.719	1.128
N.crassa	0.31	0.320	1.01	1.050
Pig mito.	0.25	0.370	0.73	1.210
Pig sol.	0.33	0.360	0.87	1.183
Chick mito.	0.27	0.347	0.824	1.133
Chick sol.	0.29	0.367	0.834	1.185

Table 4 Calculated values of polarity for MDH specimens from various sources

Pig heart soluble MDH (pMDH) has been used for comparison purposes in studies of the stability of T.aq.MDH. It was necessary to store pMDH at room temperature because it was susceptible to reversible cold inactivation at both $+4^{o}$ (Fig.5) and -18° (Fig.6). After storage at $+4^{o}$ a rise of 100% was observed in the activity of pMDH in three hours incubation in malate buffer with mercaptoethanol. Omission of mercaptoethanol precluded reactivation, and the observed activity increase was only about 30% in the case of phosphate or citrate buffers with mercaptoethanol. From

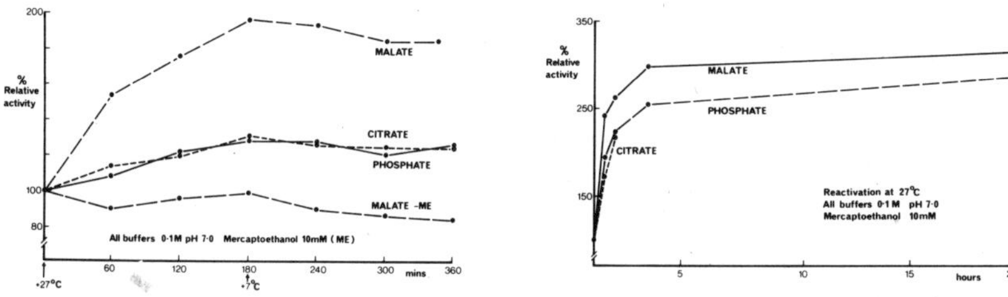

Fig.5 Reactivation of pMDH after storage at $+4^{o}$

Fig.6 Reactivation of pMDH after storage at -18°

storage at -18° the activity increase in the three buffers was two or three-fold in the presence of mercaptoethanol. No such reversible cold inactivation was observed in the case of T.aq.MDH. Nevertheless both pMDH and T.aq.MDH were routinely stored at room temperature, and saturated with chlorbutol as a preservative. The T.aq.MDH was also unaffected by preincubation at temperatures below the optimum of the organism. No increase in activity was detected when enzyme stored at room temperature was incubated at 70° prior to enzyme assay. Because T.aq.MDH is little affected by such treatment, it appears likely that the enzyme exists in a conformation, or a series of conformations, of similar catalytic potential over the temperature range -20° to 70°. If temperature dependant conformations with different properties do exist they must be rapidly interconvertible. The

Arrhenius plot for the catalysis of malate oxidation by T.aq.MDH was linear over the range of 39° to 70° and the activation energy was estimated to be 21,500 Kcal/mole (Fig.7). The criteria of Gibson for saturation with substrate and coenzyme through the temperature range in the Arrhenius plots were adhered to[19]. Sharp breaks in the Arrhenius plots have been reported for B.stearothermophilus adenosine triphosphatase[20] and fructose diphosphate aldolase[21]. There are no such breaks with T.aq.MDH which is further evidence that no conformational changes associated with changes in activity occur within the temperature range tested. B.stearothermophilus MDH is 50% inactivated in 20 min. at 75°. T.aq.MDH is clearly much more stable, being half inactivated in 1 min. at 100°, 24 min. at 95° and 120 min. at 90°.

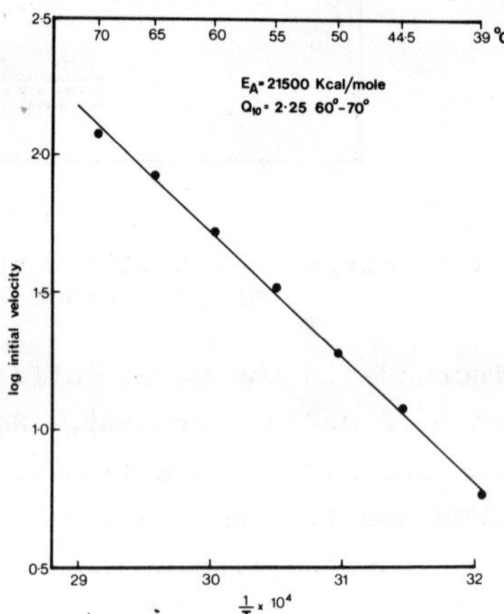

Fig. 7 Arrhenius plot for T.aq.MDH in 3.3mM NAD and 10mM malate

A series of experiments have been carried out to test the extent to which thermal stability in T.aq.MDH is paralleled by stability to inactivation by urea, guanidine and detergent. No loss of activity was detected during two hours incubation with 6M urea at 70° or 50°. The buffer used in these experiments was a mixture of tribasic weak acids that gave a linear pH gradient when titrated with sodium hydroxide. Guanidine was a much more potent inactivator of T.aq.MDH than urea, and its effect was not reversed when the guanidine was removed by dialysis. In 0.33M guanidine hydrochloride in 0.1M phosphate buffer with mercaptoethanol, pMDH is extensively inactivated. The behaviour of T.aq.MDH in the same system is peculiar (Fig.8). The experiment was carried out four times, of which two examples are illustrated. In each case the enzyme activity was reduced for the first 30 min. and then increased to 100% of the original value (or slightly more) at 40-60 min. It seems possible that in these conditions the enzyme is free to take up conformations not normally available, and that a progression takes place via forms having reduced activity to an equilibrium favouring a form of approximately the same activity as that preferred in the absence of guanidine. In 0.33M guanidine at least 90% of the original activity remained after 48 hours. A considerable proportion of the T.aq.MDH activity remained after treatment with 2M and 4M guanidine, in which pMDH activity was abolished in seconds. In 1% sodium dodecyl sulphate T.aq.MDH lost 75% of its activity within 5 seconds at 100°. At room temperature, however, it was much more stable (Fig.9), still showing more than 90% activity after 25 min., in which time pMDH activity had decayed to 5% of its original activity. The stability of this enzyme to inactivating agents will be reported in full elsewhere.

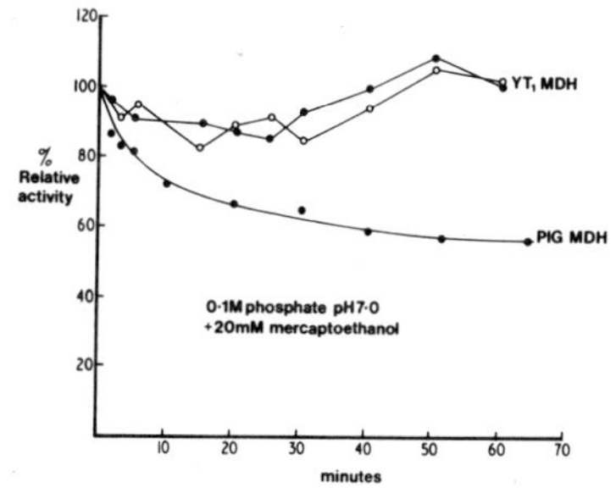

Fig.8 Effect of 0.33M guanidine upon pMDH and T.aq.MDH

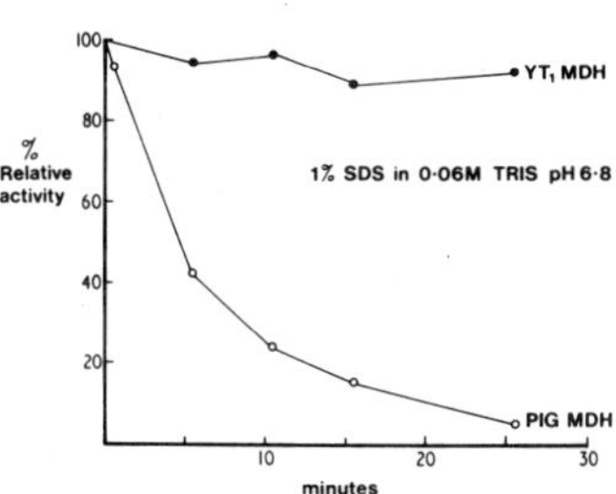

Fig.9 Effect of 1% sodium dodecyl sulphate upon T.aq.MDH

CONCLUSIONS

The malate dehydrogenase of Thermus aquaticus YT_1 seems to be similar to other bacterial MDH specimens that have been examined in all its general characteristics and in its amino acid composition. It not only catalyses the oxidation of malate with increasing rapidity as the temperature approaches 70°, but is also very stable to thermal denaturation. Furthermore, the enzyme is remarkably stable to other denaturing agents including guanidine.

This work was supported by a grant from the Science Research Council.

REFERENCES

1. Brock, T.D. and Freeze, H. (1969) J.Bact., 98, 289.
2. Williams, R.A.D. (1975) Sci.Prog., Oxford, 62, 373.
3. Hocking, D. Personal communication.
4. Craven, D.B., Harvey, M.J., Lowe, C.R. and Dean, P.D.G. (1974) Eur.J.Biochem., 41, 329.
5. Good, N.E., Winget, G.D., Winter, W., Connolly, T.N., Izawa, S. and Singh, R.M.M. (1966) Biochemistry, 5, 467.
6. Biffen, J.H.F. and Williams, R.A.D. (1975) this symposium.
7. Ornstein, L. (1964) Ann.N.Y.Acad.Sci., 121, 321.
8. Davis, B.J. (1964) Ann.N.Y.Acad.Sci., 121, 404.
9. Andrews, P. (1965) Biochem.J., 96, 595.
10. Spande, T.F. and Witkop, B. (1967) Methods in Enzymology, XI, 498.
11. Murphey, W.H., Kitto, G.B., Everse, J. and Kaplan, N.O. (1967) Biochemistry, 6, 603.
12. Murphey, W.H., Barnaby, C., Lin, F.J. and Kaplan, N.O. (1967) J.Biol.Chem., 242, 1548.
13. Thorne, C.J.R. (1962) Biochem.Biophys.Acta, 59, 624.
14. Kitto, B. and Kaplan, N.O. (1966) Biochemistry, 5, 3966.
15. Gerding, R.K. and Wolfe, R.G. (1969) J.Biol.Chem., 244, 1164.
16. Waugh, D.F. (1954) Adv.Protein Chem., 9, 326.
17. Fisher, H.F. (1964) Proc.Nat.Acad.Sci.Wash., 51, 1285.
18. Bigelow, C.C. (1967) J.Theoret.Biol., 16, 187.
19. Gibson, K.D. (1953) Biochem.Biophys.Acta, 10, 221.
20. Hachimori, A., Muramuatsu, N. and Nosoh, Y. (1970) Biochem. Biophys. Act, 206, 426.
21. Sugimoto, S. and Nosoh, Y. (1971) Biochem.Biophys.Acta, 235, 210.

PURIFICATION AND SOME PROPERTIES OF NADP$^+$-SPECIFIC ISOCITRATE DEHYDROGENASE FROM AN EXTREME THERMOPHILE, *Thermus flavus* AT-62

TAKASHI SAIKI, ISHTIAQ MAHMUD, NOBUYUKI MATSUBARA, KIYOSHI TAYA, AND KEI ARIMA
Department of Agricultural Chemistry, University of Tokyo, Tokyo, Japan

SUMMARY

Thermostable NADP$^+$-specific isocitrate dehydrogenase (EC 1.1.1.42) was purified from crude extract of an extremely thermophilic bacterium *Thermus flavus* AT-62 through DEAE-cellulose column, acetone fractionation, DEAE-Sephadex A-50 column and isoelectric focussing. The enzyme was purified about 500-folds in its specific activity and purity was found to be about 96 %.

The enzyme was not inactivated after 60 min at 70°C, but 20 and 80 % of the activity were lost after 60 min at 80° and 90°C, respectively.

Oxalacetate plus glyoxylate (each 1 mM) demonstrated 75 % inhibition of the activity in concerted manner. The degree of the inhibition and the affinity of the enzyme for isocitrate and NADP$^+$ decreased with the rise of temperature, especially above 60°C. The activation energy below and above 60°C were 14,500 and 8,000 cal per mole, respectively.

In CD spectra negative bands at 210 and 220nm were observed and α-helix content was calculated to be about 26 %. In the course of heating up to 60° practically no change in CD bands are observed, but above 60° the depth of CD bands decreased gradually and remarkably above 80°C. The effect of temperature on kinetic parameters and secondary structures of the enzyme was discussed in relation to the temperature adaptation of the organism.

INTRODUCTION

Recent studies have revealed the thermostable nature of almost all enzymes and other macromolecular structures in thermophiles. Hitherto, a number of hydrolytic enzymes, enzymes concerned with glycolytic pathway, and also enzymes involved in the biosynthetic pathway of several amino acids (such as aspartate pathway) in thermophiles have been studied (1,2,3). Some of them are purified in homogeneous state and physicochemical properties have been elucidated to a certain extent. Especially, complete X ray diffraction analysis of thermolysin (4,5) and the determination of primary structure of thermophilic ferredoxins (6) are considered to be an important step towards the elucidation of the mechanism of thermostability. However, views as to these problems are not yet clear and moreover the studies on the enzymes concerned with the TCA cycle of thermophiles are still limited (7,8,9). In this context, $NADP^+$-specific isocitrate dehydrogenase of an extremely thermophilic bacterium, *Thermus flavus* ,AT-62, has been studied.

In this paper the purification and some properties of the enzyme are described. The temperature dependent conformational change of the enzyme is also described and discussed in relation to the temperature adaptation of the organism. A part of this article is already been published in (10).

MATERIALS AND METHODS

Chemicals——DL-Isocitrate trisodium salt was obtained from Sigma Chemical Company. $NADP^+$ was purchased from Oriental Yeast Co. Ltd. (Tokyo). DEAE-Cellulose (Brown) was obtained from Seikagaku Kogyo Co. Ltd. DEAE-Sephadex A-50 was product of Pharmacia (Uppsala). Ampholine Ampholite (pH range 4.0-6.0) was obtained from AMCO. All other reagents were of reagent grade.

Determination of Protein——Protein was determined by absorbance at 280nm or by micro-biuret method described by Itzhaki and Gill (11) using bovine serum albumin as a protein standard.

Enzyme Preparation——For the study of circular dichroism the highly purified preparation described in RESULTS was used. All other experiments were performed using the partially purified enzyme described in the previous paper (10).

Enzyme Assay——$NADP^+$-specific isocitrate dehydrogenase activity was measured by using Gilford 240 recording spectrophotometer. The reaction mixture and the procedure for the activity measurement of partially purified enzyme preparation were described in the previous paper (10). One unit of enzyme activity was defined as the amount of enzyme catalyzing the reduction of 1 μmole of $NADP^+$ per min. Enzyme reaction was performed at 60°C unless noted otherwise.

Electrophoresis——Disc electrophoresis in polyacrylamide gel at pH 9.4 was performed by the method of Ornstein and Davis (12). The protein in the gel was stained with 1.0 % amide black in 7 % acetic acid and scanned by Gilford 240 recording spectrophotometer fitted with Model 2410-S linear transport.

Isoelectric focussing electrophoresis (13) was performed using 110 ml Ampholine column (LKB-Produkter). The carrier Ampholite (pH 4.0-6.0) was used at an average final concentration of 1 % and electrophoresis was continued for about 40 hr at 3-5°C.

Circular Dichroism——CD spectra were measured with a JASCO J-20 automatic recording spectropolarimeter with a temperature-controlling system employing a Haake bath. The temperature of the sample was monitored by a Takara Type SPD-1D thermistor. The mean residue weight of amino acid was assumed to be 120.

Molecular Weight Determination——Gel filtration was carried out on Sephadex G-100 column according to the method of Andrews (14). Ovalbumin and γ-globulin were used as molecular weight standards.

RESULTS

Purification of the Enzyme

Step I. Source of the enzyme——*Thermus flavus* AT-62 was aerobically tank-cultured at 75°C in a medium consisting of Kyokuto bouillon powder 0.4 %, polypeptone 0.4 %, K_2HPO_4 0.3 %, and KH_2PO_4 0.1 %, pH 7.2. The cells were harvested in late log phase and kept frozen at -20°C. Following purification procedures were carried out below 5°C.

The cell pellet (1180 g) was suspended in 3500 ml of 0.9 % NaCl solution by Potter-Elvehjem homogenizer and centrifuged at 10,000 rpm for 20 min. Supernatant was retained (Extract A, 3,500 ml). The cell pellet was resuspended in 2,500 ml of 0.02 M K_2HPO_4-KH_2PO_4 buffer, pH 7.0, containing 0.1 M KCl (Buffer A) and sonicated for 10 min employing a sonic oscillator TYPE 50-6 (Toyoriko Seisaku-sho Co. Ltd.). The homogenate was ultracentrifuged at 30,000 rpm for 60 min in a Hitachi 65P Ultracentrifuge and the supernatant obtained (2800 ml) was combined with Extract A. (Crude extract, 6,300 ml).

Step 2. DEAE-Cellulose column——The crude extract (2100 ml) of Step 1 was applied to a 10 x 50 cm DEAE-cellulose column previously equilibrated with Buffer A, and the protein having the enzyme activity was eluted with the same buffer. The same procedure was repeated for two more times and the active fractions collected were concentrated to 1600 ml by means of ultrafiltration through an Amicon PM-10 filter.

Step 3. Acetone fractionation——To 150 ml of Step 2 solution was added 100 ml of acetone with constant stirring, and the precipitate formed was removed by centrifugation. Then 50 ml of acetone was again added to the supernatant and precipitate collected by centrifugation was dissolved in minimal amount of 0.02 M K_2HPO_4-KH_2PO_4 buffer, pH 7.0 (Buffer B). The acetone fractionation was carried out at 0 to -10°C. After the acetone fractionation was repeated for necessary times, the pooled acetone 40-50 % fraction was dialyzed against 4 liters of Buffer B for 20 hr.

Step 4. DEAE-Sephadex A-50 column——The enzyme solution from Step 3 was adsorbed onto a DEAE-Sephadex A-50 column (33 x 175 mm) previously equilibrated with Buffer B and chromatography was carried out with a linear gradient of KCl to 0.3 M in the same buffer. The enzyme activity was eluted at about 0.1 M concentration of KCl. Active fractions collected were concentrated by ultrafiltration and dialyzed against 0.01 M K_2HPO_4-KH_2PO_4 buffer, pH 7.5, containing 20 mM disodium L-malate (Buffer C) overnight.

Step 5. 2nd DEAE-Sephadex A-50 column——The enzyme solution from Step 4 was charged on a DEAE-Sephadex column (16 x 90 mm) previously equilibrated with Buffer C and chromatography was carried out with a linear gradient of 0.3 M KCl in Buffer C. Active fractions collected were concentrated and dialyzed against Buffer B overnight.

Step 6. 3rd DEAE-Sephadex A-50 column——The sample from previous step was subjected to chromatography on DEAE-Sephadex A-50 column (16 x 140 mm). Equilibration and elution were performed in the same way as in Step 4. Active fractions were collected and concentrated to 14.5 ml by ultrafiltration.

Step 7. Isoelectric focussing electrophoresis——After 2.7 ml of the enzyme solution from Step 6 was dialyzed against 2 liters of distilled water overnight, the sample was subjected to isoelectric focussing electrophoresis, the procedure was described earlier. Fifty drops a fraction was collected. From the elution pattern of Ampholine column it was found that there are one main peak at pI 5.6 and two minor components at pI 5.5 and 5.4 for the isocitrate dehydrogenase activity. Two fractions of highest activity from the main peak were combined and dialyzed against 2 liters of distilled water overnight for second isoelectric focussing electrophoresis.

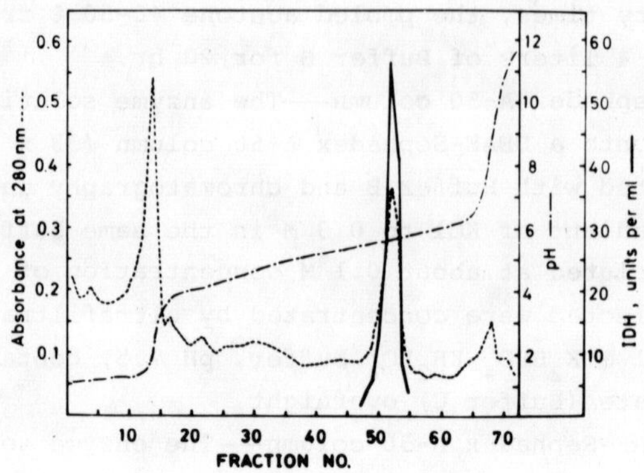

Fig. 1. Elution pattern from Ampholine column

Table I. Purification of the isocitrate dehydrogenase

	ml	Protein mg	Total unit	Sp. act. units/mg	Yield %
Crude extract	6,300	72,200	28,300	0.39	100
DEAE-cellulose	292	30,400	27,900	0.92	99
Acetone 50-60 %	235	6,600	15,800	2.39	56
DEAE-Sephadex A-50	35	593	9,800	16.5	34
2nd	98	254	6,240	24.6	22
3rd	14	172	5,600	32.5	20
Isoelectric focussing	16	9.7	1,500	155	5.3
2nd	5.2	3.5	690	198	2.4

Step 8. Second isoelectric focussing electrophoresis——The

enzyme solution from Step 7 was again subjected to isoelectric focussing electrophoresis in the same way. One of the elution pattern is shown in Fig. 1. Through these procedures the $NADP^+$-dependent isocitrate dehydrogenase was purified about 500-folds in the specific activity. The results of purification are summarized in Table I.

Homogeneity and Molecular Weight

The homogeneity of the enzyme was checked by polyacrylamide disc gel electrophoresis at pH 9.4. Fig. 2 shows that the final enzyme preparation is practically homogeneous but still contains a very small amount of impurities. The purity was calculated to be about 96 % from the peak area. The gels were also stained for the enzyme activity by the method of Reeves *et al.* (15) and activity band was observed to coincide with the major protein band.

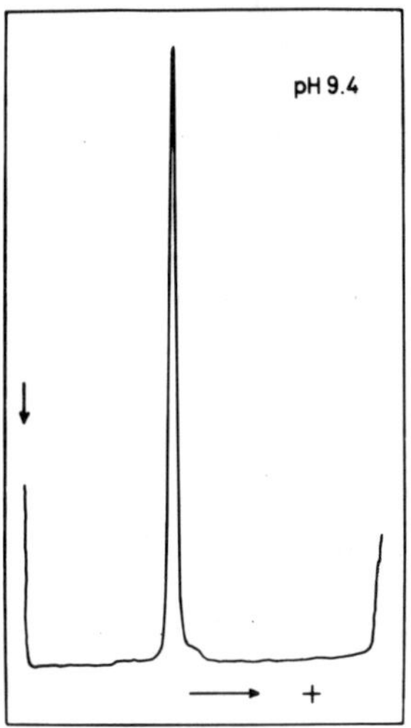

Fig. 2. Disc gel electrophoresis of the purified enzyme

The molecular weight of the dehydrogenase was estimated to be between 80,000-90,000 on Sephadex G-100 gel filtration.

Heat Stability

The enzyme solutions (10 µg per ml in 0.02 M potassium phosphate buffer, pH 7.0) were incubated at various temperatures and the residual activities were measured at intervals. As shown in Fig. 3 the enzyme has a remarkable heat stability. The enzyme was not inactivated after 60 min at 70°C and activity was lost slowly at 80°C.

Above 90°C, however, rapid inactivation was observed.

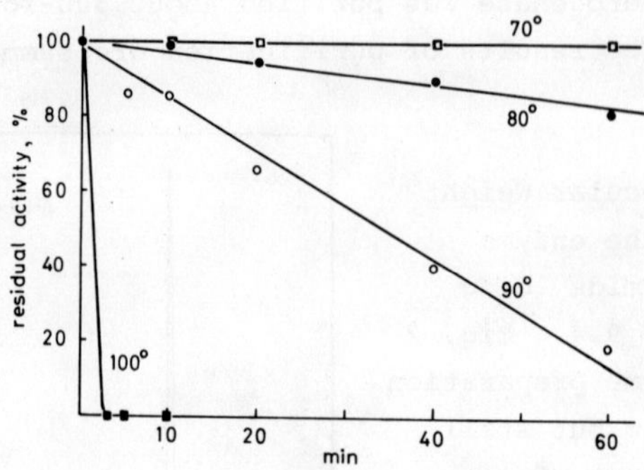

Fig. 3. Heat stability of the isocitrate dehydrogenase

Effects of Organic Acids Related to the TCA and Glyoxylate Cycles

At 1.0 mM concentration, each organic acid tested (citrate, *cis*-aconitate, α-ketoglutarate, succinate, fumarate, malate, oxaloacetate, glyoxylate, glycollate and pyruvate) showed almost no effect or only slight inhibition of the isocitrate dehydrogenase activity, but in the presence of both oxaloacetate and glyoxylate (each 1.0 mM) 75% inhibition was observed. Inhibition by 1.0 mM oxalacetate or glyoxylate was only 9 and 14 %, respectively.

From the Lineweaver-Burk plots, oxalacetate and glyoxylate were observed to inhibit the dehydrogenase activity in a mixed and a competitive manner, respectively. The type of inhibition by oxalacetate plus glyoxylate was found to be competitive (10).

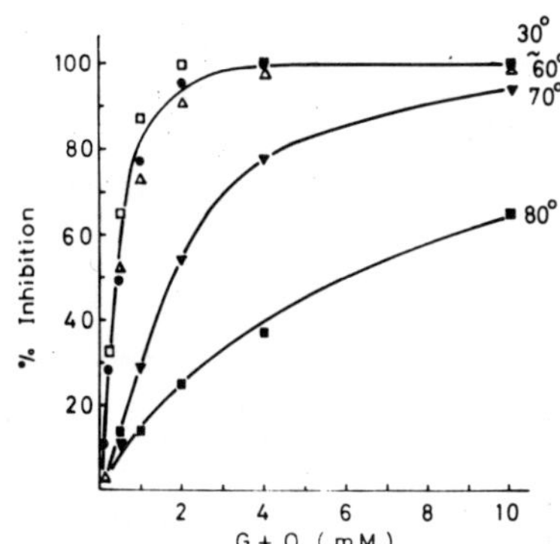

Fig. 4. The effect of temperature on the inhibition of the isocitrate dehydrogenase by oxalacetate (O) plus glyoxylate (G)

Effect of Temperature on the Inhibition by Oxalacetate plus Glyoxylate

The inhibition of the dehydrogenase activity by oxalacetate plus glyoxylate was examined as a function of their concentration in the temperature range 30-80°C (Fig. 4). The concentration of oxalacetate plus glyoxylate required for 50 % inhibition was about 0.5 mM each at 30-60°C, but 2 mM each and 6 mM each were required at 70 and 80°C, respectively. It is obvious from these results that the sensitivity of the enzyme to the inhibitors decreased markedly above 60°C. Desensitization of the enzyme to the inhibitors at 80°C was observed to be reversible (10).

Effects of Temperature on V_{max} and K_m

The values of maximum velocities (V_{max}) and Michaelis constants (K_m) for DL-isocitrate and for $NADP^+$ were estimated in the temperature range 30-80°C. The relationship between K_m for each substrate and temperature (T) is given in Fig. 5. The plots show a discontinuity of slope at 55-60°C. The temperature-V_{max} profile in the same figure indicates that this has also a discontinuity of slope at about 60°C. The activation energies calculated below and above 60°C were 14,500 and 8,000 cal per mol, respectively. These results suggest that some conformational change in the enzyme protein may occur at a transition temperature of 55-60°C.

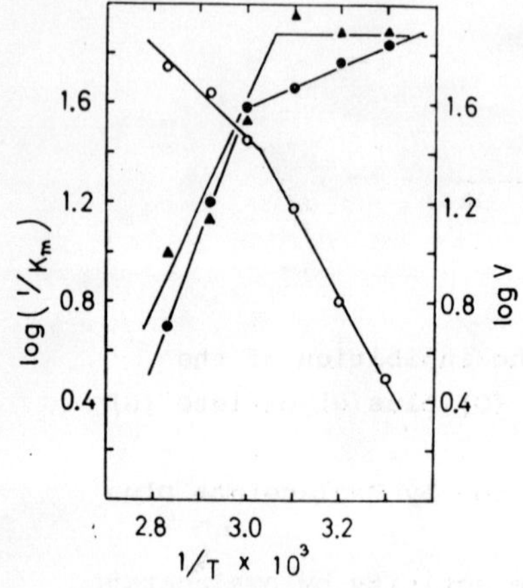

Fig. 5. Effect of temperature on K_m and V_{max} of the isocitrate dehydrogenase

● $\log(1/K_m)$ for $NADP^+$

▲ $\log(1/K_m)$ for DL-isocitrate

○ $\log(V_{max})$

Secondary Structure

The CD spectra in Fig. 6, which practically resembles that of the isocitrate dehydrogenase from *Bacillus stearothermophilus* (9), shows the two negative bands at 220 and 210 nm, demonstrating the presence of α-helical structure. The content of α-helix and β-structure at 25°C were estimated to be about 26 and 33 %, respectively, by the method of Greenfield and Fasman (16).

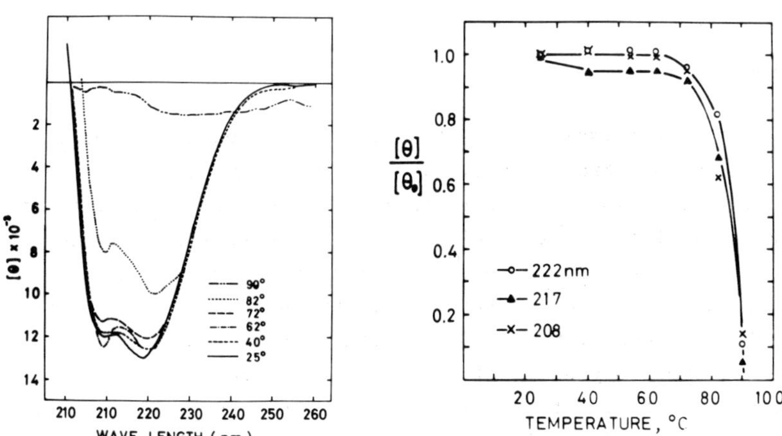

Fig. 6. (left) Effect of temperature on the CD spectra of the purified isocitrate dehydrogenase

Fig. 7 (right) Effect of temperature on CD bands at 208, 217, and 222 nm. Data are obtained from Fig. 6.

Effect of Temperature on CD Spectra

The effect of heating on the CD spectra of the enzyme solution (0.31 mg per ml in 50 mM K_2HPO_4-KH_2PO_4 buffer, pH 7.2) was examined. In this experiment, after the enzyme solution was kept at a certain temperature for about 20 min for equilibration of the temperature and scanning of CD spectra, temperature was then raised to the next higher temperature for CD scanning. The data of Fig. 7 were drawn from Fig. 6. It is evident from Fig. 6 and 7 that up to 60° practically no changes in CD bands are observed, but above 60°C the depth of CD bands decreases gradually and more remarkably above

80°C. At 90°C the enzyme was completely unfolded.

DISCUSSION

$NADP^+$-specific isoictrate dehydrogenase was purified about 500-folds from an extremely thermophilic bacterium *Thermus flavus* AT-62. Polyacrylamide gel electrophoresis showed that the preparation still contains a very small amount of impurities. As the purity was calculated to be about 96 % from peak area of gel scanning pattern, the CD spectra of this preparation are considered to reflect the inherent properties of the enzyme. It is not clear whether the minor components of isocitrate dehydrogenase are isozymes or some artifacts formed during the course of purification.

As with most enzymes from thermophiles so far studied, the dehydrogenase of *Thermus flavus* AT-62 has also remarkable heat stability, which supports the growth of the organism at temperatures up to 80°C, and seems to require no stabilizing factors. It should be noted that the homoserine dehydrogenase of the same organism is specifically stabilized against heat inactivation by monovalent cations such as Na^+ or K^+ (3).

Decreased sensitivity of the dehydrogenase to inhibition by oxalacetate plus glyoxylate was observed above 60°C. The regulatory role of inhibition by oxalacetate plus glyoxylate is described and discussed by Ozaki and Shiio (17). Similarly, the Michaelis constants of the enzyme for isocitrate and $NADP^+$ increased above 60° C and the temperature-activity profile (Arrhenius plots) showed a discontinuity of the slope at 60°, indicating a change of activation energy at 60°C. All these observations might indicate that the enzyme has rather stable properties in the range of 40-60°C (lower half of the temperature range for growth of the organism), but gradually changes its structure to show reduced affinities for substrate and reduced sensitivity to inhibitors above 60°C (upper half of the temperature range for growth of the organism). Above 90°C the enzyme protein is rapidly denatured.

As to the comparison between the CD spectra of the dehydrogenase obtained from *T. flavus* AT-62 and those of *Bacillus stearothermophilus* (9) and *Azotobacter vinelandii* (18), they resemble each other as a whole, having negative CD bands at 210 and 220 nm.

From the change of CD spectra in the course of heating, the enzyme of *T. flavus* turned out to be much more heat stable than that of *B. stearothermophilus*. It is evident from Fig. 6 and 7 that up to 60°C practically no change in CD bands are observed but above 60°C the depth of CD bands decreased gradually and more remarkably above 80°C. It is, therefore, considered that secondary structure of the enzyme is stable below 60°C but above 60°C the structure becomes a little flexible. Heat denaturation of the enzyme occurs above 80°C and the melting temperature was observed tobe at about 83°C.

The allosteric sensitivity of the aspartokinase of *T. flavus* to L-threonine decreased above 60°C (T. Saiki *et al.*, unpublished) and the homoserine dehydrogenase of the organism also lost its sensitivity to L-threonine as a feedback inhibitor. A common property of homoserine dehydrogenases of mesophiles is susceptibility to inhibition by the feedback modifier L-threonine(19,20). A decrease in allosteric sensitivity with increasing temperature is also observed in some other enzymes from thermophiles (21, 22). Increase in K_m for substrates, that is, a decrease in affinity for substrates with increasing temperature has also been observed with glucose-6-phosphate isomerase (23) and with fructose-1,6-diphosphate aldolase of *B. stearothermophilus* (24). Recently Coultate and Sundaram (25) have reported in their study about energetics of *B. stearothermophilus* growth that the molar growth yield was maximal at the lowest growth temperature employed and decreased steadily with the rise of temperature.

These observations may suggest that many, if not all, enzymes of thermophiles have adapted to elevated temperatures with success in respect to enzyme stability, but have done so less effectively as regards affinities for the enzyme substrates and regulatory function. It is suggested in certain poikilotherms that in evolutionary adaptation to temperature, the enzyme affinity for substrate is a feature more sensitive to natural selection than is molecular activity (26). Many of the thermophile enzymes, on the other hand, seem to have adapted to elevated temperatures by acquiring unusual heat stability at the sacrifice of affinity for substrate and of accuracy in metabolic regulation.

REFERENCES

1. R. Singleton Jr. and R. E. Amelunxen, Bacteriol. Rev., 37, 320 (1973)
2. T. Saiki and K. Arima, Agr. Biol. Chem., 34, 1762 (1970)
3. T. Saiki, H. Shinshi and K. Arima, J. Biochem. 74, 1239 (1973)
4. B. W. Mathews, J. N. Jansonius, P. M. Colman, B. P. Schoenborn and D. Dupourque, Nature New Biology, 238, 37 (1972)
5. B. W. Mathews, P. M. Coleman, J. N. Jansonius, K. Titani, K. A. Walsh and H. Neurath, Nature New Biology, 238, 41 (1972)
6. M. Tanaka, M. Haniu and K. T. Yasunobu, J. Biol. Chem., 248, 5215 (1973)
7. R. L. Howard and R. R. Becker, J. Biol. Chem., 245, 3186 (1970)
8. R. F. Ramaley and M.O. Hudock, Biochim. Biophys. Acta, 135, 22 (1973)
9. Y. Hibino, Y. Nosoh and T. Samejima, J. Biochem., 75, 553(1974)
10. T. Saiki and K. Arima, J. Biochem., 77, 233 (1975)

11. R. F. Itzhaki and D. M. Gill, Anal. Biochem., 9, 401 (1964)
12. L. Ornstein and B. J. Davis, Ann. New York Acad. Sci., Art. 2, 321, 404 (1964)
13. O. Vesterberg and H. Svensson, Acta Chem. Scand., 20, 820(1966)
14. P. Andrews, Biochem. J., 91, 222 (1964)
15. H. C. Reeves, G. O. Daumy, C. C. Lin and M. Houston, Biochim. Biophys. Acta, 258, 27 (1972)
16. N. Greenfield and G. D. Fasman, Biochemistry, 8, 4108 (1969)
17. H. Ozaki and I. Shiio, J. Biochem., 64, 355 (1968)
18. A. E. Chung and J. S. Franzen, Biochemistry, 8, 3175 (1969)
19. P. Datta, Science, 165, 556 (1969)
20. G. N. Cohen, R. Y. Stanier and G. LeBras, J. Bacteriol., 99, 791 (1969)
21. H. K. Kuramitsu, J. Biol. Chem., 245, 2991 (1970)
22. D. A. Thomas and H. K. Kuramitsu, Arch. Biochem. Biophys., 145, 96 (1971)
23. N. Muramatsu and Y. Nosoh, Arch. Biochem. Biophys., 144, 245 (1971)
24. S. Sugimoto and Y. Nosoh, Biochim. Biophys. Acta, 235, 210 (1971)
25. T. P. Coultate and T. K. Sundaram, J. Bacteriol., 121, 55 (1974)
26. P. W. Hochachka and G. N. Somero, Comp. Biochem. Physiol.,27, 659 (1968)

OTHER TYPES OF THERMOPHILIC ENZYMES (PROTEINS)

Maintainance of Specificity, Information, and Thermostability in Thermophilic *Bacillus sp.* Glutamine Synthetase

by: F. C. WEDLER, F. M. HOFFMANN, R. KENNEY, AND J. CARFI
Chemistry Department, Cogswell Laboratory
Rensselaer Polytechnic Inst., Troy, New York 12181, USA

ABSTRACT:

Glutamine synthetase has been purified to homogeneity from *B. subtilis* (37°) *B. stearothermophilus* (55°), and *B. caldolyticus* (75°). Those characteristics compared include size ($6.0 \pm 0.3 \times 10^5$ daltons), quaternary structure (12 SU) amino acid content, substrate K_m's and specificity for structural analogs, metal ion activation, number and kind of separate feedback modifier sites, and the complexity of modifier-substrate and modifier-modifier site interactions. Although the 37° and 55° systems are quite similar, the 75° system shows important alterations in substrate specificity and modes of modifier action. Whereas at 37° and 55° AMP inhibits synergistically with amino acids (glycine, glutamine, histidine), the 75° enzyme is inhibited directly by the products ADP, (which assumes the role of AMP) and glutamine, plus other ligands. Ligand binding domains are compared and found to be very different. Thermostabilization occurs by (a) protection by bound L-glutamate, (b) protein aggregation, (c) trends in the content of total polar residues, total Asx + Glx residues, the average hydrophobicity, and (d) disulfide bond cross-linking. Such studies provide insights to molecular evolution occurring with changes in environmental stress.

INTRODUCTION:

Because mesophilic glutamine synthetase from micro-organisms has been found to contain an extraordinary amount of functional information; namely, a large number of binding sites for metal ions, modifiers, and substrates [1], this system seemed an ideal subject for study in thermophilic organisms. The reaction catalyzed occupies a key position in nitrogen metabolism, and is therefore subject to regulation by a variety of end products. By isolation and detailed characterization of these enzymes, we have probed the question of how much complex function can be maintained in a system stable at 55° or 75°C.

Since our preliminary investigation of the B. stearothermophilus enzyme at 55° [2,3], a more complete understanding of its regulatory properties has been obtained [4]. In addition, the enzyme from B. caldolyticus has been characterized at 75° in considerable depth [5]. Comparisons of these systems with those from mesophilic sources reveal how functional complexity is maintained in the face of thermal stress. The molecular mechanisms for this thermostabilization is also a subject of investigation.

EXPERIMENTAL:

In the interest of space, detailed procedures will not be described here, except for those beyond published ones [2,3,4]. Procedures for the preparation and purification of B. caldolyticus glutamine synthetase will be published soon [6].

Kinetic experiments involving modifier pairs were carried out by procedures described previously [3]. Variation of modifier in the presence of fixed substrate levels, or vice versa, was accomplished by mixing two solutions, A and B, to the same volume (e.g., 0.2 ml) in various ratios. A and B differ only in that one of them contained the varied component at its maximal concentration, whereas the other lacked it completely. Inhibitions were expressed as *i*, fractional inhibition, calculated by published equations [7].

RESULTS AND DISCUSSION:

Size and Shape, Symmetry. The molecular weight of all glutamine synthetases isolated from micro-organisms to date is 600,000 ± 30,000. This includes that from B. caldolyticus, the data for which are shown in Fig. 1, viz., SDS-polyacrylamide gel electrophoresis, indicating a subunit MW = 50,000 ± 2,000, and an electron micrograph showing the double hexagonal arrangement of 12 subunits. Small differences do exist. The B. subtilis enzyme was found to have a subunit MW closer to 550-580,000, and its amino acid analysis showed fewer total residues than the B. caldo. and B. stearo. enzymes [2,5,8]. In contrast to the E. coli system, which has its double hexagon skewed about 7.5° from true eclipsed arrangement, the B. caldolyticus subunits are fully eclipsed [5,8].

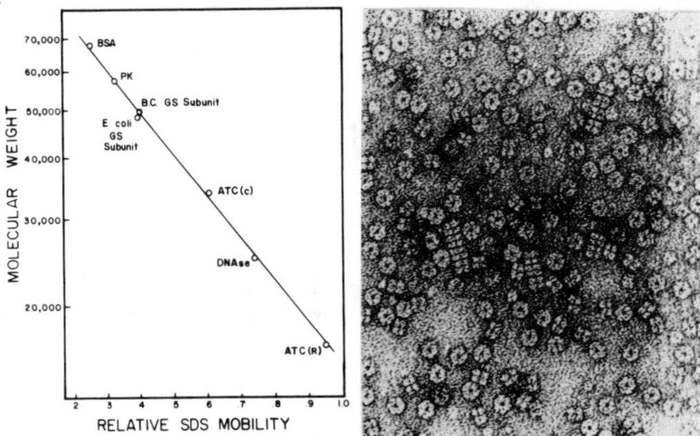

Fig. 1: Size and shape of B. caldolyticus glutamine synthetase. (Left) SDS-PAGE, showing a subunit MW of ca. 50,000. (Right) Electron micrograph, showing eclipsed double hexagonal structure, with 12 subunits per native enzyme.

Specificity: Kinetic Responses to Substrates, Metal Ions. Divalent metal ions strongly influence the response of activity to other solution parameters, especially pH and levels of ATP. As shown in Fig. 2, the biosynthetic assay for the B. caldo. enzyme is stimulated to a slightly greater extent by 50 mM $MgCl_2$ than by $MnCl_2$ equimolar with ATP. The reverse is true in the transferase assay, which has a much lower relative or specific activity. The level of Mn^{2+} relative to ATP critically alters activity, which is maximal at Mn^{2+} = ATP. Excess ATP markedly inhibits activity but excess Mn^{2+} does not, which suggests that Mn^{2+} may bridge enzyme and ATP in a productive complex, E-Mn-ATP, and that E-ATP is a non-productive, dead-end complex. The interrelationship of Mg^{2+} and ATP levels (not shown) do not show this effect, suggesting that Mn^{2+} may determine ATP conformation in the active site differently than does Mg^{2+}. At lower levels of Mg^{2+} or Mn^{2+}, say 10 mM, Mn^{2+}-stimulated activity exceeds that of Mg^{2+}, it should be noted. The relative specificity for activation by metal ions, at pH 6.5 in the biosynthetic assay, is $Mn^{2+} > Cd^{2+} > Co^{2+} > Ca^{2+} \simeq Mg^{2+} > Ni^{2+} \simeq Cu^{2+} > Ba^{2+}$, using M^{2+} = ATP = 10 mM.

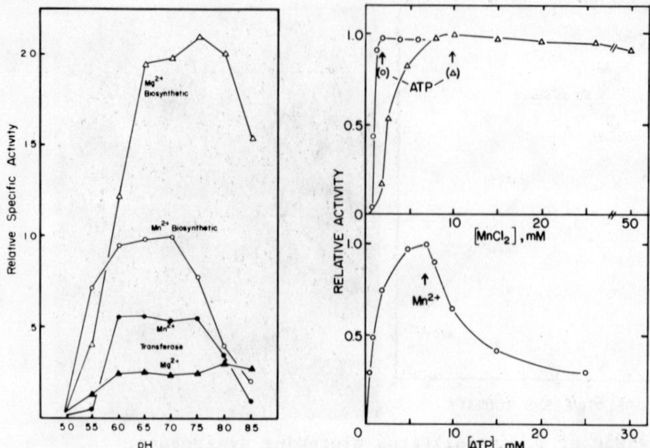

Fig. 2: Response of B. caldolyticus glutamine synthetase to divalent metal ions in the biosynthetic and transferase assays. (Left) pH-dependence. (Right) Dependence on $MnCl_2$ and ATP concentrations.

The responses of mesophilic and thermophilic enzymes to substrates, in terms of K_m values, are summarized in Table 1, part A. It is obvious that specificity for all three substrates is maintained. Especially notable, however, is that whereas MnATP binding is cooperative with the B. stearo. [3,4] and B. subtilis [9] enzymes, it is purely hyperbolic for the B. caldo. system at 72° [5].

In Table 1, part B, it is seen that the active sites of the thermophilic enzymes maintain considerable substrate specificity. Substitution of groups into ammonia, as in CH_3NH_2 and NH_2OH, results in decreased activity, probably for both steric and electronic reasons. Although NH_2OH is a much stronger nucleophile than NH_3 at pH 6.5 it is used only 1/3 as easily in the site. Specificity data for analogs of glutamate indicate that (a) the L-form is highly favored over the D-isomer, or over L-aspartate, (b) that substitution of α-OH for α-NH_2 lessens activity, suggesting a need for H-bonding, if not a charged group at this position, and that N-methylation causes no drastic decreases in activity. Steric requirements near the α-H also are not crucial. Purine nucleotides are preferred appreciably over the pyrimidines.

Table 1: Comparisons of Substrate and Substrate Analog Binding Specificities for Mesophilic and Thermophilic Glutamine Synthetases.

Source:	B. caldo.	B. stearo.	B. subtilis	B. lichen.
[Ref.]:	(5)	(3,9)	(9,10)	(11)
(A) - K_m values (mM)				
NH_3	1.3	2.0	0.4[c]	1.4[c]
L-Glu	3.1	1.5	0.8[c]	5.0
Mn-ATP	0.5 [a]	0.3 [b]	0.2 [b]	0.9
(B) - Relative Activity (conc.)				
NH_3 (50)	100	100	100	100
NH_2CH_3 (50)	11	9	30	-
NH_2OH (50)	33	28	30	-
L-glutamate (50)	100	100	100	100
D-glutamate (50)	16	11	23	20
L-aspartate (50)	3	2	0	0
L-αOH-Glut. (50)	37	10	-	-
D-αOH-Glut. (50)	9	3	-	-
D,L-glutamate (50)	85	82	-	-
D,L-αCH_3-Glu (50)	28	29	-	-
D,L-N-CH_3-Glu (50)	27	34	-	-
MnATP (10)	100	100	100	100
MnGTP (10)	47	80	29	10
MnCTP (10)	13	12	5	0
MnITP (10)	41	11	0	3
MnUTP (10)	16	18	18	0

[a] Hyperbolic binding [b] Cooperative binding [c] Substrate inhibition

Substrate-Modifier and Modifier-Modifier Interactions. Kinetic experiments with the B. stearo. and B. caldo. enzymes, using saturating substrate levels, and then with each substrate at 2 x K_m, were carried out in the presence and absence of a fixed level of modifier. These results (not shown) suggest how the mode of action of each modifier may depend on substrate concentration. These results were then used to select modifiers to test as competitors in substrate binding experiments, i.e., Lineweaver-Burke plots in the presence and absence of each modifier, as shown in Figure 3. It is seen here that on the B. caldo. enzyme neither glycine, L-alanine, nor L-glutamine compete with L-glutamate, although L-histidine apparently does. Further, ADP competes strongly and directly with ATP, but AMP has a weak noncompetitive effect. The action of amino acid modifiers is enhanced by bound ATP; GDP, CTP, and glucosamine-6-P appear to alter glutamate binding, whereas CTP competes with ATP [5]. Similar experiments with B. subtilis and B. stearo. enzymes have been published [3,10].

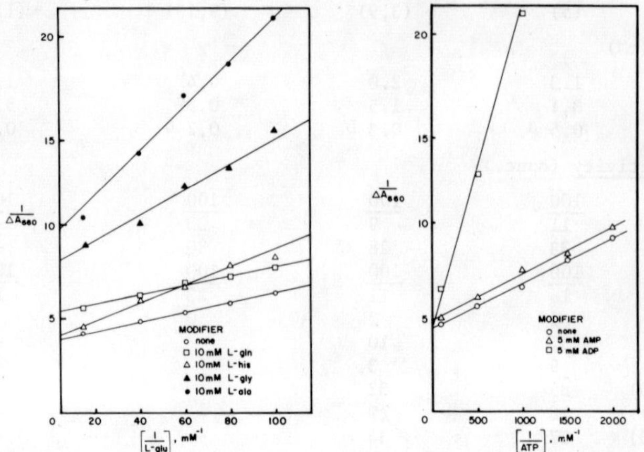

Fig. 3: Inhibition patterns for feedback modifiers of B. caldolyticus glutamine synthetase versus (Left) L-glutamate, and (Right) MnATP.

Modifier and Substrate Binding Domains. By carrying out inhibitions with pairs of these modifiers, one can determine whether they inhibit in a cumulative mechanism [1] (indicating of separate sites), in an antagonistic manner (indicating overlapping sites), in an additive manner (indicating isozymes), or synergistically. From these results one can deduce the extent of separation or overlap of modifier sites with each other, and, from the competition experiments (above), with substrate sites. Such deductions, carried out using available data for the B. subtilis [10,12], B. stearo. [3,4], and B. caldo. [5] enzymes, result in the hypothetical domain schemes shown in Figure 4. A notable feature of these schemes is the simplification of the variety of binding interactions and number of separate sites as one goes to the thermophilic systems. As a survival advantage, the basic functional information has been maintained while minimizing the surface area of the protein involved.

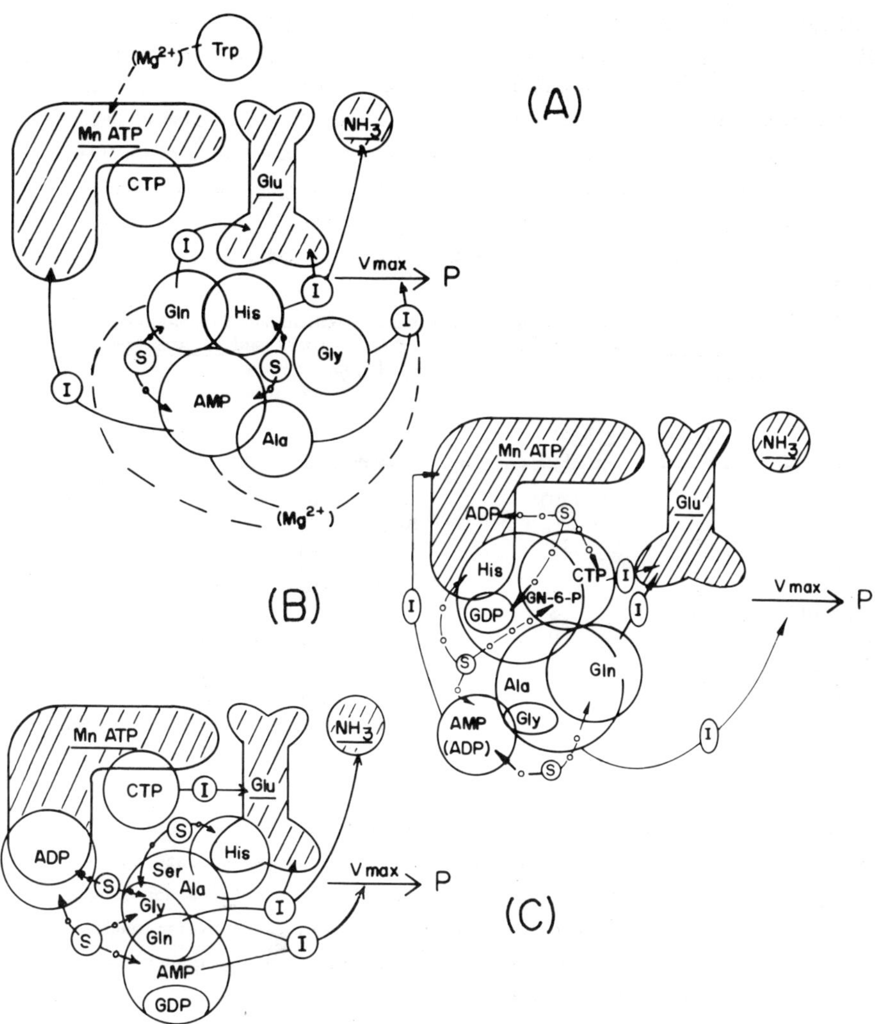

Fig. 4: Interactions of substrate and modifier binding domains in glutamine synthetases from (A) B. subtilis, (B) B. stearothermophilus, and (C) B. caldolyticus.

Thermostabilization. Bound ligands impart marked thermostability to the B. stearo. and B. caldo. enzymes, as shown in Table 2. L-glutamate stabilizes both enzymes strikingly, and MnATP (not ATP alone) has a stronger effect with the 72° enzyme than the 55° one. Glu + MnATP are powerful stabilizers, but NH_3 with either Glu or MnATP destabilizes appreciably. Bound metal ions have little effect, with the exception of Cd^{2+} with the B. stearo. enzyme. Amino acid modifiers have a stronger stabilizing effect on the 55° enzyme than the 72° one, whereas the reverse is true for the nucleotide modifiers.

Table 2: Thermostabilization by Bound Ligands with Glutamine Synthetases from B. caldolyticus (80°, 20 min., 0.075 mg/ml) and B. stearothermophilus (65°, 20 min., 0.1 mg/ml [3]).

Ligand, [mM]	% Activity Lost B. caldo.	% Activity Lost B. stearo.	Ligand [mM]	% Activity Lost B. caldo.	% Activity Lost B. stearo.
(None)	(89)	(78)	Mn^{2+}, [10]	82	69
			Mg^{2+}, [10]	92	77
NH_3, [50]	95	77	Cd^{2+}, [10]	97	43
L-Glu [50]	9	6	Ni^{2+}, [10]	94	91
ATP, [10]	97	-	Co^{2+}, [10]	92	81
MnATP, [10]	11	66			
NH_3, L-Glu	48	0	Gly [25]	38	13
NH_3, ATP	96	-	L-Ala [25]	88	5
NH_3, MnATP	33	30	L-His [25]	92	24
Glu, ATP	66	-	L-Gln [25]	92	0
Glu, MnATP	0	0	Mn-AMP [25]	11	66
			Mn-ADP [25]	44	64

Various effects of heat in relation to enzyme purity and concentration are presented in Figure 5. At the left, the Arrhenius plots indicate that as the B. caldo. enzyme is purified, it exhibits a more pronounced non-linear response above and below 56°, and that the activation energies differ much less below 56° than above. Since cooperative substrate binding is not indicated in this enzyme, in contrast to others [2,3,9], it is suspected that protein aggregation may be an explanation of these effects. This view is supported by the data in Figure 5 (right), in which higher protein concentrations thermostabilize the enzyme, as with the B. stearo. system [3]. Interestingly, the enzyme is much less stable at 75° than at 72°C.

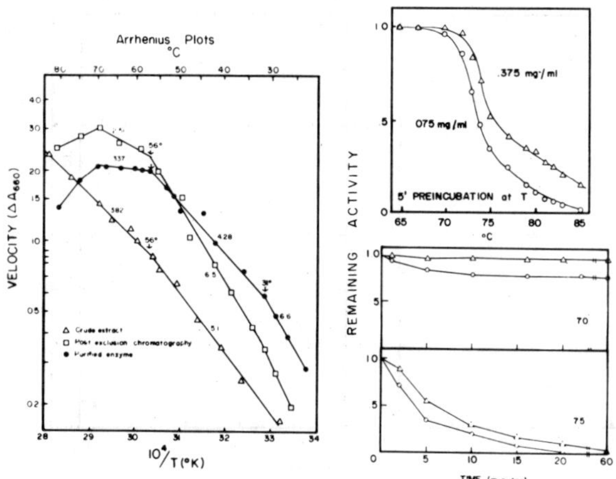

Fig. 5: Thermal responses of B. caldolyticus glutamine synthetase (Left) Arrhenius plots for enzyme at various stages of purification. (Right) Effect of protein concentration on thermal denaturation: (Top) Temperature dependence, (Bottom) Time dependence.

Finally, the possible roles played by intrinsic chemical and physical structural properties of the protein are examined in Figure 6. From the amino acid composition, potential α-helix and β-sheet contents were calculated according to Chou and Fasman, [3,13] and did not vary with growth temperature of the organism from which the enzyme was isolated. CD measurements of α-helix and β-sheet on purified proteins confirm this invariance [3]. The average hydrophobicity [14] showed a slight decrease with increasing temperature. Both the total number of polar sidechain residues and the Asx + Glx residues (Asp + Glu + Asn + Gln) increase with temperature. In this regard the thermophilic proteins resemble those from halophiles. Although the analyzed 1/2 Cys content shows considerable variation, titration of free-SH groups with specific reagents indicated that the B. subtilis enzyme has all four residues as free-SH [15], but that the B. caldo. enzyme has all six Cys as three disulfide (-S-S-) bridges [16]. Therefore disulfide bridges are likely an important stabilizing force also.

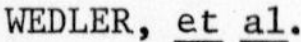

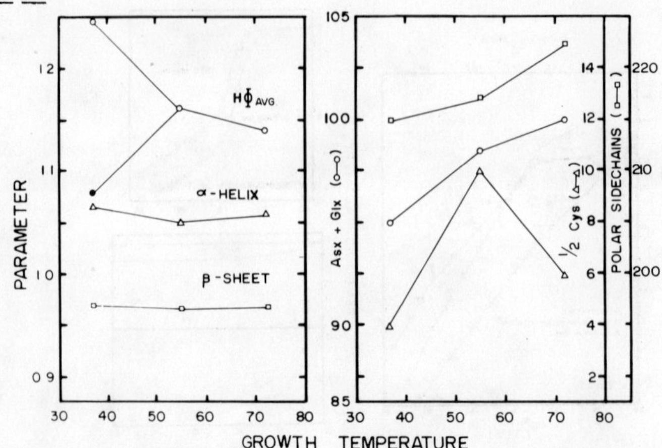

Fig. 6: Relation of physical parameters to thermostabilization of glutamine synthetase, calculated from amino acid analyses. (Left) Average hydrophobicity, HØ avg [14], and potential for α-helix and β-sheet structures [13]. (Right) Number of polar sidechains, total Asx and Glx content, and 1/2 Cys content. The 37° data are for B. subtilis, except for one point (●) for the E. coli enzyme; those at 55° are the B. stearothermophilus enzyme; those at 72° are for the B. caldolyticus system.

Conclusions. Comparing the thermophilic and extremely thermophilic glutamine synthetase enzymes to the mesophilic examples, few major differences are apparent in terms of substrate specificity, α-helix or β-sheet content, or quaternary and ternary structural features. The enzymes clearly differ in terms of their thermostability and a variety of more esoteric properties. Such differences are apparent only as the result of very detailed experimentation and data analysis. Among these, one may list: increases in polar amino acid residues, decreased average hydrophobicity, increased disulfide cross-linking, changes in the number, kind, spatial arrangements, and interactions of modifier sites relative to each other and to substrate sites. The tendency of native protein (MW 600,000) to form higher aggregates ($MW>10^6$) is virtually absent in the mesophilic enzyme [10], but becomes much more pronounced in the thermophilic ones. The extent and intensity of substrate and modifier protection against heat denaturation also shows this increasing trend. Basically, the extremely thermophilic system has minimized the protein surface area involved in ligand binding without sacrificing much in the way of functional complexity. The notable exceptions or changes in the 72° enzyme in this regard are the absence of cooperative MnATP substrate binding, and the switch from AMP to ADP as the principal nucleotide modifier acting in synergy with amino acid modifiers.

REFERENCES

[1] Shapiro, E. M., and Stadtman, E. R. (1970) Ann. Rev. Microbiol. 24, 501.

[2] Wedler, F. C., and Hoffmann, F. M. (1974) Biochemistry, 13, 3207.

[3] Wedler, F. C., and Hoffmann, F. M. (1974) Biochemistry, 13, 3215.

[4] Wedler, F. C., Carfi, J., and Ashour, A. E. (1975) Biochemistry, submitted for publication.

[5] Kenney, R. M. (1975) B. S. Thesis, Rensselaer Polytechnic Institute.

[6] Wedler, F. C., and Kenney, R. M., manuscripts in preparation.

[7] Wedler, F. C., and Boyer, P. D. (1973) J. Theor. Biol. 38, 539.

[8] Eisenberg, D., personal communication.

[9] Wedler, F. C. (1974) Biochem. Biophys. Res. Comm. 60, 737.

[10] Deuel, T. F., and Stadtman, E. R. (1970) J. Biol. Chem. 245, 5206.

[11] Hubbard, J. S., and Stadtman, E. R. (1967) J. Bacteriol. 94, 1007.

[12] Deuel, T. F., and Prusiner, S. (1974) J. Biol. Chem. 249, 257.

[13] Chou, P. Y., and Fasman, G. D. (1974) Biochemistry, 13, 222.

[14] Biglow, C. C. (1967) J. Theor. Biol. 16, 187.

[15] Deuel, T. F. (1971) J. Biol. Chem. 246, 509.

[16] Ashour, A. E., unpublished results, this laboratory.

PHOSPHOFRUCTOKINASE FROM THERMOPHILIC MICRO-ORGANISMS

H. HENGARTNER, E. KOLB and J. IEUAN HARRIS
MRC Laboratory of Molecular Biology
Hills Road, Cambridge, U.K.

INTRODUCTION

Phosphofructokinase (PFK) catalyses the interconversion of fructose 6-phosphate (F-6P) and fructose 1,6-diphosphate (FDP) in a reaction involving ATP that represents the first unique step in glycolysis (for review see (1)). The activity of PFK is regulated by a number of intracellular metabolites, and a study of its mode of action, and of the mechanisms through which it exercises control over the rate of glycolysis in the cell, is therefore of particular biochemical interest.

PFK has been isolated in pure form from a number of different sources (1). Enzymes from muscle and from yeast possess complex oligomeric structures consisting of large subunits with molecular weights of 85,000 (1) and 130,000 (2) respectively that have hitherto proved difficult to study by methods of protein chemistry and X-ray crystallography. PFKs of microbial origin have been reported (1) to possess lower molecular weights and less complex subunit structures. More specifically earlier work from this laboratory (reference (3) and unpublished results of E. Kolb and J.I. Harris) has shown that PFK from _B. stearothermophilus_ with a molecular weight of 130,000 is a thermostable tetrameric enzyme with four identical protein chains each comprising about 300 amino acid residues.

The enhanced stability of enzymes from thermophilic micro-organisms can be utilised to advantage in enzyme purification and in studies of enzyme structure and mechanism (cf. 4,5). For example, in an exploratory study of glycolytic enzymes from the extreme thermophile, _T. aquaticus_ (that grows optimally at 70-75°C), Hocking and Harris (6) showed that pure glyceraldehyde 3-phosphate dehydrogenase (GPDH) could be obtained in high yield from partially purified cell extracts by affinity chromatography on NAD-Sepharose. Other examples of the use of Sepharose-linked nucleotides in the purification of dehydrogenases and kinases have since been reported (e.g. 7,8) and we now describe a method that allows pure GPDH and PFK to be obtained from cell extracts of _T. aquaticus_ and of _B. stearothermophilus_ by the combined application of NAD-Sepharose and AMP-Sepharose. Some molecular and kinetic properties of the purified PFKs are also given together with the complete amino acid sequence of the subunit and a method for crystallising the _B. stearothermophilus_ enzyme for X-ray crystallographic analysis.

METHODS AND RESULTS

NAD-Sepharose was prepared and used to purify GPDH as previously described (6). AMP-Sepharose (N^6-(6-aminohexyl)AMP-Sepharose) was prepared by reacting CNBr activated Sepharose 4B with 6-chloropurine-5'-riboside phosphate by a method similar to that described by Guilford _et al_. (9). _B. stearothermophilus_ (NCA 1503) and _T. aquaticus_ (ATCC 25104) cells were grown (cf. 10) at 60°C and 70-75°C respectively at MRE Porton. Disruption of cells and extraction of cell paste was carried out as before (6,11). Partially purified enzyme fractions were prepared at MRE Porton by a modification (A. Atkinson and C. Bruton, to be published) of the earlier (6,11) procedures. GPDH (4,6) and PFK (12) were assayed at 20°C in thermostated cuvettes in a Guilford 222A recording spectrophotometer.

Purification of GPDH and PFK from T. aquaticus. The cell extract (6) from T. aquaticus was adsorbed batchwise onto DEAE-cellulose (2 ml settled vol per g cells). The fraction eluting between 0.1 and 0.4 M NaCl in 50 mM Tris-HCl, pH 7.5, containing 5 mM β-mercaptoethanol (ME), 1 mM $MgSO_4$ and 0.1 mM EDTA (buffer A) was dialysed against buffer A and then readsorbed onto DEAE-Sephadex (1-2 ml settled vol per g cells). Fractions rich in aldolase (a), GPDH (b) and PFK (c) were eluted with buffer A containing 180 mM NaCl, 240 mM NaCl, and 350 mM NaCl respectively.

Fraction (b) (the content of GPDH in different batches of cells varied between 50 and 200 mg per Kg frozen cells) was brought to 1.2 M ammonium sulphate, centrifuged, and the supernatant adsorbed onto NAD-Sepharose (1-2 ml settled vol/mg GPDH). The column was washed with 100 mM phosphate, pH 6.8, 5 mM β-ME, 1 mM EDTA (buffer B) containing 500 mM NaCl, and was then eluted with the same buffer containing 10 mM NAD^+ in order to obtain pure GPDH as previously described (6).

A sample of fraction (c) (100 ml containing 380 units PFK, specific activity 0.43 units/mg protein) was applied to a column of AMP-Sepharose (1-2 ml settled vol/mg PFK). The column was washed with buffer A containing 400 mM NaCl until the effluent was free of A_{280} nm absorbing material. It was then eluted with the same buffer containing 2 mM ATP/5 mM Mg^{++} and fractions containing PFK activity (50 ml) were pooled, dialysed against buffer A and concentrated by pressure dialysis. The recovery of PFK activity was 85-90%. The protein content was estimated by a colorimetric method (13) and from the results of amino acid analyses (Table 1). The specific activity was calculated to be 37 units/mg at 20°C.

Properties of T. aquaticus PFK. The material eluted from AMP-Sepharose was pure. It gave a single protein band when examined by gel electrophoresis with (14) and without (15) SDS. The pattern of protein bands given by the fractions applied to and eluted from AMP-Sepharose are shown in Fig. 1. The subunit molecular weight was estimated (from its mobility in a 0.1% SDS/12.5% acrylamide gel) to be 32,000. The molecular weight of the active enzyme (measured by gel filtration on a calibrated column of Sephadex G-200 (16) is in the range of 130,000 to 140,000 showing that T. aquaticus PFK, like its counterpart from B. stearothermophilus (3, and unpublished results from E. Kolb and J.I. Harris) is a tetramer with four probably identical protein chains each comprising approximately 310 amino acid residues (cf. amino acid analysis, Table 1). Of particular interest is the total absence of cysteine (in accord with its lack of reactivity towards DTNB and iodoacetate, even in the presence of 6 M guanidine) indicating that SH groups cannot be essential, either for the regulation or the catalytic activity of procaryotic PFK. It would appear that all but essential SH groups (such as, for example, Cys-149 in T. aquaticus GPDH (6)) are deleted in T. aquaticus enzymes (cf. 17), presumably because SH groups are possible sites of inactivation in enzymes that, of necessity, must retain activity under aerobic conditions at temperatures in excess of 70°C.

T. aquaticus PFK is remarkably thermostable (a 0.2 mg/ml solution in buffer A retains full activity after 24 hr at 80°C) and thus resembles PFK from another extreme thermophile, Flavobacterium thermophilum (18).

ATP behaves as a normal substrate with a K_M of 2.5×10^{-5} M. The K_M for the second substrate, F-6P (3×10^{-4} M) on the other hand is subject to a 5-10-fold variation, depending upon the concentration of phosphoenolpyruvate (PEP) which, in the range of 10^{-4} to 10^{-3} is an inhibitor, and of ADP which, in the same range of concentration, is a potent activator of the enzyme. PFK activity in

TABLE 1: Amino acid composition of the subunits in phosphofructokinase from *B. stearothermophilus* and *T. aquaticus*

	*B. stearothermophilus**	*T. aquaticus*†
Cysteine‡	2	0
Aspartic acid	24	19
Threonine	17	16
Serine	11	15
Glutamic acid	29	35
Proline	5	10
Glycine	42	43
Alanine	25	42
Valine	25	29
Methionine	5	7
Isoleucine	28	21
Leucine	24	26
Tyrosine	8	4
Phenylalanine	6	8
Lysine	17	10
Histidine	12	5
Arginine	18	22
Tryptophan§	1	1
TOTAL	299	313

* Obtained by sequence analysis (cf. Fig. 2)

† Nearest whole number of residues calculated from a molecular weight of 32,500

‡ Following radioactive carboxymethylation and performic oxidation

§ Spectroscopic estimation (20)

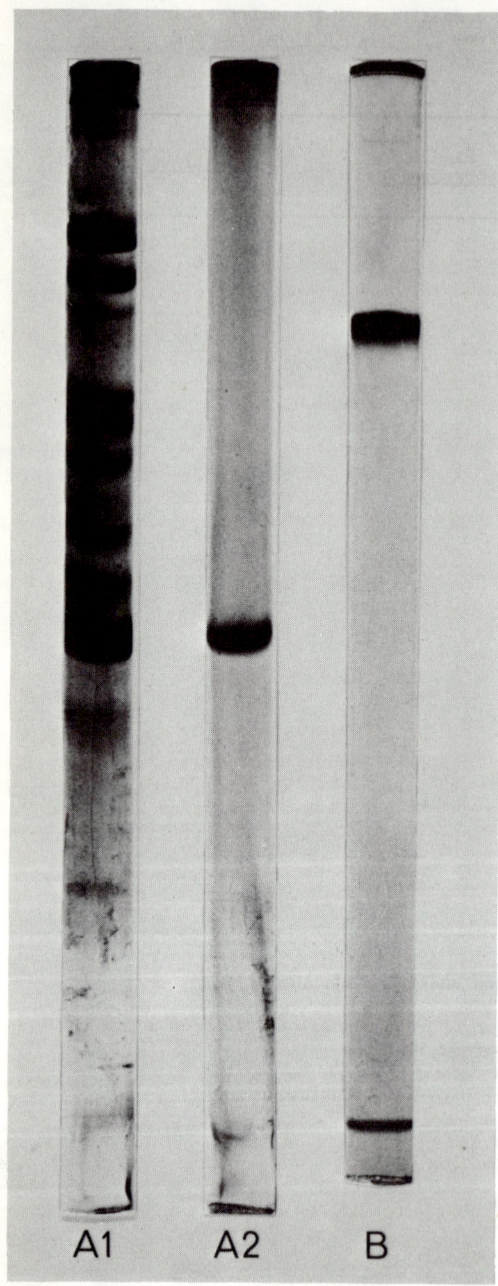

Fig. 1. Gel-electrophoresis of T. aquaticus PFK. (A) In 0.1% SDS (14); (1) before and (2) after AMP-Sepharose. (B) Native gel at pH 8.9 (15)

T. aquaticus is thus subject to regulation by the intracellular concentrations of PEP and ADP. In this respect it resembles other procaryotic PFKs (18,19) but differs from the muscle and yeast enzymes which are subject to allosteric regulation by ATP (1).

Purification of B. stearothermophilus PFK. PFK from B. stearothermophilus was obtained pure by a method similar to that already described for preparing pure T. aquaticus PFK from DEAE-Sephadex fraction (c). Cells were grown, disrupted and extracted according to published methods (10,11) and the PFK-rich fraction eluted from DEAE-Sephadex (6100 units PFK with a specific activity of 2.9 u/mg total protein) was applied to a column of AMP-Sepharose (1-2 mg settled vol/mg PFK) equilibrated with buffer A. The column was washed with several volumes of buffer A containing 400 mM NaCl and the PFK activity was eluted with the same buffer with the addition of 2 mM ATP and 5 mM Mg^{++}. The PFK prepared in this way possessed a specific activity of 105 ± 5 units/mg and was in all respects identical to enzyme prepared earlier (albeit with considerably more difficulty and in lower yield (3); E. Kolb and J.I. Harris, unpublished results) by methods not involving the use of AMP-Sepharose. It has a molecular weight of 130,000 (as determined by gel-filtration on a calibrated column of Sephadex G-200) and a subunit molecular weight of 32,000 determined by SDS-gel electrophoresis (14).

Amino acid sequence of the subunit in B. stearothermophilus PFK. Pure enzyme was carboxymethylated with $[2-^{14}C]$iodoacetate in 6 M guanidine and the S-$[2-^{14}C]$ carboxymethylated protein(containing two residues of carboxymethylcysteine per subunit) was cleaved, at methionine residues with CNBr in 70% formic acid, and at arginyl residues with trypsin following citraconylation of lysine residues. Lysine blocked fragments were subsequently deblocked and redigested with trypsin. Peptide fragments were purified by gel-filtration on Sephadex (G-75 and G-50) by ion-exchange chromatography on DEAE-cellulose (or Sephadex) and by high voltage paper electrophoresis at different pHs. For detailed references to methods see Kolb, Harris and Bridgen (ref. 21).

Purified peptide fragments were sequenced manually by the dansyl-Edman method (22) and the N-terminal sequence of the intact S-carboxymethyl-protein was determined by automated sequence analysis with a Beckman 890B sequencer (cf. 23). Overlaps were established from the sequences of peptides obtained from other (e.g. chymotrypsin) digests. In all, peptide fragments accounting for a total of 299 residues were isolated and sequenced. No other major fragments were found and it is therefore tentatively concluded that the protein subunit of B. stearothermophilus PFK comprises 299 amino acid residues and that the active enzyme contains four subunits of identical primary structure. The amino acid sequence of the subunit is given in Fig. 2. It should be noted however that overlaps were not established between residues 109 and 110, or between residues 119 and 120 so that the sequence in this part of the chain is to be regarded as provisional.

Full details of the sequence analysis (that was carried out using only 300 mg of PFK) are to be published elsewhere (E. Kolb and J.I. Harris, 1975, manuscript in preparation). As this is the first sequence to be determined for any PFK it is not as yet possible to compare the primary structures of mesophile and thermophile enzymes. Comparison of the two thermophile PFKs shows that the B. stearothermophilus and T. aquaticus enzymes possess similar molecular weights and that their N-terminal sequences are homologous (unpublished results).

Crystallisation of B. stearothermophilus PFK. Crystals (with dimensions of up to 1 mm x 1 mm x 0.3 mm) have been obtained (cf. Fig. 3) by dialysis against 2 M

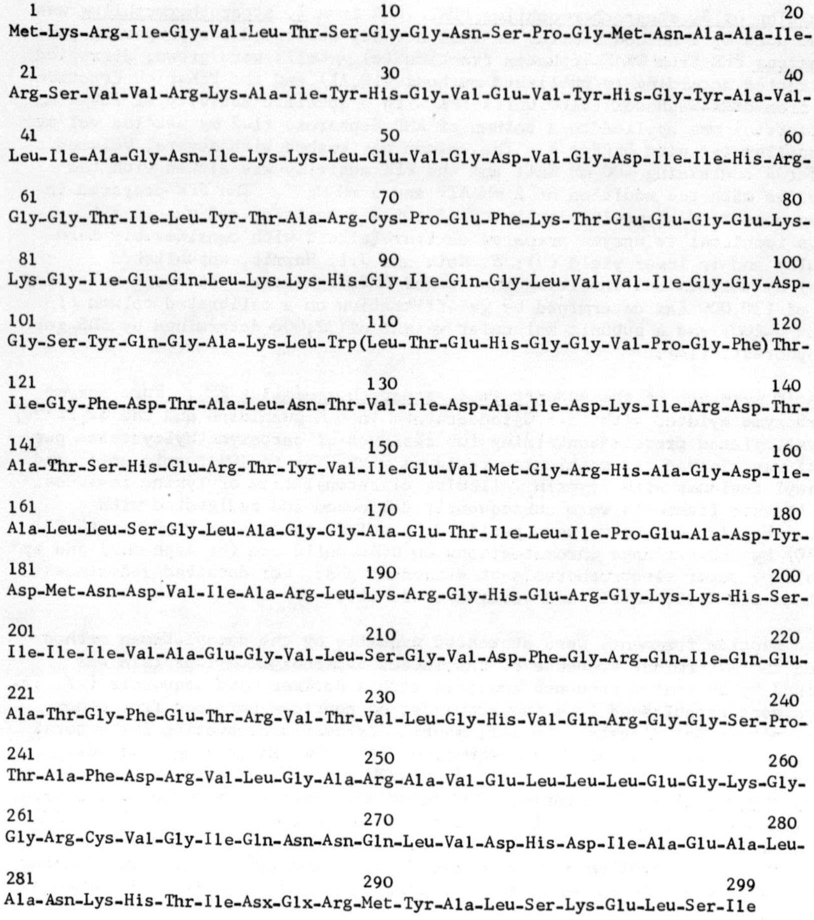

1 10 20
Met-Lys-Arg-Ile-Gly-Val-Leu-Thr-Ser-Gly-Gly-Asn-Ser-Pro-Gly-Met-Asn-Ala-Ala-Ile-

21 30 40
Arg-Ser-Val-Val-Arg-Lys-Ala-Ile-Tyr-His-Gly-Val-Glu-Val-Tyr-His-Gly-Tyr-Ala-Val-

41 50 60
Leu-Ile-Ala-Gly-Asn-Ile-Lys-Lys-Leu-Glu-Val-Gly-Asp-Val-Gly-Asp-Ile-Ile-His-Arg-

61 70 80
Gly-Gly-Thr-Ile-Leu-Tyr-Thr-Ala-Arg-Cys-Pro-Glu-Phe-Lys-Thr-Glu-Glu-Gly-Glu-Lys-

81 90 100
Lys-Gly-Ile-Glu-Gln-Leu-Lys-Lys-His-Gly-Ile-Gln-Gly-Leu-Val-Val-Ile-Gly-Gly-Asp-

101 110 120
Gly-Ser-Tyr-Gln-Gly-Ala-Lys-Leu-Trp(Leu-Thr-Glu-His-Gly-Gly-Val-Pro-Gly-Phe) Thr-

121 130 140
Ile-Gly-Phe-Asp-Thr-Ala-Leu-Asn-Thr-Val-Ile-Asp-Ala-Ile-Asp-Lys-Ile-Arg-Asp-Thr-

141 150 160
Ala-Thr-Ser-His-Glu-Arg-Thr-Tyr-Val-Ile-Glu-Val-Met-Gly-Arg-His-Ala-Gly-Asp-Ile-

161 170 180
Ala-Leu-Leu-Ser-Gly-Leu-Ala-Gly-Gly-Ala-Glu-Thr-Ile-Leu-Ile-Pro-Glu-Ala-Asp-Tyr-

181 190 200
Asp-Met-Asn-Asp-Val-Ile-Ala-Arg-Leu-Lys-Arg-Gly-His-Glu-Arg-Gly-Lys-Lys-His-Ser-

201 210 220
Ile-Ile-Ile-Val-Ala-Glu-Gly-Val-Leu-Ser-Gly-Val-Asp-Phe-Gly-Arg-Gln-Ile-Gln-Glu-

221 230 240
Ala-Thr-Gly-Phe-Glu-Thr-Arg-Val-Thr-Val-Leu-Gly-His-Val-Gln-Arg-Gly-Gly-Ser-Pro-

241 250 260
Thr-Ala-Phe-Asp-Arg-Val-Leu-Gly-Ala-Arg-Ala-Val-Glu-Leu-Leu-Leu-Glu-Gly-Lys-Gly-

261 270 280
Gly-Arg-Cys-Val-Gly-Ile-Gln-Asn-Asn-Gln-Leu-Val-Asp-His-Asp-Ile-Ala-Glu-Ala-Leu-

281 290 299
Ala-Asn-Lys-His-Thr-Ile-Asx-Glx-Arg-Met-Tyr-Ala-Leu-Ser-Lys-Glu-Leu-Ser-Ile

Fig. 2. Amino acid sequence of the subunit of B. stearothermophilus PFK. (Residues 109-110 and 119-120 not experimentally overlapped.)

sodium phosphate, pH 6, and attempts are being made to prepare a series of crystalline derivatives suitable for X-ray crystallographic analysis. One of the aims of this work is to determine the three-dimensional structure of B. stearothermophilus PFK and to seek to establish if the procaryotic enzyme with a subunit of ~32,000 and one binding site for ATP is an evolutionary precursor of the more complex ATP regulated mammalian enzyme with a subunit of ~85,000 that possesses two ATP binding sites.

Fig. 3. Crystals of B. stearothermophilus PFK from 2 M sodium phosphate (pH 6.0). The scale line represents 0.5 mm.

REFERENCES

1. Bloxham, D.P. and Lardy, H.A. (1973). In The Enzymes, 3rd Edn. Vol. 8, ed. P.D. Boyer, pp.239-278 (Academic Press, New York and London).

2. Diezel, N., Boehme, H., Nissler, K., Freyer, R., Heilmann, W., Kopperschlaeger, G. and Hoffmann, E. (1973). Eur. J. Biochem. 38, 479.

3. Kolb, E. and Harris, J.I. (1971). Biochem. J. 124, 76-77P.

4. Suzuki, K. and Harris, J.I. (1971). FEBS Letters, 13, 217.

5. Koch, G.L.E. (1974). Biochemistry 13, 2307.

6. Hocking, J.D. and Harris, J.I. (1973). FEBS Letters 34, 280.

7. Kaplan, N.O., Everse, J., Dixon, J.E., Stozenbach, F.E., Lee, C., Lee, C.T., Taylor, S.S. and Mosbach, K. (1974). Proc. Nat. Acad. Sci. USA 71, 3450.

8. Comer, M.J., Craven, D.B., Atkinson, A. and Dean, P.D.G. (1975). Eur. J. Biochem. 54, 201.

9. Guilford, H., Larsson, P.O. and Mosbach, K. (1972). Chemica Scripta 2, 165.

10. Sargeant, K., East, D.N., Whitaker, A. and Elworth, R. (1971). J. Gen. Microbiol. 65, 11.

11. Atkinson, A., Philips, B.W., Callow, D.S., Hones, W.R. and Bradford, P.A. (1972). Biochem. J. 127, 63P.

12. Ling, K.H., Marcus, F. and Lardy, H.A. (1965). J. Biol. Chem. 240, 1893.

13. Schaffner, W. and Weissmann, C. (1973). Anal. Biochem. 56, 502.

14. Laemmli, U.K. (1970). Nature 227, 680.

15. Davis, B.J. (1964). Ann. N.Y. Acad. Sci. 121, 404.

16. Andrews, P. (1965). Biochem. J. 96, 595.

17. Barnes, L.D. and Stellwagen, E. (1973). Biochemistry 12, 1559.

18. Yoshida, M. (1972). Biochemistry 11, 1087.

19. Blangy, D., Buc, H. and Monod, J. (1968). J. Mol. Biol. 31, 13.

20. Edelhoch, H. (1967). Biochemistry 6, 1948.

21. Kolb, E., Harris, J.I. and Bridgen, J. (1974). Biochem. J. 137, 185.

22. Hartley, B.S. (1970). Biochem. J. 119, 805.

23. Bridgen, J., Harris, J.I. and Northrop, F. (1975). FEBS Letters 49, 392.

PURIFICATION AND CATALYTIC PROPERTIES OF "THERMOSTABLE" FUMARASE FROM BACILLUS STEAROTHERMOPHILUS NU-10 AND THERMUS X-1

William R. Cook and Robert F. Ramaley
Department of Biochemistry, University of Nebraska Medical School
Omaha, Nebraska 68105 USA

ABSTRACT

Fumarase (L-malate hydro-lysase E.C.4.2.1.2) was purified from the thermophilic bacteria Bacillus stearothermophilus NU-10 (optimum growth temperature 62-63°C) and Thermus X-1 (optimum growth temperature 70°C). The fumarase from Thermus X-1 is slightly more thermostable and has an "optimum" catalytic reaction temperature of 83°C as compared to 81°C for the B. stearothermophilus enzyme. Increased thermostability of these fumarases permitted an examination of the properties of the enzyme catalyzed reaction at temperatures higher than had previously been possible with the fumarases from mesophilic bacteria or higher plant and animal sources. Beyond the observed thermostability of the thermophilic fumarases, the catalytic properties of thermophilic fumarases were very similar to those observed with bacterial fumarase or the well characterized pig heart fumarase (effect of temperature on substrate affinities, pH optimum, substrate inhibition by fumarate, and Haldane relationship). These similarities suggest that thermophilic enzymes may be useful in the general study of enzyme reaction mechanisms.

INTRODUCTION

Enzymes from thermophilic bacteria, because of increased thermostability (1), might provide an advantage to the study of enzyme catalytic reaction mechanisms where lability of the enzyme is a problem either during the requisite purification (2) or during prolonged incubation studies (3). However, it is necessary to demonstrate that the catalytic properties and deduced reaction mechanism are truly similar for both the thermophilic and mesophilic enzyme and that results obtained from studies with thermophilic enzymes can be generalized.

We elected to study the comparative catalytic properties of fumarases from thermophilic bacteria because of the classical studies that have been conducted on the kinetic (4-8) and physical (9) properties of the pig heart enzyme (10) and because in contrast to other hydrolyases no cofactor requirements are needed for the fumarase catalyzed reaction (10).

We selected two very dissimilar thermophilic bacteria as sources of "thermostable" fumarase. This was done not only to examine the comparative properties of fumarase from these two thermophilic bacteria and gain further information concerning the comparative physiology of these bacteria but also to provide a better indication of the common physical and catalytic properties of "thermostable" fumarase.

METHODS AND MATERIALS

Source of Thermophilic Bacteria. Bacillus stearothermophilus NU-10 was obtained from L. L. Campbell (University of Delaware). The strain was labeled as T-10 and is the same as strain 10 or Nebraska-10 (11). It was originally isolated by Marsh and Larsen (12) from Yellowstone National

Park. B. stearothermophilus Z-1 was isolated and characterized by W. Bennett from a 190 liter batch of contaminated Thermus aquaticus. B. stearothermophilus M-473 was isolated and characterized by S. Karska from University Lake at Bloomington, Indiana and is similar to B. stearothermophilus LLC which was obtained from the Indiana University Culture Collection which had received it from L. L. Campbell. Thermus X-1 was isolated by Ramaley and Hixson (13) from a thermally polluted stream on the Indiana University campus and Thermus aquaticus YT-1 was obtained from T. D. Brock and was originally isolated from Yellowstone National Park (14).

Media. 0.1% w/v yeast extract, 0.1% w/v tryptone in Casenholz salts (13) was used to grow both the B. stearothermophilus (pH 7.2) and Thermus X-1 (pH 7.6) at 60°C for B. stearothermophilus and 70°C for Thermus X-1.

Growth Studies. The bacteria were grown in a Hotpack reciprocal shaking water bath at 120 strokes per minute in side arm flasks (Bellco flasks) and the growth determined by increase in turbidity (Klett-Summerson photometer with No. 66 filter). Larger batches of B. stearothermophilus or Thermus X-1 were grown in a 14 liter Fermentation Design Fermenter.

Enzyme and Protein Assay. Fumarase was assayed by the method of Cox and Hanson (15) used with Bacillus subtilis fumarase. The assay mixture contained 0.15 M L-malate in 0.05M potassium phosphate (pH 7.6). The reaction mixture was prewarmed to 60°C and the increase in absorbancy at 240 nm followed with a Gilford 2000 recording spectrophotometer. One enzyme unit in the present report is a change of one absorbancy per min. Protein was determined by the method of Lowery et al. (16) with bovine serum albumin (Fraction V) as the standard determined.

Purification of B. stearothermophilus NU-10 Fumarase

Preparation of Cell Free Extract. B. stearothermophilus cells were harvested by centrifugation at the end of exponential growth (the level of fumarase and the other citric acid cycle enzymes is maximal in the transitional period between growth and sporulation) and stored at -20°C. The cells were thawed and suspended in 2 volumes of 0.01 M potassium phosphate (pH 7.2) containing 0.10 mM 2-mercaptoethanol (P buffer) and egg white lysozyme added to a final concentration of 50 µg per ml. After spheroplasts had formed (15 min. at 37°C), as seen by phase contrast microscopic observation, the spheroplasts were disrupted by a brief ultrasonic treatment (Branson sonifier). The broken cells were centrifuged for 2 hr at 76,000xg in a Spinco refrigerated centrifuge (#30 rotor) and the cell free supernatant fraction dialyzed overnight against four liters of P buffer.

DEAE Cellulose Chromatography. The dialyzed extract was placed on a 2.5 x 90 cm column of Whatman Microgranular DEAE cellulose previously equilibrated with P buffer. The fumarase and other proteins were eluted with a potassium chloride gradient in P buffer. The B. stearothermophilus fumarase eluted at a potassium chloride concentration of 0.03 M and before the first major protein peak.

Sephadex G-200 Gel Filtration. The fractions from the DEAE cellulose chromatography containing the largest portion of the enzyme activity were combined, the fumarase concentrated by Lypogel treatment (17) and placed on a 5 x 40 cm column of Sephadex G-200 pre-equilibrated with 0.05 M potassium phosphate (pH 7.2) containing 0.10 mM 2-mercaptoethanol and 0.05 M potassium chloride (P-KCl buffer).

Sephadex QAE-Chromatography. The fractions from the Sephadex G-200 containing the major portion of the enzyme activity were combined and placed on a 2.5 x 30 cm column of Sephadex QAE pre-equilibrated with P-KCl buffer. The fumarase and other proteins were eluted with a potassium chloride gradient in the P-Cl buffer. The fumarase eluted at a potassium chloride concentration of 0.10 M.

Brushite Chromatography. The fractions from the Sephadex QAE containing the major portion of the enzyme activity were combined, concentrated by Lypogel treatment (17) and passed through a 2.5 cm x 30 cm column of Sephadex G-25 pre-equilibrated with 0.05 M Tris-HCl (pH 7.2) containing 0.05 M KCl, 5 mM potassium phosphate (pH 7.7) and 0.1 mM 2-mercaptoethanol. The fumarase was then placed on a 1.5 x 20 cm column of brushite (a form of calcium phosphate) pre-equilibrated with the above buffer. The fumarase was eluted with a phosphate gradient in the starting buffer at a phosphate concentration of 0.01 M. The fractions containing the fumarase were concentrated with Lypogel (17) and stored at -20°C.

Stability of *B. stearothermophilus* fumarase. Although the *B. stearothermophilus* fumarase is more stable than the fumarase from *B. subtilis*, conditions have not yet been found to retain full activity upon storage at -20°C and 50% of the activity is lost per month. The enzyme is stabilized by suspension in 20% w/v glycerol and addition of 0.1 mM dithiothreitol also shows some protection, especially at pH 8.0. This instability does not appear to be caused by the presence of protease and purification of the enzyme in the presence of protease inhibitors does not give an appreciably more stable enzyme.

Analysis of Kinetic Data

The initial velocity of the fumarase catalyzed reaction as a function of substrate concentration at different temperatures was determined with the Gilford Model 2000 recording spectrophotometer. The weighted maximum velocity (Vmax) and apparent Michaelis constant (Km) at each temperature was evaluated with the aid of a Hewlett Packard Model 9830 computor. Figures 3-7 in the present report were drawn with a Hewlett Packard Model 9862A plotter attached to the Model 9830 computor.

Chemicals

Biochemicals were obtained from Sigma Chemical Company. Clemicals were obtained from Fisher Scientific Company. Sephadex G-25, G-200 and QAE Sephadex were obtained from Pharmacia Fine Chemicals. Microgranular DEAE-cellulose was obtained from Whatman. Inc. All water used to make up reagents was tap distilled water passed through a mixed bed ion exchange resin (Scientific Products).

RESULTS

Characteristics of the Thermophilic Bacteria used as Sources of the Thermostable Fumarases. We have purified fumarase from two very different thermophilic bacteria. Bacillus stearothermophilus NU-10 (12) is a gram positive, non spore-forming bacterium and Thermus X-1 is a gram-negative, non spore-forming bacterium (13). Figure 1 shows the growth rate of the two bacteria determined from increase in turbidity in a medium containing 0.1 % w/v yeast extract, 0.1% w/v tryptone in Casenholtz salts (pH 7.6) (13) at various temperatures and indicates that the "optimum" growth temperature for B. stearothermophilus NU-10 is 62-63°C and that "optimum" growth temperature for Thermus X-1 in the identical medium is 70°C. Also shown in Figure 1 for comparison purposes is a lower temperature strain of B. stearothermophilus (M-473) and Thermus aquaticus YT-1 (14).

Figure 2 shows the effect of the medium pH on the growth of B. stearothermophilus NU-10 and Thermus X-1 and indicates that only B. stearothermophilus will grow in slightly acid medium. Thermus X-1 is restricted to slightly alkaline media and this is consistent with the fact that Thermus can be isolated only from slightly alkaline, thermal, aquatic environments. Both Figures 1 and 2 show that B. stearothermophilus has a much more rapid growth rate than does Thermus X-1.

Purification and Properties of "Thermostable" Fumarase. Although B. stearothermophilus and Thermus X-1 are quite different morphologically, physiologically and presumably phylogenetically, the fumarases from these two thermophilic bacteria are quite similar in many respects. Both enzymes were eluted from ion exchange columns at the same ionic strength and similar purification schemes were used for both enzymes. Table 1 shows the purification steps used in the preparation of the B. stearothermophilus enzyme as described in the Methods and Materials section and represents an almost 200 fold purification of the enzyme. A similar purification has also been accomplished with the fumarase from Thermus X-1. However, neither enzyme has been purified to complete homogenity at the present time.

The apparent molecular weight of the fumarase from both B. stearothermophilus and Thermus X-1 is 180,000 as determined from the Stokes radius from Sephadex G-200 gel filtration. This molecular weight is still tentative pending analytical centrifugation studies with more highly purified enzymes.

Catalytic Properties of the "Thermostable" Fumarases. Both fumarases are quite thermostable demonstrating optimum catalytic reaction temperatures of 81°C for the B. stearothermophilus enzyme and 83°C for the Thermus X-1 enzyme (Figure 3) in the standard reaction mixture of 0.15 M L-malate in 0.05 M potassium phosphate (pH 7.6). Figure 4 shows the Arrhenius plot of the data in Fig. 3 and indicates an apparent activation energy of 15,900 calories for the B. stearothermophilus enzyme and 14,800 calories for the Thermus X-1 fumarase.

There was no substrate inhibition of the thermostable fumarases by high levels of malate as has been observed with the pig heart fumarase. However, high levels of fumarate did show a substrate inhibition of the reverse reaction. Both the B. stearothermophilus and Thermus X-1 fumarases have a similar pH catalytic reaction profile with L-malate as a substrate (Fig. 5) and a lower pH optimum with fumarate as the substrate.

Both fumarases had similar substrate saturation kinetics as a function of temperature but since the Thermus X-1 fumarase was slightly more thermostable and also more stable upon storage at -20°C, further studies have been conducted with this enzyme. Figure 6 shows an Arrhenius plot of the maximum velocity (Vmax) of the enzyme at pH 7.0 in 0.05 M potassium phosphate with L-malate or fumarate as the substrate. The apparent activation energy for L-malate was 18,300 calories and 3,900 calories for fumarate. Figure 7 shows the Arrhenius plot of the substrate that gives one half the Vmax (not true Michaelis constant) at pH 7.0 in 0.05 M potassium phosphate. The apparent activation energy for the Km for L-malate was 6,400 calories and -2,100 calories for fumarate. This data indicate the general reaction profile for the thermostable fumarase is quite similar to that obtained with pig heart fumarase. The Thermus X-1 shows an exothermic reaction with malate and an endothermic reaction with fumarate as has previously been demonstrated with the pig heart enzyme (5).

Determination of the Vmax and Ks as a function of temperatures also permit an examination of the Haldane relationship (6, 7) of the fumarase at temperatures up to 75°C.

$$Keq\frac{mal}{fum} = \frac{(Vmax\ \ Fum \rightarrow Mal)\ \ (Km\ \ Mal \rightarrow Fum)}{(Vmax\ \ Mal \rightarrow Fum)\ \ (Km\ \ Fum \rightarrow Mal)}$$

The $Keq\frac{mal}{fum}$ shifts from 4.4 at 25°C (7) to 1.78 at 70°C and the experimental values of $Keq\frac{mal}{fum}$ calculated from the Km and Vm follow the shift within experimental error. This is an extension of the validity of the Haldane relationship to temperatures where enzymes from non-thermophilic sources would be completely inactivated.

DISCUSSION

Comparison of B. stearothermophilus NU-10 and Thermus X-1. *B. stearothermophilus* and *Thermus* X-1 both belong to heterogeneous groups of bacteria that will require future taxanomic classification. The 8th edition of Bergey's Manual of Determinative Bacteriology states that "there has been no agreement on how classification of *B. stearothermophilus* might be improved" (18) although several groupings of the species have been proposed (19, 20). The strain 10 of *B. stearothermophilus* used in this study actually shows a gram variable to gram negative reaction upon gram staining (T. Thompson, personal communication) as do many of the high temperature strains of *B. stearothermophilus* (21) isolated from thermal aquatic environments (R. Ramaley, unpublished observations). However, the cell wall of strain 10 is a classical gram positive cell wall, as shown by the electron microscopic studies of Abram (11), and there is no evidence to suggest that *B. stearothermophilus* NU-10 is not a typical gram positive bacterium (22).

Comparison of Fumarases from B. stearothermophilus and Thermus X-1. The enzyme from *Thermus* X-1 is more thermostable than the enzyme from *B. stearothermophilus* which might be expected since *Thermus* X-1 has a higher optimum growth temperature (70°C) than *B.* stearothermophilus NU-10 (62-63°C). Both enzymes are significantly more thermostable than fumarase from mesophilic bacteria (*e.g.*, *Bacillus subtilis* (23) and R. Ramaley (unpublished observations)) or pig heart fumarase.

However, beyond this thermostability the catalytic properties of the thermophilic fumarases were extremely similar to the catalytic properties of the pig heart fumarase (4-7) and the recently purified *Pseudomonas putida* fumarase (24). It would appear that perhaps all of the fumarases have the same catalytic reaction mechanisms and thus it would be of interest to extend the "definitive isotope exchange study" (25) of Hansen, Dinovo and Boyer (26) to include the effect of temperature on the exchange rates in order to obtain the enthalpy and entropy changes involved in the different exchange rates with both the pig heart and thermophilic fumarase. Our preliminary thermodynamic balance sheet (28) for the thermophilic fumarase shows that the binding of fumarate pH 7.0 is slightly more exothermic and the binding of fumarate slightly less endothermic than is the case for the pig heart fumarase at pH 6.35 (5). However, this may be a comparative difference observed with bacterial fumarase and the temperature dependence of the Michaelis constant and maximal velocity for the *Pseudomonas* (24) or other bacterial fumarases are not available for comparison.

One of the common properties of fumarase which is also found with the thermostable fumarase is the complication of phosphate binding to the enzyme (4) and apparent dissociation and inactivation of the enzyme in the absence of phosphate (9). Recently B. M. Woodfin (27) has suggested that some of the apparent heterogenity found in pig heart fumarase (10) could be ascribed to phosphate or sulfate bound to fumarase and that bound phosphate might be the reason for the discontinuity seen in Arrhenius plots of fumarase with fumarate or with malate at high pHs (5) (B. M. Woodfin, personal communication).

ACKNOWLEDGEMENTS

We wish to thank Mr. Stanley A. Johnson for technical assistance and Dr. Jack L. Smith for the use of his Hewlett Packard computor calculator. This study was supported in part by National Science Foundation Grant GB-36831.

LITERATURE CITED

1. R. Singleton and R. E. Amelunxen. Bacteriol. Rev. 37, 320 (1973).

2. R. F. Ramaley and M. O. Hudock, Biochim. Biophys. Acta 315, 22 (1973).

3. E. H. Higa and R. F. Ramaley, J. Bacteriol. 174, 556 (1973).

4. V. Massey, Biochem. J. 53, 67 (1953).

5. V. Massey, Biochem. J. 53, 72 (1953).

6. R. M. Bock and R. A. Alberty. J. Am. Chem. Soc. 75, 1921 (1953).

7. R. A. Alberty, V. Massey, C. Frieden and A. R. Fuhlbrigge. J. Am. Chem. Soc. 76, 2485 (1954).

8. R. A. Alberty and B. M. Koerber, J. Am. Chem. Soc. 79, 6379 (1957).

9. J. W. Teipel and R. L. Hill. J. Biol. Chem. 246, 4859 (1971).

10. R. L. Hill and J. W. Teipel, in The Enzymes, 3rd. ed. (Ed. P. D. Boyer, Academic Press, New York 1971) Vol. V. p. 539.

11. D. Abram, J. Bacteriol. 89, 855 (1965).

12. C. L. Marsh and D. H. Larsen, J. Bacteriol. 65, 193 (1953).

13. R. F. Ramaley and J. Hixson, J. Bacteriol. 103 527 (1970).

14. T. D. Brock and H. Freeze, J. Bacteriol. 98, 289 (1969).

15. R. S. Hanson and D. P. Cox, J. Bacteriol. 93, 1777 (1967).

16. O. H. Lowry, N. J. Rosenbough, A. L. Farr, and R. J. Randall, J. Biol. Chem. 193, 265 (1951).

17. N. J. Fernald and R. F. Ramaley, Arch. Biochem. Biophys. 153, 95 (1972).

18. T. Gibson and R. E. Gordon, Bergey's Manual of Determinative Bacteriology, 8th ed. (R. E. Buchanan and N. E. Gibbons; Williams and Wilkins Co., Baltimore, 1974), p. 539.

19. J. Wolf and A. N. Barker in Identification Methods for Microbiologists (eds. B. M. Gibbs and D. A. Shaplan, Academic Press 1972), p. 93.

20. P. D. Walker and J. Wolf in Spore Research (Eds. A. N. Barker, G. W. Gould and J. Wolf, Academic Press, 1971), p. 247.

21. R. E. Gordon, W. C. Haynes and C. H. N. Pang in The Genus Bacillus, U. S. Dept. Agri. Handbook No. 427 (1973), p. 59.

22. M. B. Allen, Bacteriol. Rev. 17, 125 (1953).

UNIV. COLL. N.W. BANGOR LIBRARY

LITERATURE CITED (Continued)

23. M. Ohne, J. Bacteriol. 122, 224 (1975).

24. C. A. Lamartiniere, H. D. Brayner and A. D. Larson, Arch. Biochem. Biophys. 141, 239 (1970).

25. I. A. Rose, in The Enzymes, 3rd ed. (Ed. P. D. Boyer, Academic Press, New York 1971) Vol II, p. 281.

26. J. N. Hansen, E. C. Dinovo and P. D. Boyer, J. Biol. Chem. 244, 6270 (1969).

27. B. M. Woolf in Isoenzymes (Ed. C. L. Markert, Academic Press, 1975) Vol. I, p. 797.

28. M. Dixon and E. C. Webb in Enzymes, 2nd ed. (Academic Press, 1964), p. 164.

TABLE 1. Purification of B. stearothermophilus NU-10 fumarase

Purification Step	Total[1] Units	Total Protein	Specific[1] Activity	Percent Recovery
Cell free extract	3,250	13,000 mg	0.24	100%
DEAE-Cellulose (microgranular)	2,710	1,750 mg	2.2	83%
G-200 Sephadex	2,460	950 mg	2.0	75%
QAE Sephadex	1,580	99 mg	15.9	49%
Brushite	750	16 mg	46.5	23%

[1]Specific Activity is in units per milligram of protein.

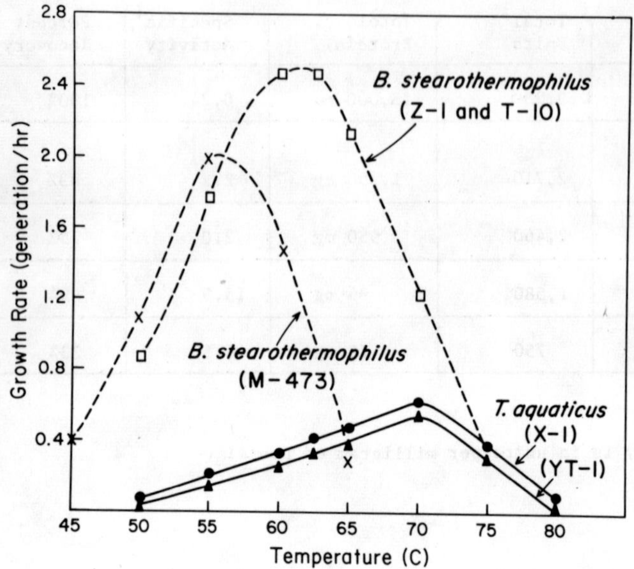

Figure 1 Effect of temperature on the growth rate of different thermophilic bacteria.

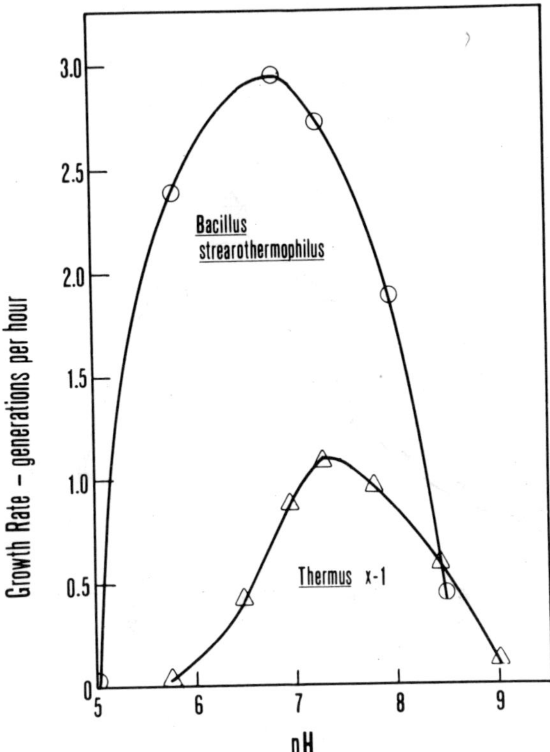

Figure 2 Effect of pH on the growth rate of B. stearothermophilus NU-10 and Thermus X-1. The pH shown is the pH of the culture during the growth rate shown (determined from semilogarthmic plots of turbidity vs. time) and not the initial pH of the medium.

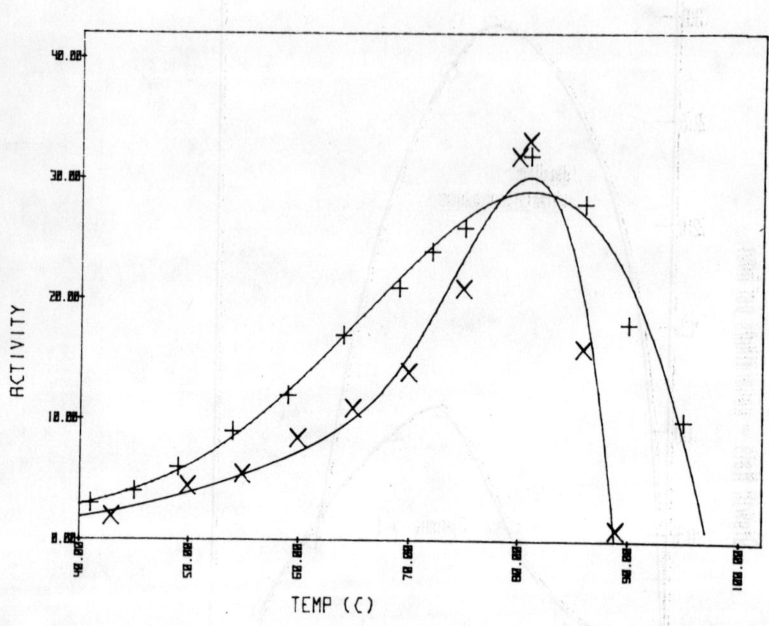

Figure 3 Effect of temperature on the initial velocity of the fumarase from B. stearothermophilus (X) and Thermus X-1 (+) under the standard assay conditions.

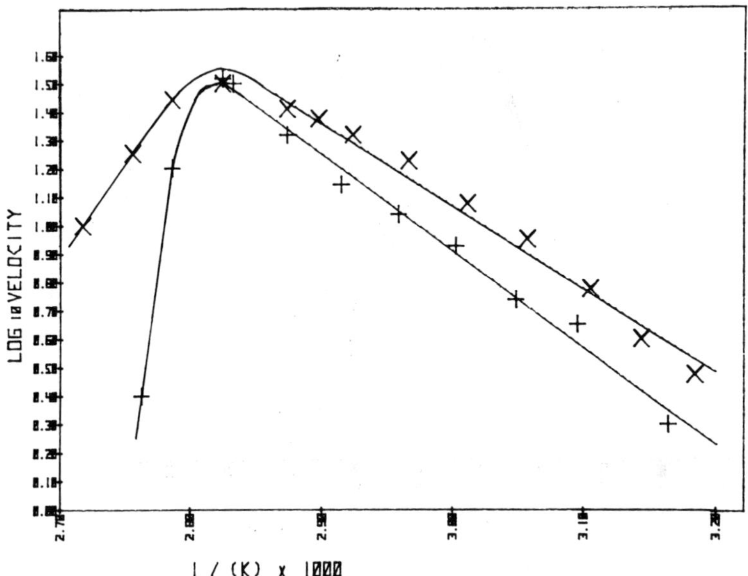

Figure 4 Arrhenius Plot of the data shown in Figure 3 with the fumarase from B. stearothermophilus (+) and Thermus X-1 (X).

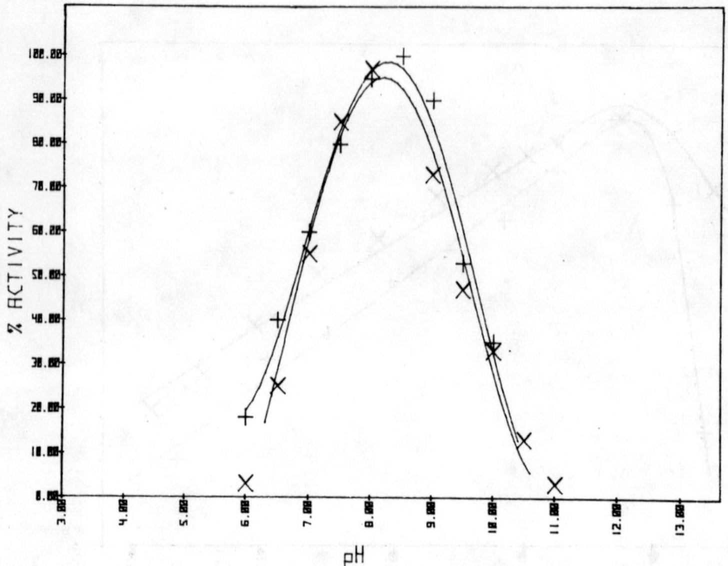

Figure 5 Effect of pH on the initial velocity of the B. stearothermophilus (X) and Thermus X-1 (+) fumarases under the standard assay conditions

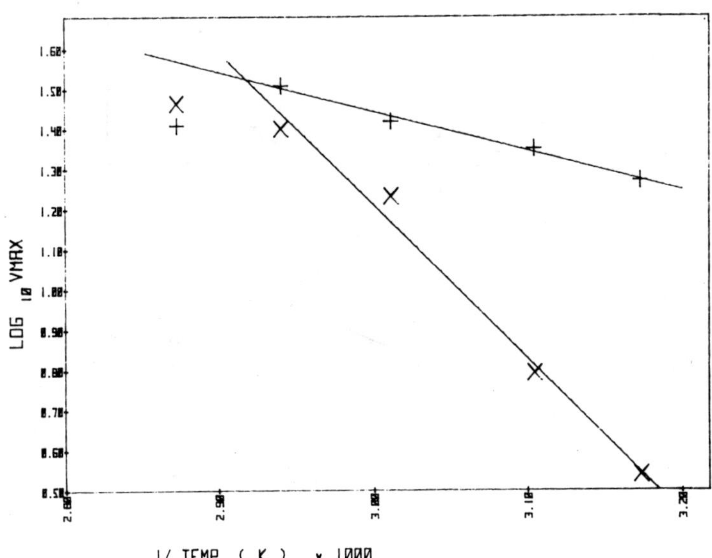

Figure 6 Arrhenius plot of the effect of temperature on the maximum velocity (Vmax) of the Thermus X-1 fumarase at pH 7.0 with malate (X0 or fumarate (+) as the substrate

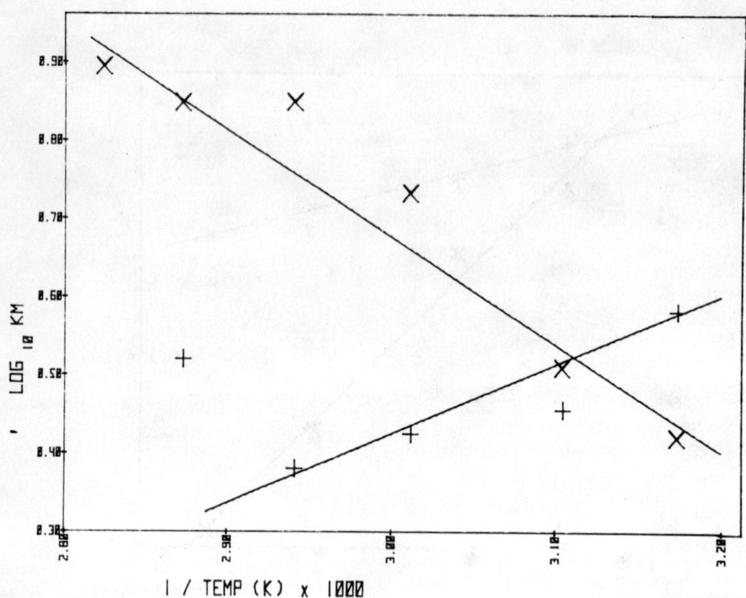

Figure 7 Arrhenius plot of the effect of temperature on the apparent Michaelis constant (Km) of the Thermus X-1 fumarase at pH 7.0 with malate (X) or fumarate (+) as the substrate

ANALYSIS OF THE THERMOSTABILITY OF ENOLASES

E. STELLWAGEN and L.D. BARNES
Department of Biochemistry, The University of Iowa,
Iowa City, Iowa 52242, USA

INTRODUCTION

We have previously (1,2) characterized the enolases purified from the thermophiles Thermus X-1 and Thermus aquaticus YT-1. In this report we wish to establish the homology of these enzymes with enolases purified from two mesophilic organisms, yeast and rabbit, to illustrate the variability of thermostability measured by a variety of procedures, and to propose a correlation between thermal stability and content of β structure.

EXPERIMENTAL PROCEDURES

Enolase was purified from Thermus X-1 and from T. aquaticus YT-1 as described previously (1,2). Rabbit muscle enolase and yeast enolase were purchased from the Sigma Chemical Company. Protein concentration and enzymic activity was measured as described previously (1,2).

Molecular weight measurements were made by equilibrium sedimentation using the meniscus depletion procedure. Frictional ratios were calculated from sedimentation and diffusion coefficients. The content of α-helix and β-structure was estimated (13) from ultraviolet circular dichroic spectra measured in 50 mM phosphate buffer, pH 7.0 between 200 and 240 nm using the three structural forms of poly-L-lysine as standards. The exposure of aromatic residues was measured by ultraviolet solvent perturbation difference spectroscopy using 20% ethylene glycol as the perturbant. The numbers of exposed protein residues was estimated using the perturbation spectra of N-acetyl ethyl ester model chromophores (4). The homology index between two proteins was obtained from their amino acid compositions (2,5) by dividing the cumulative sum of the square of the difference in mole fraction for each amino acid by the total number of different residues in the two proteins (6). Hydrophobicity was calculated from the amino acid compositions by the procedure of Bigelow (7). Melting temperatures for irreversible inactivation were obtained by preincubation of aliquots of each enzyme for exactly 5 minutes at selected temperatures followed by rapid cooling to 0° and subsequent enzymic assay at 25°, as previously described (2). Melting temperatures for conformational transitions were obtained by observing discontinuities in ΔpH/ΔT profiles, as described by Bull and Breese (8). Average residue volumes for proteins were calculated from their amino acid compositions and the residue volume for the amino acids obtained from crystallographic measurements (9). Thermal transition temperatures, T_M, were calculated from the average residue volume, ARV, using the relationship, $T_M = 204.3 + 3.258$ (ARV), H.B. Bull, personal communication.

RESULTS

A variety of properties of enolases obtained from thermophilic and mesophilic organisms are compared in Table I. Each enolase is a globular protein consisting of identical (1,2,5) subunits having a molecular weight of about 4.4×10^4 Circular dichroic measurements indicate the presence of about 30% α-helix in each enolase and a variable content of β-structure. Comparison of the amino acid composition of rabbit muscle enolase pairwise with the composition of each of the other three enolases using the homology index suggests that the primary structures of the four proteins are related. The range of homology indices within the four enolases are within the range calculated, 0.1 to 9.9, for members of the cytochrome c homologous series. The consistency of the numbers of exposed aromatic residues suggests that the tertiary and quaternary structures of the enolases are quite similar. Comparison of two catalytic parameters, the Michaelis constant for the substrate, 2-phosphoglycerate, and the activation energy for catalysis, calculated from the data shown in Fig. 1, suggests that the architecture of the catalytic sites on the four enolases are also similar. Based on these comparisons, we propose that the four enolases are members of a closely

Table I. Comparative properties of enolases

Property	Enolase Yeast	Enolase Rabbit	Enolase Thermus	Enolase T. aquaticus
Structural				
$M \times 10^{-3}$, Native	88	82	382	328
$M \times 10^{-3}$, Subunit	44	44	48	44
Frictional ratio	1.2	1.2	1.3	1.3
% α-Helix	25	30	33	32
% β-Structure	37	38	29	16
Exposed Tyr/subunit	9	8	9	-
Exposed Trp/subunit	2	2	2	-
Homology index	0.9	0.0	1.8	6.2
Hydrophobicity, cal/residue	1076	1093	1030	938
Catalytic				
K_M, 2-PGA, μM	150	90	110	110
E_{act}, Kcal/mole	11.8	16.6	15.4	15.8
Thermostability				
$T_{optimum}$, catalysis	49°	57°	70°	89°
T_M, irreversible inactivation				
50 mM Tris-HCl	40°	45°	74°	90°
plus 1 mM $MgSO_4$	58°	55°	88°	100°
T_M, $(\Delta pH/\Delta T)_{max}$				
100 mM NaCl	-	46°	64°	-
plus 1 mM $MgSO_4$	56°	41°,60°	66°,86°	-
T_M, calculated, ARV	66°	71°	69°	43°

related, if not homologous, series of proteins.

As shown in Fig. 1, the optimal temperatures for catalysis exhibited by the two

RELATIVE CATALYTIC RATE

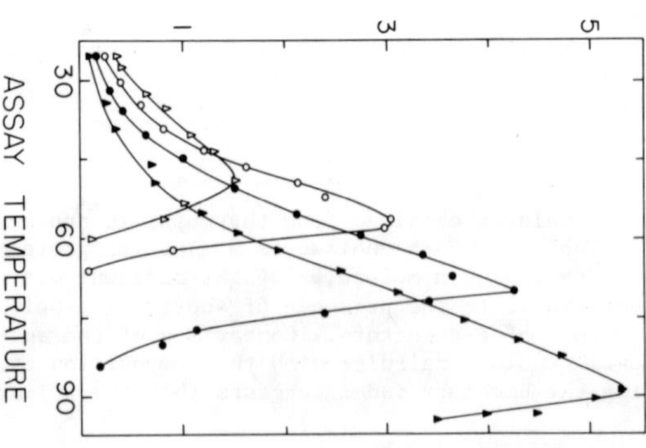

Fig. 1. Dependence of enolase catalysis on assay temperature. △, yeast enolase; ○, rabbit muscle enolase; ●, Thermus X-1 enolase; and ▲, T. aquaticus enolase. A mixture of buffer and $MgSO_4$ was placed in a cuvet and brought to the desired temperature in a thermostable cell holder of a recording spectrophotometer. An aliquot of enzyme solution was then added and equilibrated for one minute at which time the reaction was initiated by addition of substrate.

enolases purified from thermophilic organisms are distinctly higher than the two enolases obtained from mesophilic organisms. However, the optional temperatures for catalysis do not always agree with the irreversible inactivation temperatures measured by preincubation of the enzyme at elevated temperatures or with the conformational transition temperatures measured by $\Delta pH/\Delta T$ discontinuities. As shown in Table I, preincubation of the enzymes in buffer containing 1 mM $MgSO_4$, the concentration present in all assay solutions, raises the irreversible inactivation temperature of each enolase up to or above its optimal catalytic temperature. These results indicate that ligation of enzymes with one or more substrates or cofactors can increase thermostability, presumably by formation of enzyme-ligand complexes. Accordingly, the intrinsic thermostability of an enzyme is not necessarily reflected by its catalytic temperature optimum. As shown in Fig. 2, $\Delta pH/\Delta T$ measurements with rabbit muscle enolase

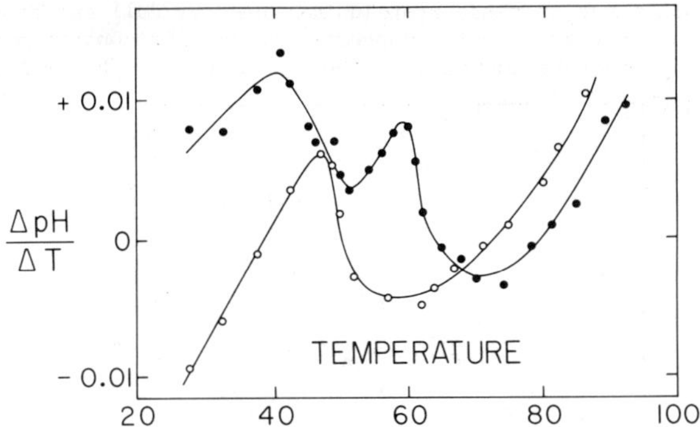

Fig. 2. Thermal transitions of rabbit muscle enolase. A solution of enolase, about 2 mg/ml, in (○) 0.1 M NaCl or in (●) 0.1 M NaCl containing 1 mM $MgSO_4$ was adjusted to pH 5.87 - 6.01 at 25° and then heated at the rate of 0.63°/minute.

reveal a single thermal conformational transition for the enzyme by itself but two thermal transitions for the enzymes in the presence of 1 mM $MgSO_4$. Similar results were obtained with Thermus enolase as indicated in Table I. In the case of rabbit muscle enolase, the loss of catalytic function correlates with the higher thermal transition while Thermus enolase catalysis appears to be lost during the lower thermal transition. We recommend that the thermostability of proteins be established by the $\Delta pH/\Delta T$ procedure since it is sensitive and simply made, capable of detecting thermal transitions which do not result in irreversible thermal inactivation and insensitive to aggregation effects which usually accompany thermal inactivation and obscure spectral procedures.

DISCUSSION

Careful compositional analysis of Thermus X-1 enolase indicated (2) that this enzyme is not a metalloenzyme and that it is devoid of nonamino acid organic moieties. Since the two mesophilic enolases are also free of such nonamino acid components (5), variations in the thermostability of the four enolases must reside in their amino acid compositions and the conformations their sequences specify. The thermal transition studies reported by Bull and Breese (10) indicate that globular proteins obtained from mesophilic organisms have thermal transitions in the range characteristic for thermophilic proteins and that their transitions correlate very well with their average residue volumes. However, the thermal transition temperatures for the four enolases predicted from their average residue volumes do not correlate well with their observed thermal transition temperatures, as shown in Table I. Similarly, the observed values do not exhibit a systematic correlation with the hydrophobicity of the four proteins. A previous correlation (2) of the content of amino acid residue capable of forming side chain hydrogen bonds with the thermal stability of enolases does not appear to pertain to other proteins (11). However, Bull and Breese (10) also report that the thermostability of 14 globular mesophilic proteins correlates very well with their glutamate content. As Chou and Fasman have noted (12), glutamate is the best helix maker and distinctly the best β-structure breaker of the twenty amino acids. It is therefore not surprising that the thermal transition temperatures of those proteins measured by Bull and Breese whose structures are known, appear directly dependent on α-helix content and inversely dependent on β-structure content, as shown in Fig. 3. The β-structure

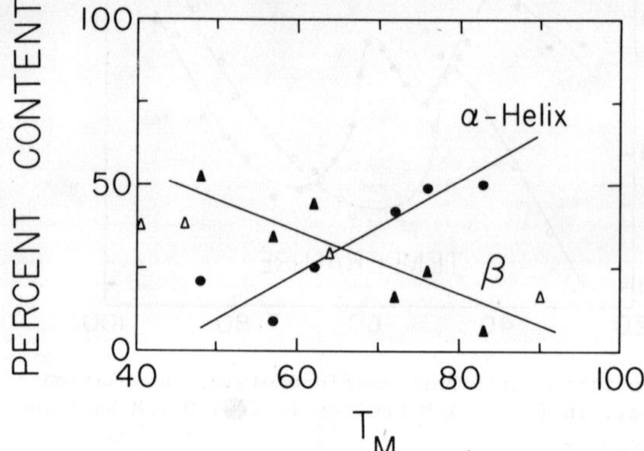

Fig. 3. Dependence of thermal transition temperature on secondary structure content. ●, helix content; △, ▲, β-structure content. Filled symbols reading from left to right represent ribonuclease-s, chymotrypsinogen, ribonuclease-A, lysozyme, insulin and α-lactalbumin. Open symbols reading from left to right represent enolases obtained from yeast, rabbit muscle, Thermus, and T. aquaticus. The secondary structure content for α-lactalbumin and the four enolases was estimated from circular dichroic measurements while the content for the other proteins was obtained from X-ray crystallographic measurements. The thermal transition temperature for ribonuclease-S was taken from (13).

content but not the α-helix content of the four enolases, estimated from circular dichroic measurements, correlates reasonably well with their thermal

UNIV. COLL. N.W. BANGOR

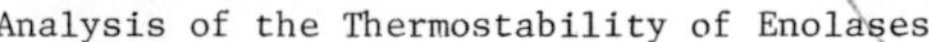

transition temperatures as also shown in Fig. 3.

Accordingly, we propose that the secondary structure content of globular proteins, particularly the β-structure content, to a first approximation determines their thermal transition temperatures. The β-sheets commonly found in a variety of protein structures serve to noncovalently link disparate portions of polypeptide sequences. Rupture of these noncovalent hydrogen bonds may cooperatively initiate the thermal unfolding of the other noncovalent interactions critical to maintainance of the native conformation. Changes in β-structure content and hence thermostabiltiy could be readily achieved in a series of homologous amino acid sequences by single base mutations, particularly in the codons for glutamate, glutamine and valine.

Inspection of the glutamate and secondary structural contents of paired proteins obtained from mesophilic and thermophilic organisms (11) reveals no systematic correlation with their catalytic temperature optima. However, these data suffer from several deficiencies. The glutamate content usually includes both glutamate and glutamine. The former is an excellent β-structure breaker while the latter is a moderately good β-structure maker. Secondly, as we have shown above catalytic temperature optima do not necessarily reflect the intrinsic thermostability of enzymes. And finally, secondary structure content can at best be only estimated from optical rotatory dispersion and circular dichroic spectra. Accordingly, we are currently measuring the conformational transition temperatures of a series of mesophilic proteins of known structure and sequence by the ΔpH/ΔT procedure to determine whether thermostability correlates with secondary structure content.

ACKNOWLEDGEMENT

This investigation was supported by a U.S. Public Health Service research grant GM-13215 and a research fellowship (to L.D.B.) grant GM-50401, both from the Institute of General Medical Sciences.

REFERENCES

1. Stellwagen, E., Cronlund, M.M., and Barnes, L.D. (1973), Biochemistry 12, 1552-1558.
2. Barnes, L.D., and Stellwagen, E. (1973), Biochemistry 12, 1559-1565.
3. Greenfield, N., and Fasman, G.D. (1969), Biochemistry 8, 4108-4115.
4. Herskovits, T.T., and Sorensen, M. (1968), Biochemistry 7, 2523-2532.
5. Wold, F. (1971), The Enzymes (3rd Ed.), 5, 499-538.
6. VanHolde, K.E., and Bruggen, E.F.J. (1971) in Subunits in Biological Systems, Part A, eds. Timasheff, S.N. and Fasman, G.D. (Dekker, N.Y.) p. 8.
7. Bigelow, C.C. (1967), J. Theoret. Biol. 16, 187-211.
8. Bull, H.B. and Breese, K. (1973), Arch. Biochem. Biophys. 156, 604-612.
9. Chothia, C. (1975), Nature 254, 304-308.
10. Bull, H.B., and Breese, K. (1973), Arch. Biochem. Biophys. 158, 681-686.
11. Singleton, R., Jr., and Amelunxen, R.E., Bact. Rev. 37, 320-342.
12. Chou, P.Y., and Fasman, G.D. (1974), Biochemistry 13, 222-245.
13. Tsong, T.Y., Hearn, R.H., Wrathall, D.P. and Sturtevant, J.M. (1970), Biochemistry 9, 2666-2677.

Analysis of the Thermostability of Proteins

transition temperatures as also shown in Fig. 3.

Accordingly, we propose that the secondary structure content of globular proteins, particularly the β-structure content, [illegible]

[illegible]

ACKNOWLEDGEMENTS

This investigation was supported by U.S. Public Health Service [illegible]

REFERENCES

[illegible]

ISOLATION AND SOME PROPERTIES OF ENOLASE FROM BACILLUS STEAROTHERMOPHILUS

E. Boccù, F.M. Veronese and A. Fontana

Institutes of Pharmaceutical Chemistry and of Organic Chemistry, University of Padova, Padova, Italy

1. INTRODUCTION

Several enzymes from the obligate thermophile *B.stearothermophilus* are under active investigation in several laboratories, in order to ascertain whether a general explanation can be given of the enhanced stability toward heat of thermophilic enzymes. We are currently involved in physico-chemical analyses of several enzymes from *B.stearothermophilus* and, in our multi-enzyme separation procedure from this bacterial source, we have also isolated enolase (EC 4.2.1.11). It seemed to us of interest to verify whether the enzyme from this moderately thermostable microorganism shows a dimeric structure similar to enolase from all mesophilic sources, from microorganisms to vertebrates (1), or an octameric structure as shown for the enzyme from two extremely thermophilic microorganisms *Thermus* X-1 and *Thermus aquaticus* YT-1 recently isolated by Stellwagen and coworkers (2,3).

2. MATERIALS AND METHODS

2.1. *Purification of the enzyme*. Frozen cells (500 g) of *B.stearothermophilus* (NCA 1503) were suspended in 1.5 l of 0.05 M Tris-HCl buffer, pH 7.5, 1 mM $MgSO_4$ (buffer A) using a blender. The suspension was treated with a Manton Gaulin homogeniser at 550 atm. The homogenate was centrifuged at 25,000 g for 60 min. The ammonium sulfate fraction between 45 and 75 % saturation, containing most of the enolase activity, was collected. The precipitate was dissolved in buffer A and, after extensive dialysis, was applied to a DEAE-cellulose column (50x60 cm) equilibrated with the same buffer. The column was washed with about 2 l of buffer A, 2 l of buffer A containing 0.15 M KCl (buffer B) and finally the enolase was eluted with a linear salt gradient (2.5 l each of buffer B and 0.8 M KCl in the same buffer). The enolase fraction was directly applied to a hydroxylapatite column (5x30 cm) equilibrated with 10 mM phosphate buffer 1 mM $MgSO_4$, pH 7.1. The column was washed with 4 column volumes of 70 mM phosphate buffer 1 mM $MgSO_4$, pH 7.1.

The enzyme was eluted with 3 l of a linear phosphate gradient between 70 and 300 mM. The active fractions were pooled and the protein precipitated by dialysis against saturated ammonium sulfate solution. The precipitate, dissolved in 0.1 M phosphate buffer, 1 mM $MgSO_4$, pH 6.8, was applied to a Bio-Gel A-1,5m (5x140 cm) column. Enolase was eluted with the same buffer and the pooled activity was concentrated to 5 ml using an Amicon ultrafiltration cell and an XM-50 membrane. The enzyme was stored as an ammonium sulfate suspension.

2.2. Enzyme assay. The activity of enolase was routinely measured at 23°C by the increase in absorbance at 230 nm using an assay solution 33 mM Tris-HCl buffer, pH 7.5, 1 mM 2-phosphoglycerate and 1 mM $MgSO_4$. The protein concentration was determined spectrophotometrically according to Layene (4).

2.3. Molecular weight evaluation. All ultracentrifuge experiments were conducted with a Spinco Model E ultracentrifuge, equipped with schlieren optics at 20°C. The rotor speed was 40,000 rev/min.

2.4. Amino acid analysis. Analyses were performed with a Jeol Model 6AH amino acid analyzer according to the procedure of Hamilton (5) for the single column analysis. Linear extrapolation to zero hydrolysis time was used for serine, threonine and tyrosine; the values obtained from the 72 hrs hydrolysis were used for valine and isoleucine. Tryptophan content was determined spectrophotometrically by the method of Goodwin and Morton (6), as modified by Beaven and Holiday (7).

3. RESULTS AND DISCUSSION

In Table 1 a summary of the purification procedure of enolase from B.stearothermophilus is reported; the enzyme was purified about 300-fold with an overall yield of 43 %. The procedure of purification here reported gives high yield of homogeneous enzyme as verified by several criteria such as single band on disc gel electrophoresis in SDS, constancy of specific activity over several purifications and sedimentation pattern with single symmetrical boundary also at high concentration of protein (Fig. 1).

The procedure here reported permits the contemporaneous purification of glyceraldehyde 3-phosphate dehydrogenase which remains in the supernatant after 75 % ammonium sulfate precipitation,and 6-phosphogluconate dehydrogenase which is eluted

TABLE 1. Typical purification of B.stearothermophilus enolase from 500 g cell paste

Fraction	Protein (mg)	Total activity	Specific activity	Yield (%)
1. Crude extract	41,000	18,000	0.43	
2. $(NH_4)_2SO_4$ fractionation	14,000	17,000	1.2	94
3. DEAE-cellulose	1,400	14,000	10.0	77
4. Hydroxylapatite	115	9,775	85	54
5. Bio-Gel A-1.5m	60	7,800	130	43

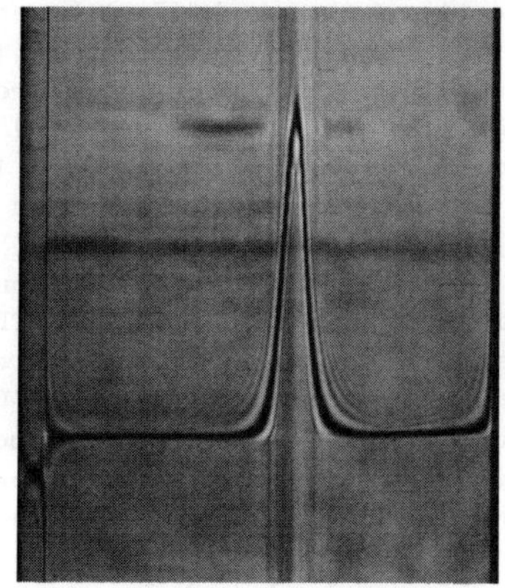

FIG. 1. Sedimentation pattern of purified enolase. The protein concentration was 7 mg/ml in 0.033 M Tris-HCl buffer, pH 7.5, 1 mM $MgSO_4$. Migration is from the left to the right. The photograph was taken 80 min after reaching a speed of 40,000 rev/min at 20°C.

from the DEAE-cellulose column before the enolase activity. The possibility of a multienzyme preparation is of particular interest since these enzymes are under current investigation in our laboratory.

The pH optimum for activity was found to be 7.8. This figure is similar to that observed for enolase from bacteria such as E. coli (8) and Th.aquaticus YT-1 (2) and considerably more alkaline than the pH optimum found for vertebrate enolases (1).

The optimum concentration of Mg^{++} was found to be 10^{-3} M which is of the same order as that found for the enzyme from other sources; furthermore the enzyme exibits the drop in activity at high Mg^{++} concentration typical for all enolases so far studied (1). The Km for 2-phosphoglycerate was found $1x10^{-4}$M.

On the basis of these parameters, the *B.stearothermophilus* enolase does not show particularly unusual features. This is an additional case in which an enzyme from thermophiles is similar to those extracted from mesophilic sources.

A study of the effect of the assay temperature on the rate of enolase catalysis at pH 8 indicates that the temperature optimum of activity is about 60°C. This value corresponds to the growth temperature of the microorganism. This correlation of temperature with the optimum of activity has been already observed with other mesophilic as well as thermophilic microorganisms, as for instance with 6-phosphogluconate dehydrogenase from *B.stearothermophilus* and *E.coli* (9), neutral proteases from *B.stearothermophilus* grown at different temperatures (10) and phosphoglycerate kynase from *B.stearothermophilus* (11). A linear Arrhenius plot up to 65°C is obtained corresponding with an activation energy of -7845 cal/mole.

The enzyme is rather thermostable the temperature of half inactivation after 15 minutes heating being 67°C. The Mg^{++} shows a remarkable stabilizing effect on the enzyme, since in its absence the temperature of half inactivation is 47°C (Fig. 2). This stabilizing effect of ions on the thermal properties of enzymes has been frequently observed, and is in general explained by assuming that the ions bring the structure of the enzyme into the correct position of a stable conformation.

The amino acid composition calculated for a subunit of 47,000 mol wt is reported in Table II. The enzyme showed the absence of cysteine, a residue which is present in all mesophilic enolases. This property appears related more to the philogenetic position of the organism than to its thermal properties, since cysteine is present in high level in rabbit muscle (12), salmon (13) and potato (14) enolase, in low level in yeast (15) and *E.coli* (8), but absent in *Thermus* X-1 (2) and *Th.aquaticus* YT-1 (3).

Our data on amino acid composition and thermostability of *B.stearothermophilus* and *E.coli* enzymes fit properly in a graph where the number of arginine, glycine, acidic amino acids and amino acids generating hydrogen bonds were plotted against the degree of thermostability of four other enolases by Barnes *et al.* (3) (Fig. 3). The meaning of this correlation for the understanding

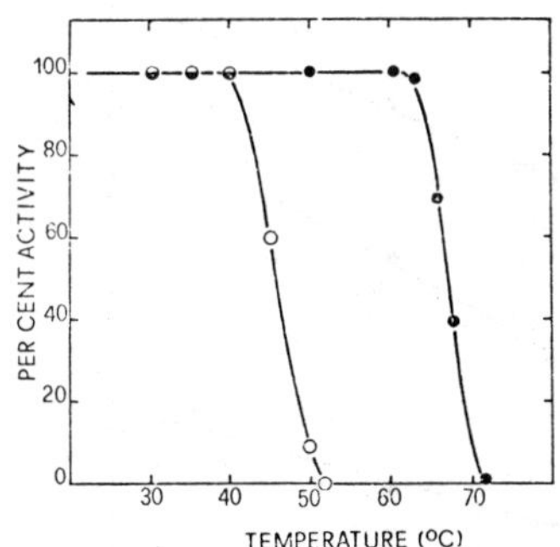

FIG. 2. Heat inactivation of enolase. The enzyme solution (0.1 mg/ml) was incubated at the indicated temperature for 15 min, cooled and assayed for residual activity. The enzyme preparation was depleted of Mg^{++} by dialysis against 0.1 M phosphate buffer pH 7.5, 0.1 M EDTA followed by dialysis against phosphate buffer. (●—●) enzyme incubated in buffer containing 1 mM $MgSO_4$. (o—o) enzyme incubated in buffer without Mg^{++}.

TABLE II. Amino acid composition of B.stearothermophilus enolase

Amino acid	Residue/ Polypeptide	Amino acid	Residue/ Polypeptide
Aspartic acid	45	Methionine	9
Threonine	24	Isoleucine	27
Serine	22	Leucine	38
Glutamic acid	55	Tyrosine	15
Proline	16	Phenylalanine	9
Glycine	43	Lysine	28
Alanine	42	Histidine	7
Cysteine	0	Arginine	16
Valine	35	Tryptophan	4

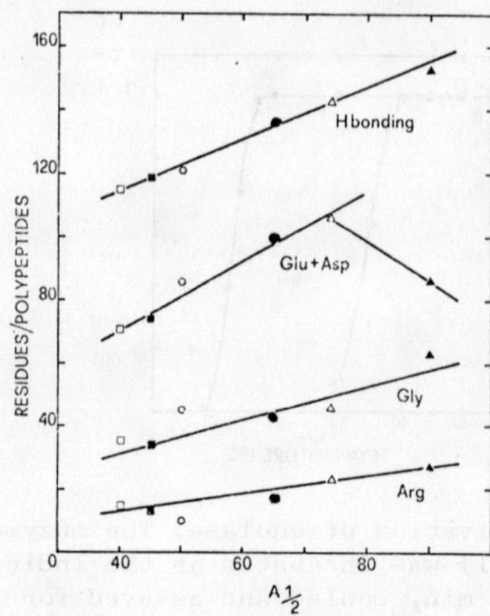

FIG. 3. Correlation of the content of various amino acids with $A_{1/2}$ (temperature at which 50 % of the initial catalytic activity of enzyme sample devoid of Mg^{++} is lost in five minutes). Enolase from rabbit muscle, □ ; yeast, ■ ; E.coli, ○ ; B.stearothermophilus, ● ; Thermus X-1, △ ; Thermus aquaticus YT-1, ▲ . The points for E.coli and B.stearothermophilus enzymes are obtained from our experiments.

of thermostability of proteins, although impressive, is difficult to evaluate since contradictory observations are also reported (11).

The B.stearothermophilus enolase sediments as a single symmetrical boundary in 50 mM Tris-HCl buffer 1 mM $MgSO_4$, pH 7.2, at a protein concentration range of 2-7 mg/ml. The $S^o_{20,w}$ at infinite dilution is 14.63 ± 0.3, a value obtained by least-squares analysis of the experimental data. A value for $D^o_{20,w}$ of $3.64 \pm 0.2 \times 10^{-7}$ cm^2/sec was estimated by the height-area method at eight different concentrations of proteins. Combination of the values derived for $D^o_{20,w}$, $S^o_{20,w}$, and $\bar{V}$ in the Svedberg equation gives a mol wt of 376,000. In making these calculations a $\bar{V}$ of 0.74 ml/g derived from the amino acid composition was used. The mol wt of the dissociated protein was determined by SDS disc gel electrophoresis according to Weber and Osborn (16). The protein migrated as a single sharp band with a mobility corresponding to a mol wt of 48,000. These data indicate that the enzyme is composed of eight apparently identical subunits.

This would imply that enolase, extracted from a microorganism grown at 60°C, only 10 degrees lower than growth temperature of Th.aquaticus X-1 and only 20 degrees above that of E.coli, also posesses an octameric structure like enolase from Th.aquaticus X-1, whereas enolase from E.coli, like those from all other mesophilic organisms, posesses a dimeric structure. Such change in quaternary structure in the enolase series is not accompanied by significant differences in the properties of the enzymes. This great difference in the subunit arrangement between enolase from organisms grown above or below 50°C remains so far an unique and unexplained phenomenon. In this context it is worth recalling that the octameric structure of enolase from Th.aquaticus YT-1 has been shown to be stable over the temperature range 25-90°C (2).

In conclusion, enolase from B.stearothermophilus exhibits kinetic and physical properties in common with enolase from other mesophilic species, the main unusual feature being its octameric structure.

Preliminary experiments seem to indicate that the octameric structure can be related to the temperature of folding of the protein. In fact, samples of enolase denaturated in 8 M urea renature to octameric enzyme on dialysis at 56°C, whereas at 23°C dimeric and octameric species are obtained. These investigations are feasible, since the enzyme can be denatured in 8 M urea while both structure and activity are restored by removal of the denaturant at low as well as at high temperature. This aspect of enolase chemistry is presently under current investigation and will be reported elsewhere.

Acknowledgement

The authors wish to thank Dr. B.Salvato for his advice in ultracentrifugation analysis and Mrs. R.Largajolli and Mr. V. Carbone for skilful technical assistance. The work was carried out with the financial support of the Italian C.N.R.

REFERENCES

(1) Wold, F. (1971), Enzymes, 3rd, 5, 499-538.

(2) Stellwagen, E., Cronlund, M.M., and Barnes, L.D. (1973), Biochemistry, 12, 1552-1558.

(3) Barnes, L.D., and Stellwagen, E. (1973), Biochemistry, 12, 1559-1565.

(4) Layene, E. (1967), Methods in Enzymology, 3, 447-454.
(5) Hamilton, B.P. (1963), Anal.Chem., 35, 2055-2063.
(6) Goodwin, T.W., and Morton, R.A. (1946), Biochem.J., 40, 628-632.
(7) Beaven, G.H.,and Holiday, E.R. (1952), Adv.Protein Chem., 7, 319-385.
(8) Spring, T.G. (1970), Ph.D.Thesis, Univ. of Minnesota, U.S.A.
(9) Veronese, F.M., Grandi, C., Boccù, E.,and Fontana, A., these proceedings.
(10) Suzuki, K.,and Imahori, K.(1974), J.Biochem., 76, 771-782.
(11) Sidler, W.,and Zuber, H. (1972), FEBS Letters, 25, 292-294.
(12) Holt, A.,and Wold, F. (1961), J.Biol.Chem., 236, 3227-3235.
(13) Ruth, R.C., Soja, D.M., and Wold, F. (1970), Arch.Biochem. Biophys., 140, 1-10.
(14) Boser, H. (1959), Z.Physiol.Chem., 315, 163-172 .
(15) Brewer, J.M., Fairwell, T., Travis, J., and Lovins, R.E. (1970), Biochemistry, 9, 4002-4012.
(16) Weber, R., and Osborn, M. (1969), J.Biol.Chem., 244, 4406-4412.

PROPERTIES OF ENZYMES FROM CLOSTRIDIUM THERMOACETICUM AND CLOSTRIDIUM FORMICOACETICUM*

LARS G. LJUNGDAHL+, DAVID W. SHEROD, MICHAEL R. MOORE, and JAN R. ANDREESEN
Department of Biochemistry, University of Georgia, Athens, Georgia 30602 (U.S.A.) and Institut für Mikrobiologie der Universität Göttingen, D-3400 Göttingen, Federal Republic of Germany

Clostridium thermoaceticum and *Clostridium formicoaceticum* are obligate anaerobic organisms, with terminal spores and a morphology and metabolism that are very similar. They carry out a homoacetate fermentation of sugars with acetate as the only product, and almost three moles of acetate are formed per mole of hexose (1,2,3). The fermentation proceeds via the Embden-Meyerhof pathway to pyruvate, which is further metabolized to yield two moles of acetate. The third mole of acetate is formed by reduction of CO_2 to a methyl group, which is carboxylated to yield acetate in a transcarboxylation reaction involving the carboxyl group of pyruvate (4). An outline is shown in Figure 1 of the reactions involved in the homoacetate fermentation by *C. thermoaceticum*.

Of particular interest to us has been the synthesis of acetate from CO_2. It is now well established that in *C. thermoaceticum* CO_2 is first converted to formate, which is further reduced via inter= mediates of tetrahydrofolate to methyltetrahydrofolate, and that methylcorrinoids are involved in the last steps (3,5,6,7). A similar series of reactions has also been demonstrated in *C. formicoaceticum* (8).

[1] Fontaine, F., Peterson, W. H., McCoy, E., Johnson, M. J. and Ritter, G. *J. Bacteriol.*, *43*, 701 (1942).
[2] Andreesen, J. R., Gottschalk, G. and Schlegel, H. G. *Arch. Mikrobiol.*, *72*, 154 (1970).
[3] Ljungdahl, L. G. and Wood, H. G. *Annu. Rev. Microbiol.*, *23*, 515 (1969).
[4] Schulman, M., Ghambeer, R. K., Ljungdahl, L. G. and Wood, H. G. *J. Biol. Chem.*, *248*, 6255 (1973)
[5] Thauer, R. K. *FEBS Letters*, 27, 111 (1972).
[6] Parker, D. J., Wu, T. F. and Wood, H. G. *J. Bacteriol.*, *108*, 770 (1971).
[7] Andreesen, J. R., Schaupp, A., Neurauter, C., Brown, A. and Ljungdahl, L. G. *J. Bacteriol.*, 114, 743 (1973).
[8] O'Brien, W. E. and Ljungdahl, L. G. *J. Bacteriol.*, *109*, 626 (1972).

*Supported by grants from National Science Foundation GB13031 and National Institute of Health AM12912 and "Forschungsmittel des Landes Niedersachsen".
+A Senior U.S. Scientist Award for research and teaching in Göttingen, Germany during 1974/75 from Alexander von Humboldt-Stiftung, Bonn-Bad Godesberg is gratefully acknowledged.

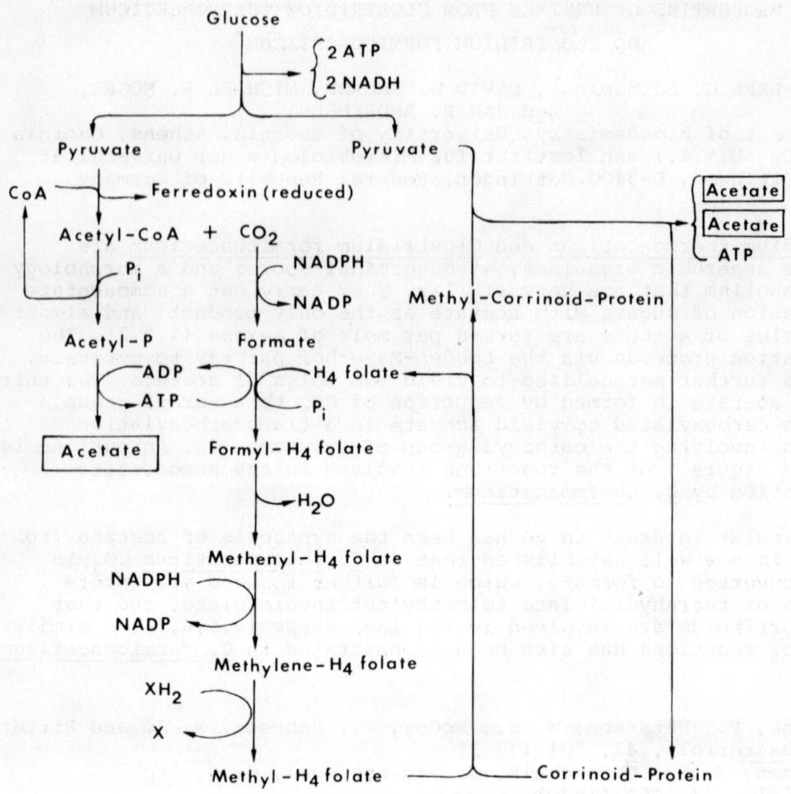

Fig. 1. Homoacetate fermentation by Clostridium thermoaceticum

Although the two organisms appear almost identical there are important differences. For instance the electron donor for both formate dehydrogenase and methylenetetrahydrofolate dehydrogenase in C. thermoaceticum is NADPH (9,10). In C. formicoaceticum the latter enzyme is NADH specific (11), while the natural electron donor for

[9] Li, L.-F., Ljungdahl, L. G. and Wood, H. G. J. Bacteriol., 92, 405 (1966).

[10] O'Brien, W. E., Brewer, J. M. and Ljungdahl, L. G. J. Biol. Chem., 248, 403 (1973).

[11] Moore, M. R., O'Brien, W. E. and Ljungdahl, L. G. J. Biol. Chem., 249, 5250 (1974).

the formate dehydrogenase has not been established (12). Another important difference between the two organisms is found in their requirements of growth temperature. C. formicoaceticum is mesophilic and grows optimally around 37°C, grows slowly at 44°C, and fails to grow at 52°C (2). C. thermoaceticum has optimum growth between 55 to 60°C, a maximum growth at about 65°C and does not grow below 45°C. The spores of C. thermoaceticum are quite heat resistant and survive for 8 hours at 100°C or 15 min at 120°C, but not for 17 hours at 100°C or 30 min at 120°C (1).

The difference in growth temperatures between the two organisms is reflected also in the ability of extracts to catalyze the synthesis of acetate from CO_2 and pyruvate. Thus extracts from C. thermoaceticum catalyze the acetate synthesis at 57°C but not or very little below 40°C (13,14). The optimum temperatures of several enzymes from C. thermoaceticum are at 60°C or above and some of these enzymes have very low activities below 40°C (15). The enzymes in extracts from C. formicoaceticum, on the other hand, are generally rapidly inactivated at temperatures above 45°C. These observations show that the enzymes from C. thermoaceticum compared with the enzymes from C. formicoaceticum are in general more heat stable and that this higher heat stability is not dependent upon their being contained in the intact cell.

The similarities between C. thermoaceticum and C. formicoaceticum in most aspects except in the requirement for growth temperature indicate that these organisms are closely related. To find the basis for the higher thermostability of the C. thermoaceticum enzymes, we have started a program involving the isolation of similar enzymes from the two microorganisms, and the comparison of the properties of these enzymes. Thus far we have succeeded in the purification of two pairs of enzymes, methylenetetrahydrofolate dehydrogenase and formyltetrahydrofolate synthetase. A comparative study of formyltetrahydrofolate synthetase enzymes from Clostridia will be presented at this symposium by O'Brien et al. (16). Now we wish to discuss the properties of methylenetetrahydrofolate dehydrogenase enzymes from C. formicoaceticum and C. thermoaceticum as well as some properties of formate dehydrogenase and acetate kinase from the latter organism.

PROPERTIES OF METHYLENETETRAHYDROFOLATE DEHYDROGENASE

Methylenetetrahydrofolate reductase (E. C. 1.5.1.5) catalyzes the reaction:

$$CH_2 = H_4\text{folate} + NADP^+ \rightleftharpoons CH \equiv H_4\text{folate}^+ + NADPH$$

[12] Andreesen, J. R., El Ghazzawi, E. and Gottschalk, G. Arch. Microbiol., 96, 103 (1974).
[13] Ljungdahl, L. G., Irion, E. and Wood, H. G. Biochemistry, 4, 2771 (1965).
[14] Poston, J. M., Kuratomi, K. and Stadtman, E. R. J. Biol. Chem., 241, 4209 (1966).
[15] Ljungdahl, L. G. and Wood, H. G. Bacteriol. Proc., 109 (1963).
[16] O'Brien, W. E., Brewer, J. M. and Ljungdahl, L. G. (This symposium).

In the reaction, there is a direct transfer of hydrogen from 5,10-methylenetetrahydrofolate to NADP (17) and a hydride ion mechanism has been proposed (18). This enzyme has been demonstrated in different organisms and various tissues and it has been purified from several sources (19) including Clostridium cylindrosporum (20), C. thermoaceticum (10) and C. formicoaceticum (11). The enzymes from C. cylindrosporum and C. thermoaceticum are specific for NADP as electron acceptor. The enzyme from C. formicoaceticum uses NAD and with NADP there is less than one one hundreth of the activity with NAD, if any activity at all. The specific activities (μmoles min^{-1} mg^{-1} of protein) of the purified enzymes from the three Clostridia differ and are 1,400 and 360 from C. formicoaceticum and C. thermoaceticum, respectively, both measured at 37°C (10,11), and 95 from C. cylindrosporum measured at 25°C (20). At 60°C, the growth temperature of C. thermoaceticum, the enzyme from this organism has a specific activity of about 1,000.

The physical parameters of methylenetetrahydrofolate dehydrogenase enzymes from C. thermoaceticum and C. formicoaceticum are listed in Table I.

TABLE I. Physical Properties of Methylenetetrahydrofolate Dehydrogenase Enzymes

Property	C. thermoaceticum enzyme	C. formicoaceticum enzyme
MW	55,000 ± 5,000	60,000
$S^{o}_{20,w}$	3.8	3.8
$\bar{V}$, cc/g	0.752	0.74
Stokes Radius Å	31.7	33.2
Subunits, number	2	2

Both enzymes have a molecular weight of about 60,000 and consist of 2, probably equal, subunits. The molecular weight for the C. cylindrosporum enzyme was reported to be approximately 70,000 (20). It is evident from Table I that the enzymes have almost identical physical properties.

[17] Ramasastri, B. V. and Blakley, R. L. J. Biol. Chem., 239, 112 (1964),

[18] Huennekens, F. M. and Scrimgeour, K. G. in Pteridine Chemistry (Pfleiderer, W., and Taylor, E. C., eds.) Pergamon Press, The MacMillan Comp., N.Y., p. 355 (1964).

[19] Blakley, R. L. in The Biochemistry of Folic Acid and Related Pteridines (Neuberger, A., and Tatum, E. L., eds.) North-Holland Publishing Co., Amsterdam, Netherland, p. 190 (1969).

[20] Uyeda, K. and Rabinowitz, J. C. J. Biol. Chem., 242, 4378 (1967).

Some enzymatic properties of methylenetetrahydrofolate dehydrogenase enzymes are found in Table II.

TABLE II. Enzymatic Properties of Methylenetetrahydrofolate Dehydrogenase Enzymes

Property	*C. thermoaceticum* enzyme	*C. formicoaceticum* enzyme	*C. cylindrosporum* enzyme*
Co-enzyme	NADP	NAD	NADP
Spec. activity (μmoles min^{-1} mg^{-1})	360 (37°C)	1,400 (37°C)	95 (25°C)
Km (cofactor)	9×10^{-5} M	7.9×10^{-4} M	3.4×10^{-5} M
Km (CH_2=H_4folate)	3.5×10^{-5} M	6.6×10^{-5} M	2.6×10^{-5} M
Temperature optimum	>64°C	45°C	-
Activation energy (cal/mole)	6,500 (35-65°C) 9,630 (10-35°C)	8,500	-
pH-optimum+	6.2	6.4	6.8-8

* Data from Uyeda and Rabinowitz (20).
+ Measured in the direction forming NADPH (NADH).

It is apparent from Table II that the enzymes from *C. thermoaceticum* and *C. formicoaceticum* differ enzymatically. The most noticeable difference is of course, as already has been pointed out, the specificity for NADP by the *C. thermoaceticum* enzyme but NAD by the *C. formicoaceticum* enzyme. The latter enzyme has a temperature optimum around 45°C and above that temperature it losses activity rapidly due to denaturation. The *C. thermoaceticum* enzyme has a temperature optimum over 64°C. The enzymes could not be assayed at higher temperatures due to a rapid breakdown of the substrate 5,10-methylenetetrahydrofolate.

The activation energies were calculated from plots of the logarithm of the velocity at different temperatures versus the reciprocal of the temperature (°K). (Arrhenius plots). The Arrhenius plot from the *C. formicoaceticum* enzyme was linear over the temperature range it was assayed and an activation energy of 8,500 cal/mole was calculated. The line in the plot from the *C. thermoaceticum* enzyme was broken with the break at about 35°C. The activation energy below this temperature is 9,630 cal/mole and above is 6,500 cal/mole. The finding of a broken line in an Arrhenius plot is not unique to the methylenetetrahydrofolate dehydrogenase from *C. thermoaceticum*. Such breaks have been reported for many types of enzymes from different sources and the breaks occur over a wide range of temperatures from 0° to 55°C.

The broken Arrhenius plot obtained with methylenetetrahydrofolate dehydrogenase from C. thermoaceticum may suggest that the enzyme undergoes a conformational change when heated, with a transition temperature around 35°C. Evidence for this is found in observations indicating that the enzyme exists in two forms. During electrophoresis the enzyme divides up in two active forms both of which after separation, again give rise to the same two forms. During ultracentrifugation or chromatography on Sephadex the enzyme behaves homogeneously, indicating that the two forms are not obtained by association or dissociation of subunits. Results from heating experiments also indicate at least two forms. When the enzyme is exposed to temperatures over 50°C in the absence of substrate, and then assayed at 37°C, a rapid partial inactivation is observed. After this initial inactivation the enzyme is stable at the lower level of activity for an extended period of time; as if it had undergone a conversion into a more stable, but less active form.

The amino acid compositions of methylenetetrahydrofolate dehydrogenase enzymes from C. thermoaceticum and C. formicoaceticum are listed in Table III.

TABLE III. Amino Acid Compositions of Methylenetetrahydrofolate Dehydrogenase Enzymes, Assuming a MW of 60,000

Amino Acid	C. thermoaceticum enzyme	C. formicoaceticum enzyme
Lysine	39	54
Histidine	13	5
Arginine	13	10
Aspartic acid	49	64
Glutamic acid	52	67
Threonine	20	25
Serine	20	24
Proline	31	29
Glycine	49	47
Alanine	63	47
Half-cystine	9	7
Valine	63	63
Methionine	9	12
Isoleucine	44	36
Leucine	47	44
Tyrosine	4	9
Phenylalanine	8	17
Tryptophan	+	10

Although the amino acid composition of the two enzymes are similar, some differences are noticeable. The C. thermoaceticum enzyme has less lysine than the C. formicoaceticum enzyme, but more histidine and arginine. The C. formicoaceticum enzyme has more aspartic and glutamic acid residues compared with the C. thermoaceticum enzyme; this is reflected in a stronger binding of the C. formicoaceticum enzyme to DEAE-columns, and in a faster migration during electrophoresis. The hydrophobic amino acid residues are almost equal in the enzymes except that there is more phenylalanine and less isoleucine in the C. formicoaceticum enzyme compared with the C. thermoaceticum enzyme. We have found no evidence for the presence of non-proteinacious material in the two dehydrogenases.

The methylenetetrahydrofolate dehydrogenase enzyme from both C. thermoaceticum and C. formicoaceticum have been investigated using initial velocity and product inhibition studies to establish the mechanisms of the two enzymes. Although these studies have not been completed, it is apparent that the two enzymes have a similar if not identical mechanism. The initial velocity experiments clearly indicate a sequential mechanism in which both substrates must add before release of product. The product inhibition patterns suggest a rapid equilibrium random bibi mechanism for both enzymes. However, some of our data are not in complete agreement with such a mechanism and further experiments are needed to establish the true mechanism of the methylenetetrahydrofolate dehydrogenase reaction.

The results of our investigation of methylenetetrahydrofolate dehydrogenase enzymes from C. thermoaceticum and C. formicoaceticum indicate that the enzymes have very similar physical, chemical, and kinetic properties. However, the enzyme from C. thermoaceticum is much more thermostable than the enzyme from C. formicoaceticum. There is no evidence that the higher thermostability of the C. thermoaceticum enzyme results from non-proteinacious material, unusual protein structure, or an external factor. We have concluded, as has been with so many enzymes from thermophilic organisms (21), that higher thermostability depends on small differences in the amino acid composition or sequence.

PROPERTIES OF FORMATE DEHYDROGENASE FROM C. THERMOACETICUM

Formate dehydrogenase from C. thermoaceticum catalyzes the reversible reaction:

$$CO_2 + NADPH \rightleftharpoons HCOO^- + NADP^+$$

The enzyme is best determined in the direction of CO_2 formation and besides NADP as electron acceptor methyl viologen or benzyl viologen can also be used. It is inactive with NAD and ferredoxin (22). The C. thermoaceticum enzyme is the only known NADP-dependent formate dehydrogenase. Its level in the cell is greatly increased by the addition of tungstate and selenite to the growth medium. Molybdate replaces tungstate but a higher enzyme activity is obtained with the latter. Recently using ^{75}Se-selenite and ^{185}W-tungstate it has

[21] Singleton, R., Jr. and Amelunxen, R. E. Bacteriol. Rev., 37, 320 (1973).
[22] Andreesen, J. R. and Ljungdahl, L. G. J. Bacteriol., 120, 6 (1974).

been shown that both selenium and tungsten are incorporated into the active enzyme, suggesting that it is a tungsto-seleno-protein (23,24).

The formate dehydrogenase from C. thermoaceticum is very sensitive to oxygen, and any work with it must be done under strictly anaerobic conditions. Stored under oxygen free nitrogen in presence of a thioglycolate-iron complex functioning as an oxygen scavenger the enzyme is stable for several days. By including the thiol-iron complex in buffer solutions, it has been possible to partly purify the enzyme using precipitation with ammonium sulfate, chromatography on DEAE-cellulose, and gel-filtration (22,24). The highest specific activity (μmoles min^{-1} mg^{-1}) obtained is 50, measured with NADP as electron acceptor at 45°C. Using data, obtained from the

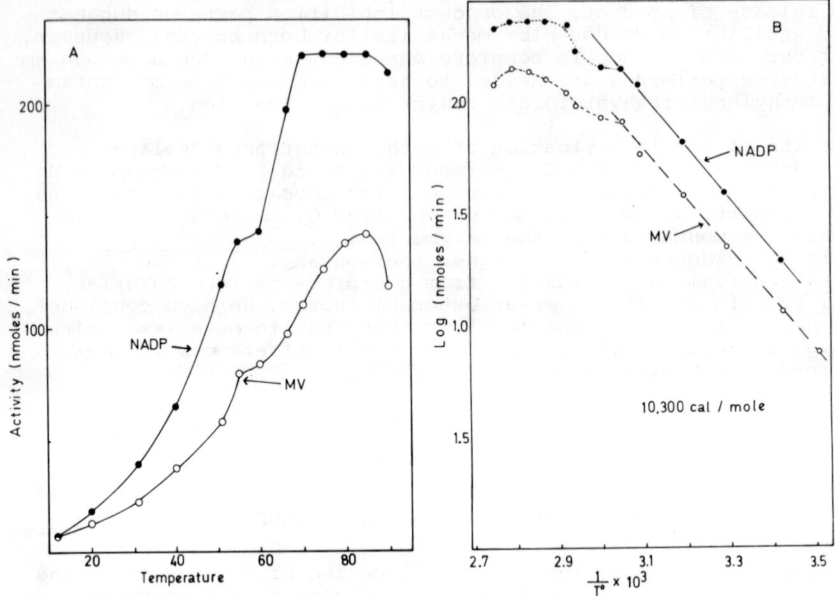

Fig. 2. A. Effect of temperature on the activity of a partly purified formate dehydrogenase from C. thermoaceticum using NADP and methyl viologen (MV) as electron acceptors. With NADP 40 μg of protein and with MV 8 μg of protein was used.
B. Arrhenius plot of the data shown in graph A.

[23] Andreesen, J. R. and Ljungdahl, L. G. J. Bacteriol., 116, 867 (1973).

[24] Ljungdahl, L. G. and Andreesen, J. R. FEBS Letters, 54, 279 (1975).

incorporation studies with ^{185}W-tungstate, it is possible to calculate that the pure enzyme, containing 1 g-at. W per mole, should have a specific activity of about 200. Thus it is apparent that the enzyme is far from homogeneous. Elution profiles from Sepharose-6B, Sephadex G-200 and Biol-Gel A-5 m and A-0.5 m columns indicate that the enzyme has a molecular weight of about 300,000 (22,24). However, during ultracentrifugations in surose gradients two peaks having formate dehydrogenase activity with methyl viologen were obtained, corresponding to apparent molecular weights of 320,000 and 270,000.

The activity of formate dehydrogenase from C. thermoaceticum has been measured at different temperatures using both NADP and methyl viologen as electron acceptors (22). The results of these experiments are shown in Figure 2. The optimum temperature, using NADP as electron acceptor, is approximately 70°C; with methyl viologen it is about 85°C. The increase in activity with temperature follows a smooth curve to about 55°C at which temperature a definite break occurs regardless of electron acceptor (Figure 2, A). This behavior has been noticed repeatedly and is also observed when crude extract is used. At temperatures below 55°C straight lines are obtained in the Arrhenius plot and the slope of the lines are the same with both electron acceptors, indicating the same activation energy of approximately 10,500 cal/mole (Figure 2, B). When the C. thermoaceticum formate dehydrogenase is incubated at 70°C it is activated as shown in Figure 3, A. The activation is especially evident when the incubation is in the presence of the substrate formate, but it occurs also in the presence of the inhibitor sodium azide or in the absence of substrate and inhibitor. After the activation period, which is completed in about 1 hour, the enzyme slowly loses activity. However, after 4 hours a substantial activity remains, especially in the presence of sodium azide, which apparently protects the enzyme from denaturation. At 80°C (Figure 3, B) the enzyme is rather rapidly inactivated, but again some protection by both sodium azide and sodium formate is seen.

C. formicoaceticum contains a very oxygen sensitive formate dehydrogenase which appears to be similar to that of C. thermoaceticum. Its formation is also dependent upon the presence of selenite and tungstate in the growth medium (12). Molybdate again can replace tungstate but less effectively. To obtain high formate dehydrogenase activity both selenite and tungstate must be present during the growth of cells; in other words during synthesis of the enzyme. Addition of the metals to extracts from cells grown in the absence of the metals does not result in increased enzyme activity. Therefore, it is likely that the metals are incorporated into the active protein, as has been shown with the formate dehydrogenase of C. thermoaceticum. The formate dehydrogenase from C. formicoaceticum, like that of C. thermoaceticum, is activated when incubated anaerobically. This activation is observed at 50°C and is followed by rather rapid denaturation at that temperature.

Lack of data prohibits an extensive comparison of the formate dehydrogenase enzymes. However, it is safe to conclude that the presence of selenium and tungsten in the C. thermoaceticum enzyme does not *per se* contribute to its thermostability, since the

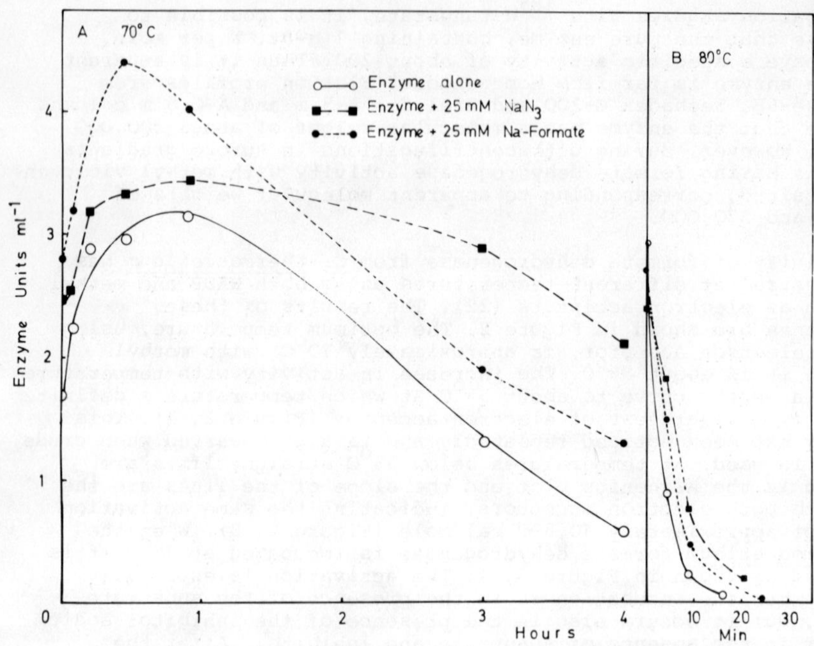

Fig. 3. Activation and denaturation of formate dehydrogenase from C. thermoaceticum.
A. Incubation at 70°C of extract containing 15 mg of protein/ml in 0.1 M triethanolamine-maleate buffer pH 7.5, 0.025 M sodium thioglycolate, 2 mM ferrous ammonium sulfate and 0.1 mM EDTA under nitrogen, and with 0.025 M sodium azide or 0.025 M sodium formate as indicated. Formate dehydrogenase activity was assayed as previously published (22) at 45°C with NADP.
B. Same as A., but with incubations at 80°C.

C. formicoaceticum enzyme also appears to contain these metals. We would like to suggest that the thermostability resides in the protein portion of the enzyme and depends upon the amino acid composition and sequence.

PROPERTIES OF ACETATE KINASE FROM C. THERMOACETICUM

Acetate kinase has been purified about 40 fold to a specific activity (µmoles min^{-1} mg^{-1}) of 282, determined at 50°C from extracts of C. thermoaceticum (25). The enzyme appears homogeneous

[25] Schaupp, A. and Ljungdahl, L. G. Arch. Microbiol., 100, 121 (1974).

in the ultracentrifuge and during gel filtration, but is heterogeneous when examined by electrophoresis in polyacrylamide gels at pH 9.0 and 9.5. The molecular weight of the enzyme is approximately 60,000.

When the velocity of the enzyme reaction was studied, as a function of the temperature, a temperature optimum of about 60°C was observed. The Arrhenius plot of the data exhibits a broken line with the break around 40°C. The activation energies calculated from the Arrhenius graph were 10,500 cal/mole and 3,110 cal/mole between 20 and 40°C and 40 and 60°C, respectively. During incubations in 0.1 M potassium phosphate pH 7.0, in the absence of substrate, the enzyme is stable at 50°C but loses about 70 % of its activity at 60°C in a 30 min period.

The C. thermoaceticum acetate kinase exhibits normal Michaelis-Menten kinetics with acetate as the variable substrate. However, with Mg-ATP as the variable substrate sigmoidal kinetics were observed, indicating a cooperative binding of the Mg-ATP complex. A Hill coefficient (n) of 1.78 was obtained from the kinetic data, suggesting the presence of two interacting substrate sites. Acetate kinase from C. thermoaceticum may thus be an example of an enzyme from a thermophilic organism exhibiting allosteric control.

DISCUSSION

The existence of thermophilic organisms raises a number of interesting questions: how did these organisms evolve; how are they able to function at high temperatures; and why are obligate thermophiles unable to grow at the lower "normal" temperatures, at which mesophilic organisms grow?

C. thermoaceticum grows optimally at 60°C and C. formicoaceticum at 37°C. They have very similar morphology and metabolism and seem to be closely related (1,2). One might, therefore, predict that their cellular components, including enzymes, are similar. However, the cell free extract experiments clearly show that the enzymes from C. thermoaceticum are generally more heat stable than the corresponding enzymes from C. formicoaceticum. It is apparent that part of the reason, why C. thermoaceticum is able to grow at high temperatures, is because it can synthesize enzymes stable at its growth temperature. The results reported here, concerning methylenetetrahydrofolate dehydrogenase and formate dehydrogenase and those reported by O'Brien et al. (16) dealing with formyltetrahydrofolate synthetase, demonstrate that the enzymes in C. thermoaceticum are as predicted very similar to corresponding enzymes in C. formicoaceticum. The purified enzymes have almost identical chemical, physical and kinetic properties, yet the enzymes from C. thermoaceticum are more thermostable. The purified C. thermoaceticum enzymes do not contain any non-proteinacious material, which might stabilize them at high temperatures. We have thus concluded that the higher thermostability is an intrinsic property of the protein, and that small changes in amino acid composition or sequence might explain the difference in thermo=stability between similar enzymes from the two Clostridia. If this is so, it would be necessary to establish the exact structure of the proteins to find to what extent individual amino acid residues may contribute to stability by forming hydrophobic interactions, hydrogen bonds or salt bridges. Such an analysis was recently made by Perutz and Raidt with ferredoxin and haemoglobin A_2 (26).

Results reported here and elsewhere show that enzymes from C. thermoaceticum, although they are more thermostable, have properties similar to the corresponding enzymes from mesophilic organisms. For example, the sigmoidal kinetics observed for acetate kinase from C. thermoaceticum indicate that this enzyme may be under metabolic control. The acetate kinase from Veillonella alcalescens has similar properties (27). The formate dehydrogenase enzymes from C. formicoaceticum and C. thermoaceticum have identical metal requirements and probably similar catalytic mechanisms.

The rates of enzyme reactions normally doubles with an increase in temperature of 10°C. If enzymes from thermophiles would have the same catalytic activity as enzymes from mesophiles, the thermophilic enzymes would have extremely high activities at thermophilic temperatures. This is not observed. Generally enzymes from thermophiles have at the optimum growth of the thermophiles about the same catalytic activity as the corresponding enzymes from mesophiles assayed at their growth temperature. For instance, the methylenetetrahydrofolate dehydrogenase from C. thermoaceticum has a specific activity of 1,000 at 60°C and the enzyme from C. formicoaceticum about 1,400 at 37°C. It thus logically follows that enzymes from thermophilic organisms often have rather low activities at temperatures below 40°C. The metabolic rates of the thermophiles may thus be too low to sustain growth. This would explain why obligate thermophiles are unable to grow at low temperatures.

Several of the enzymes from C. thermoaceticum show broken lines in Arrhenius plots and the activation energies calculated from these plots are generally higher at lower temperatures than at the higher optimum growth temperatures. This phenomenon may also explain why C. thermoaceticum is unable to grow below 40°C. Clearly a higher activation energy for an enzyme reaction at lower temperature would decrease the rate of the reaction.

SUMMARY

Methylenetetrahydrofolate dehydrogenase from C. thermoaceticum and C. formicoaceticum have been purified to homogeneity and compared. The two enzymes are very similar physically, chemically, and kinetically, but the C. thermoaceticum enzyme has a higher thermostability, which is an intrinsic property of the protein. Formate dehydrogenase enzymes from both bacteria require selenite and tungstate for formation and these enzymes also appear to have similar properties, although the C. thermoaceticum is stable at 70°C for more than one hour. Acetate kinase from C. thermoaceticum appears to be under metabolic control. It can be concluded that enzymes from C. thermoaceticum, although they are more thermostable, are very similar to corresponding enzymes from mesophilic organisms.

[26] Perutz, M. F. and Raidt, H. Nature, 255, 256 (1975).
[27] Pelroy, R. A. and Whiteley, H. R. J. Bacteriol., 105, 259 (1971).

CHEMICAL, PHYSICAL AND ENZYMATIC COMPARISONS OF FORMYLTETRAHYDROFOLATE SYNTHETASES FROM THERMO- AND MESOPHILIC CLOSTRIDIA*

W. E. O'BRIEN[†], J. M. BREWER AND LARS G. LJUNGDAHL
Department of Biochemistry, University of Georgia
Athens, Georgia 30602 (U.S.A.)

INTRODUCTION

10-Formyltetrahydrofolate synthetase (Formate:tetrahydrofolate ligase (ADP), E. C. 6.3.4.3) catalyzes the synthesis of 10-formyltetrahydrofolate:

$$HCOOH + ATP + H_4\text{-folate} \xrightarrow[Mg^{++}]{} 10\text{-Formyl-}H_4\text{-folate} + ADP + Pi$$

10-Formyltetrahydrofolate synthetase has been purified from pigeon liver[1], *Micrococcus aerogenes*[2], *Clostridium acidiurici*[3], *C. cylindrosporum*[3] and *C. thermoaceticum*[4,5]. The *C. cylindrosporum* enzyme has been crystallized[3] and extensively studied[6-10].

The enzyme in *C. thermoaceticum* is one of several enzymes which convert CO_2 into the 5-methyl group of 5-methyltetrahydrofolate which is subsequently carboxylated to yield acetate[9,11]. The enzyme appears to serve the same function

[1]Greenberg, G. R., Jaenicke, L. and Silverman, M. Biochim. Biophys. Acta, 17, 589 (1955).
[2]Whiteley, H. R., Osborn, M. J. and Huennekens, F. M. J. Biol. Chem., 234, 1538 (1959).
[3]Rabinowitz, J. C. and Pricer, W. E., Jr. J. Biol. Chem., 237, 2898 (1962).
[4]Sun, A. Y., Ljungdahl, L. G. and Wood, H. G. J. Bacteriol., 98, 842 (1969).
[5]Ljungdahl, L. G., Brewer, J. M., Neece, S. H. and Fairwell, T. J. Biol. Chem., 245, 4791 (1970).
[6]Himes, R. H. and Rabinowitz, J. C. J. Biol. Chem., 237, 2903 (1962).
[7]Scott, J. M. and Rabinowitz, J. C. Biochem. Biophys. Res. Comm., 29, 418 (1967).
[8]Himes, R. H. and Wilder, T. Arch. Biochem. Biophys., 124, 230, (1968).
[9]Rabinowitz, J. C. and Pricer, W. E., Jr. J. Biol. Chem., 229, 321 (1957).
[10]Himes, R. H. and Harmony, J. A. K. CRC Critical Reviews in Biochemistry, 501 (1973).
[11]Ljungdahl, L. G. and Wood, H. G. Ann. Rev. Microbiol., 23, 515 (1969).

*Supported by NSF GB13031 and NIH AM12913.

[†]Present address: Department of Biochemistry, Case Western Reserve University, Cleveland, Ohio 44106.

in C. formicoaceticum[12].

C. thermoaceticum and C. formicoaceticum are identical, physiologically speaking, except that the former is an obligate thermophile whereas the latter is an obligate mesophile. Enzymes isolated from C. thermoaceticum have higher optimum temperatures than the same enzymes isolated from mesophilic organisms[4,13].

In this paper we report the purification, stability and composition of the enzyme isolated from C. formicoaceticum and discuss these data in comparison with the properties of the enzymes from C. cylindrosporum, C. acidiurici and C. thermoaceticum.

MATERIALS AND METHODS

Cell Material. C. formicoaceticum was grown in 40° in the medium described by O'Brien and Ljungdahl[12]. Cells were harvested by centrifugation and stored at -15° until they were to be used. Enzyme activity was stable in such cells for at least six months.

Analytical Methods. Assays for 10-formyltetrahydrofolate synthetase were done according to Rabinowitz and Pricer[3]. Protein was determined by the biuret method[14] using bovine serum albumin as a standard and by measurement of 280/260 ratios[15].

The amino acid composition of the enzyme was determined after hydrolysis in 6 N HCl. The enzyme was dialyzed in dilute potassium maleate buffer, pH 7.0, before hydrolysis since dialysis against distilled water led to irreversible precipitation of the enzyme. Aliquot portions were made 6 N in HCl, evacuated and sealed.

The hydrolysis tubes were incubated at 110° for 24, 48, 72 and 96 hrs. After the allotted time the samples were repeatedly taken to dryness to remove residual HCl. The samples were then applied to a Beckman model 120C automatic amino acid analyzer. Values for serine and threonine were corrected for destruction by extrapolation to zero time and valine and isoleucine were determined by extrapolation to infinite time. Half-cystine and methionine contents were determined by performic acid oxidation[16].

Disc gel electrophoresis was performed as described by Brewer and Ashworth[17]. Sedimentation velocity centrifugations were made in a Beckman Model E analytical ultracentrifuge equipped with schlieren optics as described by Schachman[18].

[12]O'Brien, W. E. and Ljungdahl, L. G. J. Bacteriol., 109, 626 (1972).
[13]Li, L. F., Ljungdahl, L. G. and Wood, H. G. J. Bacteriol., 92, 405 (1966).
[14]Lowry, O. H., Rosebrough, H. J., Farr, A. L. and Randall, R. J. J. Biol. Chem., 193, 265 (1951).
[15]Warburg, O. and Christian, W. Biochem. Z., 310, 384 (1941).
[16]Hirs, C. H. W., in C. H. W. Hirs (Editor), Methods in Enzymology, Vol. XI, Academic Press, New York, 1967, page 197.
[17]Brewer, J. M. and Ashworth, R. B. J. Chem. Educ., 46, 41 (1969).
[18]Schachman, H. K. Ultracentrifugation in Biochemistry, Academic Press, New York (1959).

Sedimentation equilibrium centrifugation was performed by the method of Yphantis[19] Measurements of pH were done with a Radiometer pH meter 26 with an internal temperature compensator. Absorption measurements were made on a Gilford model 204 spectrophotometer.

Chemicals. ATP and Trisma base were purchased from Sigma. Tetrahydrofolate was prepared by a modification of the method of Blair and Saunders[12]. Ammonium sulfate was the "enzyme grade" quality from Schwarz/Mann. Diethylaminoethyl cellulose (DE-23) was obtained from Whatman, and Sephadex G-200 from Pharmacia.

PURIFICATION OF ENZYME

The purification of formyltetrahydrofolate synthetase from *C. formicoaceticum* was carried out at 0-4° in potassium maleate buffer, pH 7.0, or potassium phosphate buffer at pH 7.0.

Preparation of Cell-Free Extract. Frozen cells (60 g wet weight) were suspended in 180 ml of 0.05 M maleate buffer. The suspension was passed through a French pressure cell (American Instrument Co., Silver Spring, Md.) and centrifuged at 37,000 x g for 1 hr. The supernatant solution was decanted and the residue washed with 170 ml of 0.05 M maleate buffer, centrifuged again and the two supernatant solutions combined.

DEAE Column Chromatography. The cell-free extract (350 ml) was dialyzed against 0.002 M maleate buffer, pH 7.0, overnight and applied to a 5 x 35 cm DEAE-cellulose column. Batchwise elution with buffer solutions of increasing molarity was employed. The enzyme eluted in 0.08 M maleate buffer in a volume of 1845 ml and was precipitated by addition of crystalline $(NH_4)_2SO_4$ to 80% saturation[20].

Chromatography on Sephadex G-200. Sephadex G-200 was swollen in boiling water and then washed with several volumes of 0.10 M phosphate buffer. A 10 x 95 cm column was prepared for ascending flow and washed with at least 10 volumes of 0.10 M phosphate buffer. The precipitate from the DEAE column was dissolved in 36 ml, applied to the column, and eluted with the same buffer. Fractions of approximately 18 ml were collected and those with specific activities between 100 and 134 were pooled.

Chromatography on DEAE-cellulose. The pooled fractions from the Sephadex G-200 column were dialyzed against 2 changes of 4 liters each of 0.002 M maleate buffer. The enzyme was applied to a 2 x 20 cm DEAE-cellulose column equilibrated with the 0.002 M maleate buffer. The column was washed successively with 100 ml of 0.002, 0.01 and 0.02 M maleate buffer. The enzyme was eluted with a 0.02 to 0.20 M 1 liter linear gradient. Fractions of 3.5 ml were collected during the gradient, and those with specific activities between 160 and 260 were pooled and precipitated by addition of solid ammonium sulfate to 80% saturation.

Back Extraction with Ammonium Sulfate. The ammonium sulfate precipitate was successively washed at pH 7.2 with 20 ml of 65%, 55% and 45% saturated solutions of ammonium sulfate in which the pH had been adjusted with 4N NH_4OH. Very little activity was found in the 65% and 55% washes. The enzyme dissolved in the 45% wash. The purification is shown summarized in Table I.

[19]Yphantis, D. A. Biochemistry, 3, 297 (1964).
[20]DiJeso, F. J. Biol. Chem., 243, 2022 (1968).

TABLE I. Purification of Formyltetrahydrofolate Synthetase from C. formicoaceticum

Fraction	Activity, i.u.	Protein, mg	Specific Activity i.u./mg	% Recovery	Purification
Crude	82,500	8,120	10	100	1
1st DEAE	49,000	1,470	32	60	3
G-200	22,600	192	118	27	11
2nd DEAE	5,093	30.5	166	6	16
$(NH_4)_2SO_4$	3,500	13.8	254	4	25

RESULTS

Enzyme Stability. The purified enzyme has been stored for periods of up to one month in 0.10 M potassium maleate or tris-HCl buffers, pH 7.0, without significant loss of activity.

Purity of Enzyme. Figure 1 shows a schlieren pattern obtained with purified formyltetrahydrofolate synthetase. A single symmetrical peak was obtained.

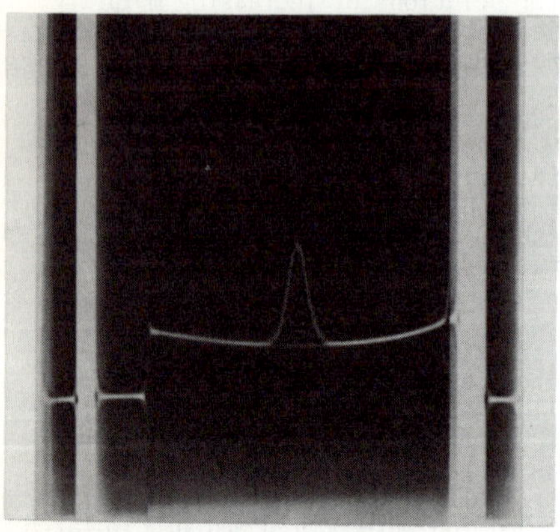

Figure 1. Schlieren Pattern of Formyltetrahydrofolate Synthetase from C. formicoaceticum. Formyltetrahydrofolate synthetase of specific activity 254 i.u./mg at a concentration of 6.8 mg/ml was centrifuged at a rotor speed of 60,000 rpm in 0.1 M potassium maleate, pH 7.0. The temperature was 20°.

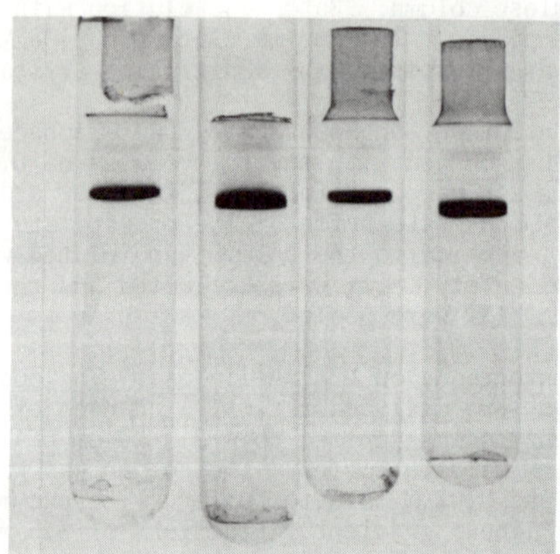

Figure 2. Analytical Disc Acrylamide Gel Electrophoresis of Formyltetrahydrofolate Synthetase. The first two gels were run at pH 9.5 and the two right-hand gels at pH 9.0 in tris-glycine buffers[17]. The first and third contained 30 ug of protein and the second and fourth contained 60 ug of protein. 7 1/2% gels were used.

The molecular weight of the enzyme was determined by short column Yphantis sedimentation equilibrium (see below). The logarithm of protein concentration was a linear function of the square of the distance from the center of rotation. This suggests that the protein preparation is homogeneous in terms of mass and density. However, the length of the column limits the value of this method as a test for homogeneity.

Figure 2 shows patterns obtained with enzyme of specific activity 254 on electrophoresis in acrylamide gels at pH 9.0 and 9.5. At these pH values two thin sharp bands are observed which move more slowly than the major component. Ljungdahl et al.[5] observed this phenomenon with the enzyme from C. thermoaceticum. These minor bands are believed to be aggregates of the enzyme[5].

These criteria indicate that our preparations of formyltetrahydrofolate synthetase from C. formicoaceticum exhibit a high degree of homogeneity, and are at least 95% pure.

ENZYMATIC PROPERTIES

Effect of Temperature on the Formyltetrahydrofolate Synthetase Reaction. To determine the optimum assay temperature, the enzyme was assayed under the standard assay conditions[3] at various temperatures (Figure 3).

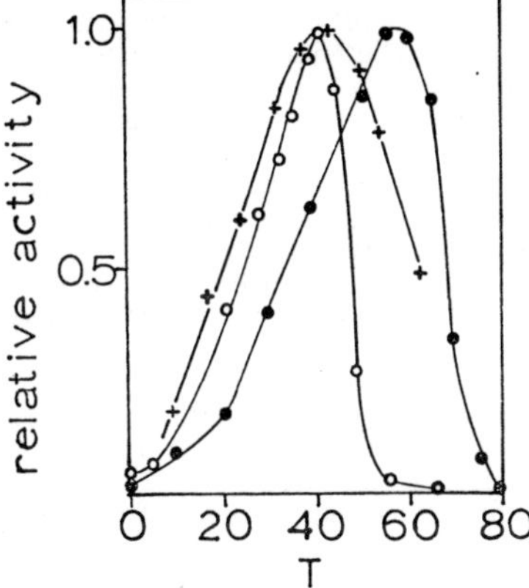

Figure 3. The Effect of Temperature on the Reaction Rate of Formyltetrahydrofolate Synthetases. Reaction mixtures (1.5 ml) were incubated at the desired temperature for five minutes to allow for temperature equilibration. Then 0.022 i.u. (0.08 ug) of C. formicoaceticum enzyme was added to initiate the reaction. The C. formicoaceticum data (+) are shown compared with data obtained using the enzymes from C. thermoaceticum (•)[4,5] and C. cylindrosporum (o)[6].

A broad temperature profile was obtained with an optimum at 45°. An Arrhenius plot of this data gives an activation energy, Ea, of 7500 cal/mole. All subsequent investigations were performed at 37° unless otherwise noted. The temperature optimum of the C. formicoaceticum enzyme is thus close to that of the C. cylindrosporum enzyme (42°) but lower than that from C. thermoaceticum (55°)[4].

Effect of pH. The optimum pH for the reaction was determined by measuring the rate of the enzymatic reaction at various pH values in maleate, triethanolamine, and glycine buffers. The enzyme exhibits maximum activity between the pH values of 8.0 and 8.5 (data not shown) in triethanolamine or glycine buffers. The enzyme was slightly more active in the glycine-KOH buffers.

Effect of Monovalent Cations. Whiteley and Huennekens[21] first reported that formyltetrahydrofolate synthetase from M. aerogenes was stimulated by monovalent cations. Himes and Wilder[8] determined that the increase in activity of the C. cylindrosporum enzyme that is observed when Na^+ is replaced with K^+ in the assay is due to a decrease in the Km for formate from approximately 50 mM to 7 mM. Also, the Km for tetrahydrofolate is decreased slightly. The enzyme from C. thermoaceticum has recently been shown to require monovalent cations for maximum thermal stability[22]. These ions also lower the Km for formate with the C. thermoaceticum enzyme, but do not affect V_{max}.

The monovalent cation requirement was investigated with the enzyme from C. formicoaceticum. NH_4^+ (10 mM) gave the greatest stimulation (almost 8-fold over that obtained using tris$^+$) with an equal concentration of K^+ giving somewhat lower rates. Li^+, Rb^+, and Na^+ all gave less than a two-fold stimulation of activity. These results are very similar to those obtained by Himes and Wilder[8] except in the case of Rb^+ which could activate the enzyme from C. cylindrosporum.

To determine if a decrease in the Km for formate was responsible for the activation of the enzyme from C. formicoaceticum, the Michaelis constant was calculated from the results of experiments in which formate was the variable substrate[23] and sodium or potassium was the cation employed to neutralize the assay reagents (except for tetrahydrofolate which was in tris$^+$). When Na^+ is the monovalent cation in the assay, the apparent Km for formate was 38 mM. This is significantly higher than the value of 7.2 mM obtained with K^+ as the cation. This change in Km is sufficient to cause the difference in activity observed. The Michaelis constants for this and the other Clostridial enzymes are presented in Table II. It is clear that the values, and the effect of the activating monovalent cation, are very similar.

The maximum velocity obtained with the C. formicoaceticum enzyme was 0.575 µmoles/min/µg. Using this value, a turnover number of 1.4×10^5 moles of product formed per minute per mole of enzyme is obtained at 37°, assuming a molecular weight of 240,000 daltons. The corresponding values[10] for the other enzymes are given in Table II.

Thermal Stability of Formyltetrahydrofolate Synthetase. The C. thermoaceticum enzyme has been shown to be stable at elevated temperatures and also to be stabilized against heat inactivation by NH_4^+ and K^+[22]. The enzyme from C. formicoaceticum was investigated in regard to its thermal stability and also the protection of the enzyme against heat denaturation by monovalent cations (Figure 4).

[21] Whiteley, H. R. and Huennekens, F. M. J. Biol. Chem., 237, 1290 (1962).
[22] Shoaf, W. T., Neece, S. H. and Ljungdahl, L. G. Biochim. Biophys. Acta, 334, 448 (1974).
[23] Lineweaver, H. and Burk, D. J. Am. Chem. Soc., 56, 638 (1934).

TABLE II. Comparison of Enzymatic Parameters[5,10]

Property	Enzyme Source			
	C. acidi-urici	C. cylindro-sporum	C. formico-aceticum	C. thermo-aceticum
Turnover number moles/minute/ mole enzyme	$1.1x10^5$	$1.1x10^5$	$1.4x10^5$	$0.9x10^5$
Temperature Optimum	-	42°	45°	55°
Activation Energy (cal/mole)	-	11,000	7,500	11,000(20-40°) 8,600(40-55°)
Km(Formate) + K^+:	$4.8x10^{-3}M$	$6.7x10^{-3}M$	$7.2x10^{-3}M$	$5x10^{-3}M$
- K^+:	-	$50x10^{-3}M$	$38x10^{-3}M$	$50x10^{-3}M$
Km(Mg-ATP)	$1.2x10^{-4}M$	$1.4x10^{-4}M$	$1.9x10^{-4}M$	$1x10^{-4}M$
Km(H_4-folate)(+K^+)	$2.8x10^{-4}M$	$5.2x10^{-4}M$	$6.2x10^{-4}M$	$2x10^{-4}M$

Data from the first three enzymes were obtained at 37°; the C. thermoaceticum enzyme was assayed at 50°.

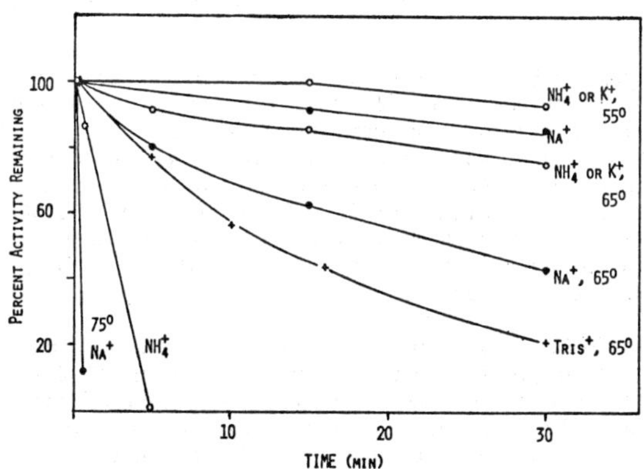

Figure 4. The Effect of Monovalent Cations on the Thermal Stability of C. formicoaceticum Formyltetrahydrofolate Synthetase. Samples (one ml) of 0.1 M $tris^+$ maleate, pH 7.0, containing one of the various cations at a final concentration of 0.01 M, were placed in a water bath at the desired temperature and incubated for five minutes. 0.84 i.u. (3.2 ug) of enzyme was added to the buffer and 0.01 ml aliquot portions removed on a time schedule and assayed at 37° in the standard reaction mixture[3].

The formyltetrahydrofolate synthetase from C. *formicoaceticum* is stable at temperatures up to 55° for at least 30 minutes under the conditions of the experiment. At 65° NH_4^+ and K^+ afford a definite protection against thermal denaturation compared to Na^+ and $tris^+$. The activity loss observed is irreversible since cooling the thermally denatured enzyme under various conditions produced no increase in activity. Overall, the thermal stability of the C. *formicoaceticum* enzyme is considerably higher than those of the C. *cylindrosporum* or C. *acidiurici* enzymes, and is comparable to that of the C. *thermoaceticum* enzyme. Similar experiments demonstrated no protection of the enzyme from thermal inactivation by substrates.

In 1967 Scott and Rabinowitz[7] reported that the C. *cylindrosporum* enzyme dissociates in the absence of activating monovalent cations into four subunits with a corresponding loss of activity. The C. *acidiurici* enzyme behaves similarly[7,10]. However, we have never been able to observe such an effect of monovalent cation on the subunit structure of the enzyme from C. *thermoaceticum*. An examination of the sedimentation behavior of the C. *formicoaceticum* enzyme also showed no effect of replacement of Na^+ or $tris^+$ by NH_4^+ either at 20° or at 42°. Clearly, the strength of the interactions between subunits seems greater in the more thermostable formyltetrahydrofolate synthetases from C. *thermoaceticum* and C. *formicoaceticum* than in the enzymes from C. *cylindrosporum* and C. *acidiurici*.

PHYSICAL PROPERTIES

Determination of the Molecular Weight of Formyltetrahydrofolate Synthetase from C. formicoaceticum. The molecular weight of the enzyme was determined by short column Yphantis sedimentation equilibrium experiments. Assuming a partial specific volume of 0.749 cc/g (see below), these experiments gave weight-average molecular weights of 252,000, 237,000, and 240,000 to give an average M_w of 243,000.

Determination of the Subunit Molecular Weight. The molecular weight of the subunits of this enzyme was determined by equilibrium centrifugation of enzyme which had been dialyzed extensively against 4 M guanidine-HCl containing 0.01 M dithiothreitol. The dependence of the logarithm of the protein concentration on the square of the distance from the center of rotation was linear in each case (not shown). Weight average molecular weights of 53,000, 63,900, and 71,000 were obtained to give an average molecular weight of 62,900 for the subunit. No dependence of apparent molecular weight on rotor speed, like that observed with the enzyme from C. *thermoaceticum*, was found[24,25].

The protein was also electrophoresed in polyacrylamide gels containing sodium dodecylsulfate after incubation in 1% sodium dodecylsulfate and 1% 2-mercaptoethanol[26]. Only one band was observed (not shown). Comparison of the R_f value for the enzyme with the R_f values of known standards gave a molecular weight of 60,000 for the subunit. We conclude that this enzyme, like that from C. *thermoaceticum*, is composed of four subunits of equal molecular weight.

[24]Brewer, J. M., Ljungdahl, L., Spencer, T. E. and Neece, S. H. *J. Biol. Chem.*, *245*, 4798 (1970).
[25]Rawitch, A. B. *Fed. Proc.*, *30*, 1321 Abs. (1971).
[26]Weber, K. and Osborne, M. *J. Biol. Chem.*, *244*, 4406 (1967).

Determination of the Sedimentation Constant, $S^{\circ}_{20,w}$. The dependence of the observed sedimentation constant of the enzyme on protein concentration was identical to that of the C. thermoaceticum enzyme (not shown)[24]. A value for $S^{\circ}_{20,w}$ of 9.9 S was calculated from the data, which is also identical to that of the C. thermoaceticum enzyme[24].

Determination of the Stokes Radius. The Stokes radius of formyltetrahydrofolate synthetase from C. formicoaceticum was found to be identical with that of the enzyme from C. thermoaceticum, since the enzymes elute from a Sephadex G-200 column in the same volume (not shown)[27]. The measured Stokes radius is in good agreement with the Stokes radius calculated from the sedimentation constant, molecular weight and partial specific volume.

Absorption Spectrum. The C. formicoaceticum enzyme has an absorption maximum at 277 nm and a 280/260 ratio of 1.5 (not shown). The presence of tryptophan is clearly indicated.

Comparison of Physical Properties of Formyltetrahydrofolate Synthetases. A comparison of some physical properties of the enzymes from C. formicoaceticum, C. thermoaceticum, C. acidiurici and C. cylindrosporum is given in Table III. It is evident that these proteins are similar in size, density and shape.

CHEMICAL PROPERTIES

Amino Acid Composition of Formyltetrahydrofolate Synthetase from C. formicoaceticum. Formyltetrahydrofolate synthetase of specific activity 260 i.u./mg was hydrolyzed and its amino acid composition determined. The composition, assuming 100% recovery of total residues, is shown in Table IV. From these data a partial specific volume of 0.749 cc/g was obtained[28]. The average standard deviation in the C. formicoaceticum data is ±5% of the residue values. Table IV also gives the amino acid compositions of the enzymes from C. cylindrosporum, C. acidiurici and C. thermoaceticum[10]. They are clearly similar (see Discussion).

Presence of Non-Proteinaceous Material in the Enzyme. We have no direct evidence for or against the presence of substances such as carbohydrate or lipid in the enzyme. However, the physical parameters such as Stokes radius, molecular weight, subunit molecular weight, absorption spectrum and sedimentation constant are very similar if not identical to those parameters of the enzyme from C. thermoaceticum, which has been shown to have little or no non-proteinaceous material present[24].

The internal agreement between the parameters of the enzyme from C. formicoaceticum also argues against the presence of large amounts of carbohydrate, lipid or nucleic acids[5,10].

DISCUSSION

Purity. Our data, though less extensive than presented for the enzyme from C. thermoaceticum[5,10], nevertheless indicates a high degree of homogeneity for our preparations of formyltetrahydrofolate synthetase from C. formicoaceticum. We are confident that our comparisons of properties of this enzyme with the same enzyme from other sources is meaningful.

[27] Ackers, G. K. Biochemistry, 3, 723 (1964).
[28] Cohn, E. J. and Edsall, J. T. Proteins, Amino Acids and Peptides, Reinhold, New York, 1943, p. 373.

TABLE III. Comparison of Physical Parameters[5,10].

Property	Enzyme Source: C. acidi-urici	C. cylindro-sporum	C. formico-aceticum	C. thermo-aceticum
M_W	240,000	238,000	243,000	244,000
$S^{\circ}_{20,w}$, S	9.3	9.3	9.9	9.9
$\bar{V}$, cc/g	0.742	0.740	0.749	0.747
f/fo	1.42	1.43	1.30	1.27
Stokes Radius, Å	-	-	55.1	55.1
M_W subunit	60,000	60,000	60,000	60,000
$\varepsilon^{0.1\%}_{280}$, 1 cm	3.7	5.3	-	7.32
$\lambda^{20^{\circ}}_{max,abs}$ (nm)	-	279	277	277
fluorescence emission maximum, nm λ_{ex} = 280 nm	340*	338*	323	325

*Corrected values (J.A.K. Harmony, personal communication).

Comparison of Enzymatic Properties. The enzymes appear quite similar in their Michaelis constants. The pattern of activation by potassium or ammonium ion is also very similar. However, the dissociation of the enzymes from C. acidiurici and C. cylindrosporum upon replacement of activating cations with $tris^+$ or Na^+ has no apparent parallel in the other two enzymes.

The turnover number of the enzyme from C. thermoaceticum is lower than that of the other three enzymes, when the difference in assay temperatures is allowed for. We do not know if this is a consistent pattern: if thermophilic organisms have effectively traded high turnover for high thermal stability.

Comparison of Physical Properties. Brandts[29] has concluded that the temperature of maximum stability of a protein could be increased by substituting hydrophobic amino acids for more polar ones. Most proteins from thermophiles, including formyltetrahydrofolate synthetase from C. thermoaceticum, are not unusually rich in hydrophobic amino acids. However, a logical extension of Brandts ideas is that the thermal stability of a protein would also be enhanced by excluding solvent from access to the interior of a protein.

[29]Brandts, J. F., in A. H. Rose (Editor), Thermobiology, Academic Press, New York, 1967, page 25.

TABLE IV. Comparison of Amino Acid Compositions[5,10].

Residue	Enzyme Source				Correlation with temperature optimum**
	C. acidiurici	*C. cylindrosporum*	*C. formicoaceticum*	*C. thermoaceticum*	
Tryptophan	9	12	*	20	+?
Lysine	186	181	186	133	0
Histidine	45	32	35	32	0
Ammonia	-	(177)	(181)	(100)	?
Arginine	66	61	72	97	+
Aspartate	277	260	248	210	-
Threonine	110	123	127	121	0
Serine	62	76	57	69	0
Glutamate	180	197	186	179	0
Proline	76	96	88	96	0
Glycine	204	215	186	196	0
Alanine	253	280	236	239	0
Half-Cystine	24	24	16	24	0
Valine	152	184	155	167	0
Methionine	45	65	32	52	0
Isoleucine	124	129	148	148	+
Leucine	210	217	226	206	0
Tyrosine	37	36	33	57	0
Phenylalanine	75	73	66	63	-

*Present but not quantitated.
**Minus signs mean a negative correlation; plus signs a positive correlation.

We examined the fluorescence emission spectra of the *C. thermoaceticum* and *C. formicoaceticum* enzymes. We expected to find the tryptophan emission of the less thermostable enzymes occurring at predominantly longer wavelengths than that of the *C. thermoaceticum* enzyme, since the environment of the tryptophans should be "dryer", more excluded from the solvent, in the protein from the thermophile[24]. In fact, the maximum of the *C. formicoaceticum* enzyme occurs within 2 nm of that of the *C. thermoaceticum* enzyme. On the other hand, the *C. formicoaceticum* enzyme is relatively thermostable, and the emission maxima of the *C. cylindrosporum* and *C. acidiurici* enzymes are at wavelengths about 10-15 nm longer than those of the two more thermostable enzymes[10], when the different characteristics of the instruments are allowed for. So the fluorescence emission maximum wavelength may indeed provide some indication of relative thermostability.

When we speak of the "hydrophobicity" of the "interior" of a protein, we are speaking of a time average property. Recently, the idea that proteins in aqueous solution undergo constant conformational fluctuations has gained acceptance[30], and it seems reasonable, extending the ideas mentioned above, that the "dryness" of the interior of a protein and hence its thermostability would be enhanced if the protein were made more rigid. This is provided the conformations still allowed were the most restrictive of solvent access to the "interior" of the protein, i.e. were the most compact ones. In this connection, the sedimentation constants of the two less thermstable formyltetrahydrofolate synthetases are significantly lower (by 6-10%) than those of the more thermostable proteins, although all four enzymes have the same molecular weight, within experimental error[10]. So the fluorescence data are at least consistent with the sedimentation data in rationalizing the observed thermostabilities.

Another difference we have found between the two more thermostable enzymes and the two less thermostable ones is the greater strength of the interactions between subunits in the case of the former. This seems to persist even in 4 M guanidine hydrochloride in the case of the C. thermoaceticum enzyme. This is also consistent with the arguments given above, and a survey of the literature has turned up some apparent parallels. While most enzymes from thermophiles have the same molecular weight as their counterparts from mesophiles, the enolases from two extreme thermophiles are octomers, while enolases from all mesophiles are dimers[31,32]. Wedler and Hoffman[33] have shown that glutamine synthetase from B. stearothermophilus forms higher molecular weight forms at higher temperatures which stabilize the enzyme. In this connection, Donovan and Beardslee[34] found that complex formation between trypsin and ovomucoid and soybean trypsin inhibitor stabilized all three proteins to thermal denaturation. Stronger interactions or more interactions between proteins - or different parts of a single protein - of any kind would tend to have a stabilizing effect.

In order to get some idea of the nature of the source of additional stabilization, we examined the effect of deuterium oxide on the relative thermal stabilities of the enzymes from C. formicoaceticum and C. thermoaceticum. Deuterium oxide, because it forms stronger hydrogen bonds, would tend to strengthen hydrophobic interactions in proteins and to weaken intraprotein hydrogen bonding[35].

We determined the stability of the C. thermoaceticum and C. formicoaceticum enzymes in 0, 25 and 50% deuterium oxide at 60° (Figure 5). Deuterium oxide definitely destabilized the mesophilic enzyme while having no initial effect on the thermophilic. We do not believe these results are an artifact of the small effect of deuterium oxide on the pH (less than 0.3 pH units). No destabilization was seen at 4°. This result agrees with the idea that the C. thermoaceticum enzyme is stabilized by hydrophobic interactions, but more work must be done to confirm and extend these results, particularly to the other synthetases. Of course, it is quite possible that different thermophilic proteins will turn out to have different sources of additional thermostabilization.

[30]Lakowicz, J. R. and Weber, G. Biochemistry, 12, 4171 (1973).
[31]Stellwagen, E., Cronlund, M. M. and Barnes, L. D. Biochemistry, 12, 1552 (1973).
[32]Barnes, L. D. and Stellwagen, E. Biochemistry, 12, 1559 (1973).
[33]Wedler, F. C. and Hoffman, F. M. Biochemistry, 13, 3215 (1974).
[34]Donovan, J. W. and Beardslee, R. A. J. Biol. Chem., 250, 1966 (1975).
[35]Lewin, S. Displacement of Water and Its Control of Biochemical Reactions, Academic Press, London (1974).

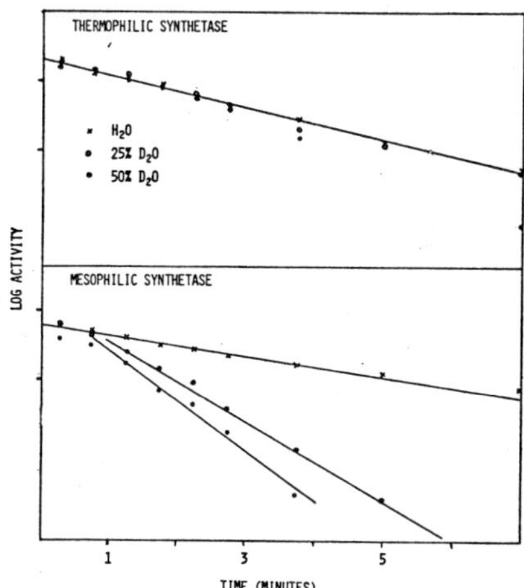

Figure 5. Effect of Deuterium Oxide on the Thermal Stability of Formyltetrahydrofolate Synthetase from C. formicoaceticum (lower) and C. thermoaceticum (upper). The buffer was 0.1 M Tris-HCl, pH 7.4 and the temperature was 60°. The experiment was carried out otherwise as described in the legend to Figure 4.

Comparison of Chemical Properties. The comparative amino acid composition data (Table IV) show correlations with thermal stability and an increased content of isoleucine, arginine and possibly tryptophan and a lower content of phenylalanine and aspartic acid. However, Singleton and Amelunxen, in an extensive survey of amino acid compositions, found no general correlation between composition and thermal stability[36]. Again, it is possible that different proteins from thermophiles will have different "mixes" of sources of thermal stability.

SUMMARY

Formyltetrahydrofolate synthetase from the mesophiles Clostridium cylindrosporum, C. acidiurici, and C. formicoaceticum and the thermophile C. thermoaceticum have been purified to homogeneity, the first two by Rabinowitz and Himes and their coworkers and the latter two by ourselves. The physical properties of these proteins are very similar. All four enzymes are tetrameric and all are activated by NH_4^+ or K^+, and the mechanism of this activation always involves a decrease in K_m for formate. The enzyme from C. formicoaceticum is more thermostable and has a higher temperature optimum than the C. cylindrosporum or C. acidiurici enzymes and the C. thermoaceticum enzyme is the most thermostable. Comparisons of the amino acid compositions indicate possible correlations between thermostability and an increased content of isoleucine, arginine and tryptophan and a decreased content of phenylalanine and aspartic acid. Both of the less thermostable enzymes dissociate above 37° if NH_4^+ or K^+ are removed, but neither of the more stable

[36]Singleton, R., Jr. and Amelunxen, R. E. Bacteriol. Rev., 37, 320 (1973).

enzymes do. In 25% or 50% D_2O, the C. formicoaceticum enzyme is destabilized at elevated temperatures, relative to the C. thermoaceticum enzyme. It is possible that stronger hydrophobic interactions between subunits are responsible for increased thermostability in these enzymes.

ANAPLEROTIC ENZYMES OF ACETATE AND PYRUVATE METABOLISM: DISTINCTIVE CHARACTERISTICS IN BACILLUS STEAROTHERMOPHILUS

T.K. SUNDARAM, SUSAN LIBOR and R.M. CHELL
Department of Biochemistry,
University of Manchester Institute of Science and Technology,
Manchester M60 1QD,
ENGLAND.

INTRODUCTION

Until recently, the unusual thermostability of proteins from thermophilic organisms was expected to be the consequence of gross peculiarities in their structure, not shared by their mesophilic counterparts (11). However, in a large number of instances, enzymes from thermophiles have appeared similar to cognate mesophile enzymes in most properties. This has led to the recognition that the structural peculiarity underlying the stability of thermophile proteins may be subtle (11). Indeed it has been argued on theoretical grounds that very small structural changes could account for the observed differences in stability between thermophile and mesophile proteins (14). Moreover the degree of thermal resistance can vary from one thermophile protein to another, and it is probable that in the evolution of thermophiles the balancing of the two requirements of thermostability and catalytic efficiency of their proteins, possibly mutually conflicting, was a key factor. Thus it is likely that no single general mechanism will explain the stability of all thermophile proteins. There is a case therefore, for examining a wide variety of thermophile enzymes for unusual characteristics. Our study of isocitrate lyase (L_s-isocitrate glyoxylate-lyase, EC 4.1.3.1), malate synthase (L-malate glyoxylate-lyase [CoA-acetylating], EC 4.1.3.2) and pyruvate carboxylase (pyruvate: CO_2 ligase (ADP), EC 6.4.1.1), all enzymes involved in the generation of C_4 compounds to meet the anaplerotic (6) needs of metabolism, reveals distinctive features in the first two enzymes and a possibly distinctive feature in the carboxylase, other than thermostability, that are not apparently possessed by their counterparts in mesophiles.

MATERIALS

A wild-type prototrophic strain of *Bacillus stearothermophilus* var *nondiastaticus* (4) and a mutant derivative, PC2 NG35, which is devoid of pyruvate carboxylase activity and derepressed for isocitrate lyase and malate synthase (12), were the organisms used in this work.

RESULTS AND DISCUSSION

Isocitrate lyase

The enzyme was purified to the same specific activity from both the wild-type strain and mutant PC2 NG35 (Table 1) and was judged to be homogeneous by the following criteria. It sedimented in the ultracentrifuge yielding a single, symmetrical peak of protein (Fig. 1), yielded a single band upon disc electrophoresis in polyacrylamide gel (Fig. 2) - a similar result was obtained upon electrophoresis on cellulose acetate strips at several pHs - and gave rise to a single polypeptide band when denatured with sodium dodecyl sulphate (SDS)-mercaptoethanol and then electrophoresed in SDS-polyacrylamide gel and also when carboxymethylated in guanidinium chloride and electrophoresed in the SDS-gel (Fig.2). It may be inferred from the results of the electrophoretic analysis of the denatured enzyme that the native enzyme contains only one kind of subunit. The data on the purification of the enzyme (Table 1) suggest that the lyase forms about 16% of the proteins in the cell-free extract of mutant PC2 NG35.

Table 2 summarizes several molecular and kinetic properties of the thermophile isocitrate lyase and provides a comparison between it and some mesophile isocitrate lyases. The thermophilic species is quite similar to the mesophilic ones in molecular size, tetrameric structure and kinetic parameters such as pH optimum and K_m for isocitrate but differs from them in possessing an appreciably lower catalytic efficiency (turnover number). Another distinctive feature of the *B. stearothermophilus* lyase not seen in several mesophile lyases examined is that it is strongly activated by salt (Fig. 3). A variety of salts can bring about this activation, although the maximum extent of activation or the salt concentration required to elicit the maximal activation is not necessarily the same for every salt (5). Potassium chloride, sodium chloride and magnesium chloride are three of the best activators, increasing the enzyme activity about three-fold. Isocitrate lyase from another strain of *B. stearothermophilus* (NCA1518 Ra2) is also strongly activated (Fig. 3), suggesting that this property may be a characteristic of thermophile isocitrate lyases. The activation becomes progressively weaker as the pH is raised (5). More interestingly, it is also temperature-dependent. As the temperature of the enzyme reaction is raised, the activation becomes less strong and at 55°C, the optimum temperature for the growth of the thermophilic bacterium, it is only about 20% (Fig.4). As a result the Arrhenius plot becomes non-linear in the presence of potassium chloride; the slope of the plot also decreases under these conditions indicating a drop in the temperature coefficient of the enzyme reaction (Fig. 5). Potassium chloride also stabilizes the enzyme against thermal inactivation at 55°C. However, it is likely that the effects of salt on the activity and on the stability are independent because the effect on activity, unlike that on stability, is quite weak at temperatures around 55°C.

The lower turnover number of the thermophile lyase suggests that the enzyme molecule, in its evolution, may have sacrificed some of the catalytic efficiency in order to acquire thermostability. It is not clear whether this is the general rule with all thermophile enzymes. Since the salt activation is observed with the purified enzyme as well as in cell-free extracts, it is apparently an intrinsic characteristic of the thermophile isocitrate lyase molecule rather than due to the release of the enzyme by salt from an inhibitor. A probable explanation of this effect is that the electrolyte promotes a change in conformation of the enzyme by interacting with charged groups on the protein or/and by enhancing the interaction between hydrophobic side chains. The reduced activation at the higher temperatures may be interpreted as signifying a change in the conformation of the enzyme with temperature. The property of salt activation is seen in some mesophile

Table 1. Purification of isocitrate lyase and malate synthase

Specific activity = enzyme units/mg protein. Enzyme units: Isocitrate lyase - 1 unit catalyzes the formation of 1 μmole of glyoxylate per min at pH 6.8 and 30°C; Malate synthase - 1 unit catalyzes the cleavage of 1 μmole of acetyl coenzyme A per min at pH 8 and 30°C, in the presence of glyoxylate.

Purification schedule	Wild type			Mutant PC2 NG35		
	Protein mg	Specific activity of isocitrate lyase	Specific activity of malate synthase	Protein mg	Specific activity of isocitrate lyase	Specific activity of malate synthase
Crude cell-free extract	3640	0.2	0.6	1450	0.49	1.6
Protamine sulphate treatment: supernatant fraction	3165	0.23	0.69	1347	0.53	1.7
Ammonium sulphate fractionation: precipitate at 50-65% saturation	1152	0.82	1.3	720	0.82	2.2
Effluent from phosphocellulose column at pH 6	479	1.5	2.4	402	1.5	3.7
DEAE cellulose chromatography at pH 7.5: gradient elution with NaCl	122	2.5	-	130	2.3	-
	22.8	-	24.0	31.8	-	26.6
Adsorption to phosphocellulose at pH 5.5, elution with 1M NaCl at pH 7	69	2.9	-	70	2.95	-

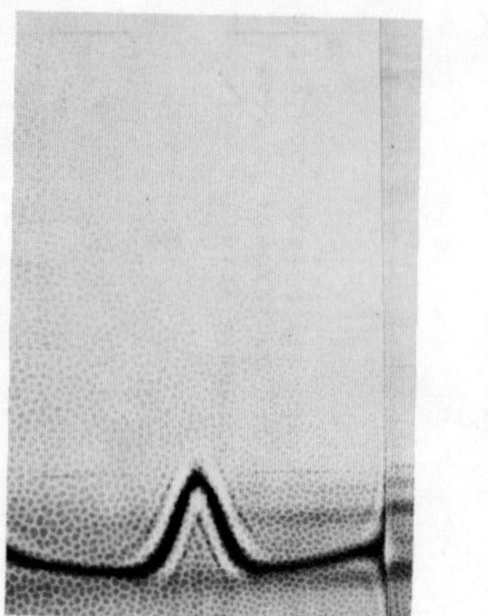

Fig. 1. Sedimentation of isocitrate lyase in the ultracentrifuge. Purified isocitrate lyase (3.2 mg/ml) in 50 mM phosphate buffer, pH 7, 1 mM EDTA, was used. Photograph on the right was taken when top speed (50740 rev./min) was attained and that on the left was taken 40 min later.

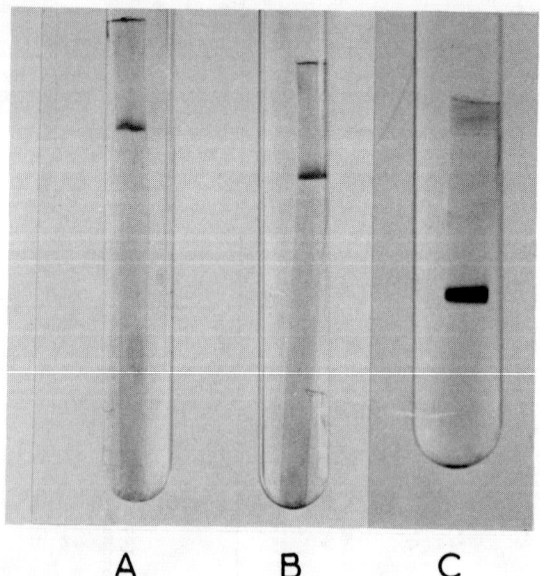

Fig. 2. Electrophoretic analysis of isocitrate lyase. A: Enzyme carboxymethylated in guanidinium chloride; B: Enzyme denatured in boiling SDS-mercaptoethanol; C: Native enzyme. A and B: Electrophoresis in polyacrylamide (12.5% acrylamide) containing 0.1% SDS; C: Electrophoresis

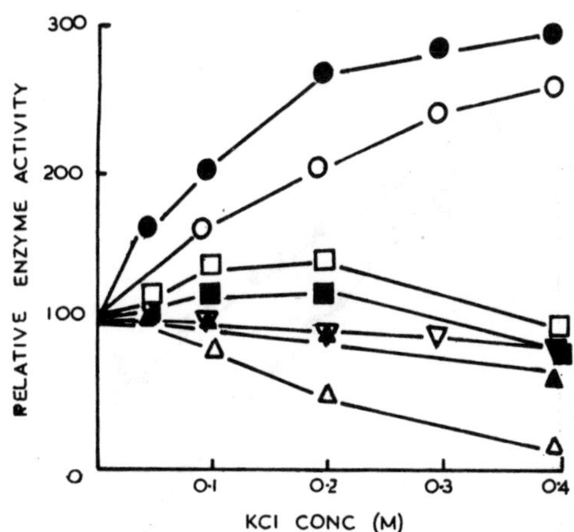

Fig. 3. Effect of KCl on the activity of thermophile and mesophile isocitrate lyases. Cell-free extracts were used. ●, B. stearothermophilus var nondiastaticus; ○, B. stearothermophilus NCA1518 Ra2; □, Bacillus megaterium; ■, Bacillus licheniformis; ▲, Escherichia coli; ▽, Pseudomonas indigofera; △, Aspergillus nidulans.

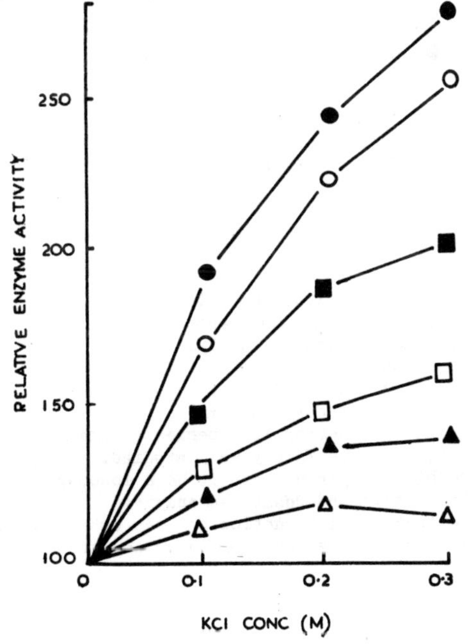

Fig. 4. Effect of KCl on B. stearothermophilus var nondiastaticus isocitrate lyase at different temperatures. A dialyzed cell-free extract was used.
●, 22°C; ○, 30°C; ■, 40°C; □, 45°C; ▲, 50°C; △, 55°C.

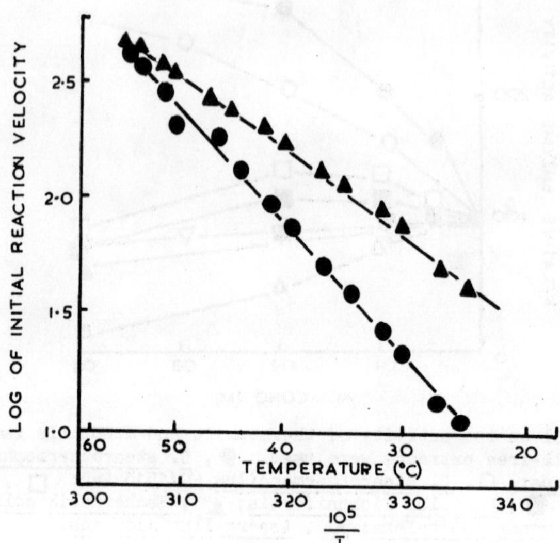

Fig. 5. Arrhenius plots of thermophile isocitrate lyase activity assayed in the presence and absence of KCl. ●, no KCl; ▲, 0.4 M KCl.

isocitrate lyases in an attenuated, incipient form (Fig. 3). It is likely that the events in evolution that conferred thermal resistance on the isocitrate lyase molecule concomitantly caused an accentuation of the salt activation characteristic.

Malate synthase

This enzyme, like isocitrate lyase, was isolated pure from the wild-type strain and mutant PC2 NG35 (Table 1). The final preparation was homogeneous by the electrophoretic and ultracentrifugation criteria applied to the lyase (Figs. 6 and 7). In the mutant malate synthase appears to constitute about 6% of the extractable proteins. Some molecular and kinetic properties of this enzyme are presented in Table 3. Since no mesophile malate synthase has been characterized in great detail, comparison with the thermophile species becomes difficult. However, the synthase from yeast, which is fairly well studied, is an oligomer (10); by contrast the thermophile enzyme appears to be a monomer as suggested by the molecular weights of the native and denatured enzymes (Table 3), unless it is considered possible that the interaction between subunits in the enzyme molecule is too strong to be disrupted by the denaturing conditions of

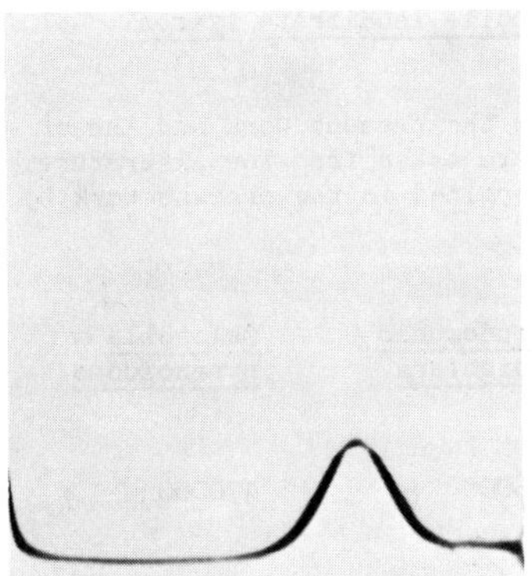

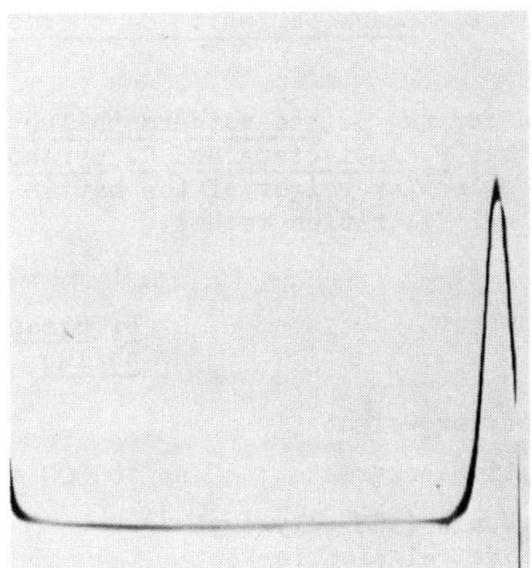

Fig. 6. Sedimentation of malate synthase in the ultracentrifuge. Purified malate synthase (6 mg/ml) in 0.1 M Tris-HCl buffer, pH 8, was used. Photograph on the right was taken when top speed (50740 rev./min) was attained and that on the left was taken 80 min later.

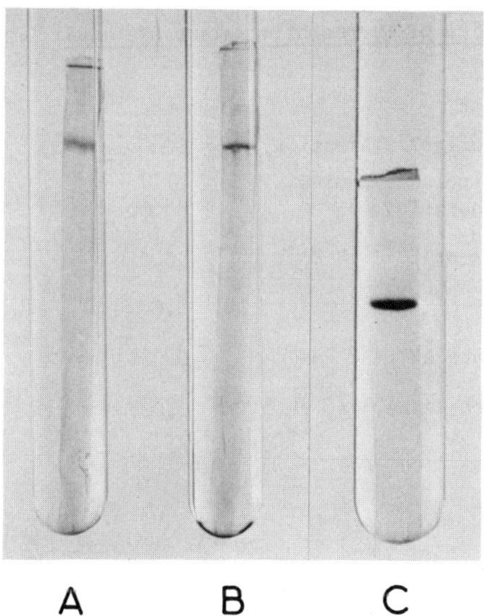

Fig. 7. Electrophoretic analysis of malate synthase. A: Enzyme carboxymethylated in guanidinium chloride; B: Enzyme denatured in boiling SDS-mercaptoethanol: C: Native enzyme. A and B: Electrophoresis in polyacrylamide (12.5% acrylamide) containing 0.1% SDS; C: Electrophoresis in polyacrylamide (7.5% acrylamide) pH 9.2.

Table 2. Characteristics of thermophile and mesophile isocitrate lyases

Data for the B. stearothermophilus enzyme are from the present work and those for the P. indigofera and C. pyrenoidosa enzymes are taken from the literature. The molecular weight of the native enzyme was determined in the present work by the gel filtration method.

Parameter	B. stearothermophilus	Pseudomonas indigofera	Chlorella pyrenoidosa
Molecular weight			
Native enzyme	180000 ± 18000	206000	170000
Subunit (SDS-polyacrylamide gel electrophoresis)	48000 ± 3300	48200	
Turnover number at 30°C	1440 at pH 8	7300 at pH 7.7	5950 at pH 7.6
pH optimum	8	7.7	7.6
K_m for isocitrate at pH 6.8 (mM)	0.021	0.04	0.023

Table 3. Characteristics of thermophile malate synthase

Molecular weight	
Native enzyme (gel filtration)	62000 ± 6200
Subunit (SDS-polyacrylamide gel electrophoresis)	58000 ± 4000
Turnover number at 30°C pH 8	1600
pH optimum	8.6
K_m for glyoxylate at pH 8 (mM)	0.088
K_m for acetyl coenzyme A at pH 8 (mM)	0.008

our experiments (see Fig. 7).

The thermophile malate synthase possesses a high degree of resistance to thermal denaturation, losing less than 5% of its activity in 1 hour at 60°C (Fig. 8). This unusual stability may derive, at least in part, from its monomeric structure. In a study of bacterial ferredoxin Perutz and Raidt (9) suggest, on the basis of an atomic model of this protein, which contains a single polypeptide chain, that in monomeric enzymes endowed with extraordinary thermal stability the extra energy of stabilization can be provided without disturbance of the tertiary structure by a few extra salt bridges on the molecular surface. A prediction from this conclusion is that increased ionic strength would weaken the salt bridges and labilize the protein structure. This prediction is borne out with the thermophile malate synthase: as shown in Fig. 8, the addition of potassium chloride results in a pronounced acceleration of the thermal decay of this enzyme.

Pyruvate carboxylase

The procedure for purifying this enzyme from the wild-type strain consisted of the following steps: (i) treatment of the cell-free extract with protamine sulphate to remove nucleic acids, (ii) precipitation of the enzyme with ammonium sulphate, (iii) ion-exchange chromatography on phosphocellulose, (iv) chromatography on DEAE-cellulose and (v) gel filtration through Sephadex G-200 or Sepharose 6B. It yielded a preparation that was homogeneous by the criteria applied to isocitrate lyase and malate synthase (Fig. 9); an additional criterion was that gel filtration of the carboxymethylated enzyme through Sepharose 6B in 6 M guanidinium chloride in acetate buffer (pH 4.6) revealed a single polypeptide component. More recently we have developed a specific affinity chromatography step on Sepharose-avidin for the purification of the enzyme (7).

Some molecular parameters of the thermophile pyruvate carboxylase are given in Table 4. The enzyme generally resembles several mesophile pyruvate carboxylases in molecular size, tetrameric structure, biotin content (1 molecule/subunit), metal content (1 atom/subunit) (Table 4) and being strongly activated by the effector, acetyl coenzyme A (3). The electrophoretic and gel filtration analyses (see above) of the denatured enzyme suggest that the native enzyme contains only one kind of subunit. Apart from its higher resistance to thermal denaturation vis-a-vis its mesophilic counterparts, the thermophile carboxylase is distinguished by its greater insensitivity to other denaturing and inhibitory agents, as shown by the comparative data presented in Table 5.

Other biotin-containing carboxylases such as acetyl coenzyme A carboxylase of *Escherichia coli* (1) and methylmalonyl coenzyme A: pyruvate transcarboxylase of *Propionibacterium shermanii* (15) have been dissociated into three kinds of subunit, each fulfilling a distinct function. One subunit, containing the biotin, serves as carrier for CO_2, another catalyzes the attachment of CO_2 to the carrier protein and the third mediates the transfer of CO_2 from the carrier to the eventual acceptor substrate. In view of the similarity in the mechanism of action of the biotin-containing carboxylases, it has been

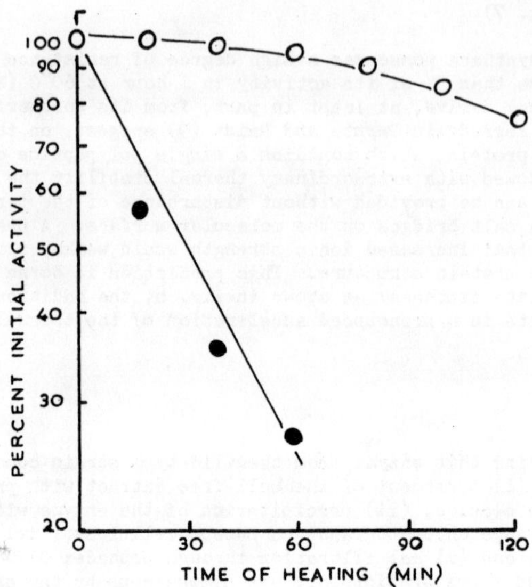

Fig. 8. Effect of KCl on the stability of malate synthase at 60°C. Purified malate synthase (28 µg/ml) in 50 mM Tris-HCl, pH 8, was used. O, no KCl; ●, 0.2M KCl.

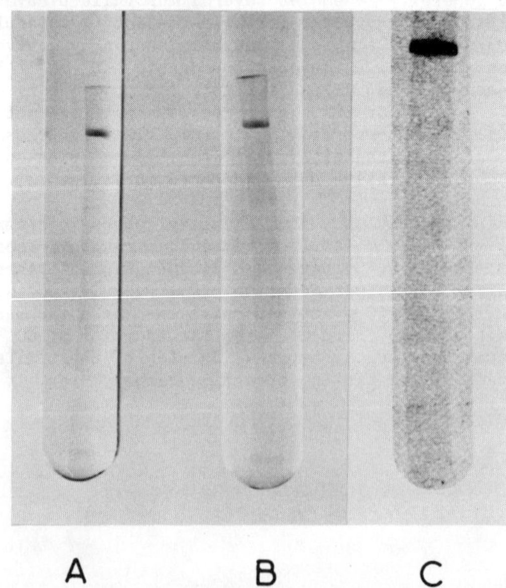

Fig. 9. Electrophoretic analysis of pyruvate carboxylase. A: Enzyme denatured in boiling SDS-mercaptoethanol; B: Enzyme carboxymethylated in urea; C: Native enzyme. A and B: Electrophoresis in polyacrylamide (12.5% acrylamide) containing 0.1% SDS; C: Electrophoresis in polyacrylamide (acrylamide 6%) pH 8.9.

Table 4. Molecular parameters of thermophile pyruvate carboxylase

The molecular weight of the native enzyme was determined by gel filtration through Sepharose 6B and confirmed independently by determining the sedimentation coefficient in the ultracentrifuge. The biotin content was obtained by measuring the radioactivity of the enzyme labelled with ^{14}C - biotin; the result was confirmed by microbiological assay after hydrolysis of the enzyme. The zinc content was determined by atomic absorption spectral analysis.

Molecular weight	
Native enzyme	600000 ± 40000
Subunit (SDS-polyacrylamide gel electrophoresis)	150000 ± 15000
Biotin content	1 mole per 160000 g
Zinc content	1 g-atom per 145000 g

suggested (1) that all these carboxylases might have a similar subunit constitution. Our results suggest that the thermophile pyruvate carboxylase is made up of but one type of subunit. The situation in regard to mesophile pyruvate carboxylases seems unclear. The report of Barden and Taylor (2) suggests that they contain only one type of polypeptide; the results of McClure *et al.* (8) and of Warren and Tipton (13) indicate, on the other hand, that they may be composed of more than one type of subunit. It is tempting to speculate that, in the more hardy thermophile system, the three functions in the carboxylation reaction reside in a large single polypeptide represented by the quarter-molecular subunit identified in our experiments; this subunit may comprise the three putative components held together by covalent linkage.

In conclusion, in the thermophile system the salt activation property of isocitrate lyase, the monomeric constitution of malate synthase and possibly the dissociability of pyruvate carboxylase into only one kind of subunit appear to be unique features related to the unusual stability of these enzymes.

ACKNOWLEDGEMENTS

This investigation was supported by the Science Research Council, Great Britain through grant No. B/RG/3594/6 and by the Royal Society through an equipment grant.

Table 5. Effect of denaturing agents and sulphhydryl reagents on the activity of pyruvate carboxylase

Treatment	Effect on pyruvate carboxylase from	
	Chicken liver	*B. stearothermophilus*
Exposure to low temperature	Inactivation	No inactivation
Urea	Inactivation accompanied by dissociation of enzyme with 0.4 M urea at 23°C; acetyl coenzyme A protects against the inactivation. Reversibility of the inactivation has not been demonstrated.	No inactivation with 0.4 M urea. 7.5 M urea at pH 7 inactivates; inactivation is completely reversed upon dilution of the urea.
Sodium dodecyl sulphate (0.025%)	Irreversible inactivation	Irreversible inactivation
pH	Reversible inactivation at pH 6.2 and 7.9; acetyl coenzyme A protects against the inactivation.	Little inactivation at pH 6.2 and 7.5. 85% of activity is retained after exposure to pH 5 and pH 11 for 2 min.
Sulphhydryl reagents [p-chloromercuribenzoate or 5,5' dithiobis (2-nitrobenzoic acid)]	10 to 20-fold molar excess of reagent inactivates reversibly. Irreversible inactivation with over 100-fold excess of reagent.	No inactivation with 10 to 20-fold molar excess of reagent. With over 100-fold excess of reagent, inactivation which is completely reversed upon treatment with excess of dithiothreitol.

REFERENCES

1. Alberts, A.W., Nervi, A.M. and Vagelos, P.R. (1969) Proc. Natl. Acad. Sci. U.S.A. 63, 1319.

2. Barden, R.E. and Taylor, B.L. (1973) Fed. Proc. 32, 510.

3. Cazzulo, J.J., Sundaram, T.K. and Kornberg, H.L. (1970) Proc. Roy. Soc. Lond. B. 176, 1.

4. Epstein, I. and Grossowicz, N. (1969) J. Bacteriol. 99, 414.

5. Griffiths, M.W. and Sundaram, T.K. (1973) J. Bacteriol. 116, 1160.

6. Kornberg, H.L. (1966) Essays in Biochemistry 2, 1.

7. Libor, S., Warwick, R. and Sundaram, T.K. (1975) FEBS Letters (in Press).

8. McClure, W.R., Lardy, H.A. and Kneifel, H.P. (1971) J. Biol. Chem. 246, 3569.

9. Perutz, M.F. and Raidt, H. (1975) Nature 255, 256.

10. Schmid, G., Durchschlag, H., Biederman, G., Eggerer, H. and Jaenicke, R. (1974) Biochem. Biophys. Res Commun, 58, 419.

11. Singleton, Jr., R. and Amelunxen, R.E. (1973) Bacteriol. Revs. 37, 320.

12. Sundaram, T.K. (1973) J. Bacteriol. 113, 549.

13. Warren, G.B. and Tipton, K.F. (1974) Biochem. J. 139, 297.

14. Wilkinson, S. and Knowles, J.R. (1974) Biochem. J. 139, 391.

15. Wood, H.G. Ahmad, F., Jacobson, B., Chuang, M. and Brattin, W. (1975) J. Biol. Chem. 250, 918.

OXYGEN INDUCED TRANSFORMATION OF THE NADH-NITRATE OXIDOREDUCTASE FROM BACILLUS STEAROTHERMOPHILUS

R.J. Downey* and M.P. Stambaugh
Department of Zoology and Microbiology
Ohio University, Athens, Ohio 45701

Introduction

The elusive character of the basis for heat stability of proteins in thermophilic organisms is universally acknowledged. Thermophiles appear to beat the high cost of faster metabolic rates by high turnover of many intracellular components including certain enzymes (2). A lower level of dissolved oxygen is also a fact of life at elevated temperatures. In many types of procaryotes, denitrification is a useful alternative to limited atmospheric oxygen. There may be appreciable selective pressure for retention of nitrate respiration among thermophilic microorganisms as a prime energetic expedient. The postulated mechanisms whereby bacteria continually adjust their enzymatic composition to cope with a variable environment are increasing in number and complexity. Among the most important of these is the manner in which the cell recognizes and responds to continually varying levels of available terminal oxidant whether it be oxygen, nitrate, nitrite, thiosulfate, sulfite or tetrathionate.

Oxygen is a strong inhibitor of denitrification in the *Bacillus* having a significant effect on the proportion of NADH-nitrate reductase enzyme that is lodged in the cell membrane (5,6). We have examined the behavior of induced and uninduced cultures of *Bacillus stearothermophilus* NCA 2184 as well as isolated enzyme in response to a carefully regulated input of oxygen under rigidly controlled conditions of growth or in vitro assay. We report here some evidence for the discreet nature of the NADH-nitrate reductase in the membrane and its reversible transformation from active to inactive states in the presence of controlled and continuously measured oxygen tension.

Methods and Materials

The strain of thermophile used in this study was *Bacillus stearothermophilus* NCA 2184. The growth medium, manner of cultivation, enzyme assay and protein assay procedures have been described previously (5,6,7). In the experiments reported here, the growth medium (CHSN) contained 10.0 g casein hydrolysate, 5.0 g sucrose, 10.0 g yeast extract, 1.0 g KNO_3, 7.0 mg $FeCl_3 \cdot 6H_2O$, 15.0 mg $MnCl_2 \cdot 4H_2O$ and water to 1 liter. Where indicated KCL replaced KNO_3 in medium lacking the inducer (CHSCl). The temperature utilized throughout these experiments was 65 C.

Oxygen partial pressure (pO_2) was monitored with a Clark electrode in combination with a Beckman Model 160 Physiological Gas Analyzer. The oxygen partial pressure was maintained in cultures or test suspensions at a fixed input level by use of a New Brunswick Model DO-60 Controller. Antifoam (HODAG Type KG, Hodag Chemical Co., Skokie, Ill) was used only when necessary and was added to the medium prior to inoculation.

Electrophoresis of Triton dissociated membrane proteins was performed on 12.5 % running gels and 3.0 % stacking gels according to the procedure reported by Beard and Connolly (1). Gels were scanned with an IsCo U.V. ultraviolet analyzer (Instrumentation Specialties Co., Lincoln, Nebraska). Turbidometric determination of cell density was performed on a Bausch-Lomb Spectronic 20 at 625 nm. An absorbance of 0.1 represented a viable cell density of approximately 2×10^7 cells/ml. A unit of enzyme activity is defined as the production of one nanomole of nitrite produced per min except where indicated. The enzyme specific activity is given in units per mg protein. The protein concentration was measured by the method of Lowry *et al* (12).

Results and Discussion

Since nitrate reductase can be a factor in its own synthesis, the minimum inducing nitrate level is difficult to determine. The induced synthesis of nitrate reductase may be restricted by the low basal level of the enzyme or an insufficient concentration of nitrate. We asked whether or not synthesis of the enzyme can be initiated and sustained by exposure to a small limited amount of nitrate for a short period. The accumulated biosynthetic capacity for nitrate reductase was examined after withdrawal of the inducer nitrate which is a non-replenishable oxidant under the conditions of the experiment. We did this by a method in which cells were retrieved by rapid filtration following exposure to the inducer for 5 min. The induced formation and continued synthesis of the NADH-nitrate reductase required the presence of nitrate as expected (Fig. 1). Little, if any, enzyme was formed in the presence of 10 mM nitrate if it was removed after a 5 min pulse. As a result of this, there is insufficient nitrate respiration for

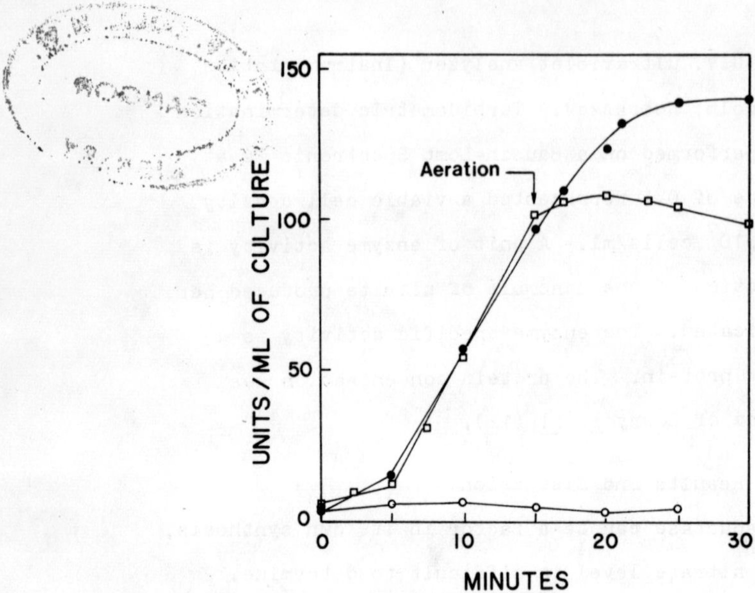

Fig. 1. Induction of the NADH-nitrate reductase in a 2 liter culture of Bacillus grown aerobically in the absence of nitrate and shifted to anaerobic conditions by gassing with a nitrogen:CO_2 mixture. At zero time, KNO_3 was added and enzyme was assayed in samples obtained at the indicated intervals (●). A similar culture (○) was afforded KNO_3 (10 mM) for a period of 5 min and then filtered free of inducer on millipore membranes (0.45 um) with pre-warmed, deoxygenated medium prior to resuspension in fresh CHSCl medium. Vigorous aeration (input pO_2 125 mm Hg) was initiated where indicated in a companion culture growing on regular CHSN medium (□).

the culture's growth. However in the case where nitrate is kept available for possible transport and induction, the synthesis is initiated quickly after addition of the inducer. It typically lagged briefly and then increased at a linear rate to a steady state, the level of which was determined principally by the available nitrate. Aeration at a rate sufficient to maintain the pO_2 at an input partial pressure of approximately 125 mm Hg resulted in the repression of further enzyme synthesis (Fig. 1 open squares). It should be noted that these cells were fully induced

UNIV. COLL. N.W. BANGOR LIBRARY

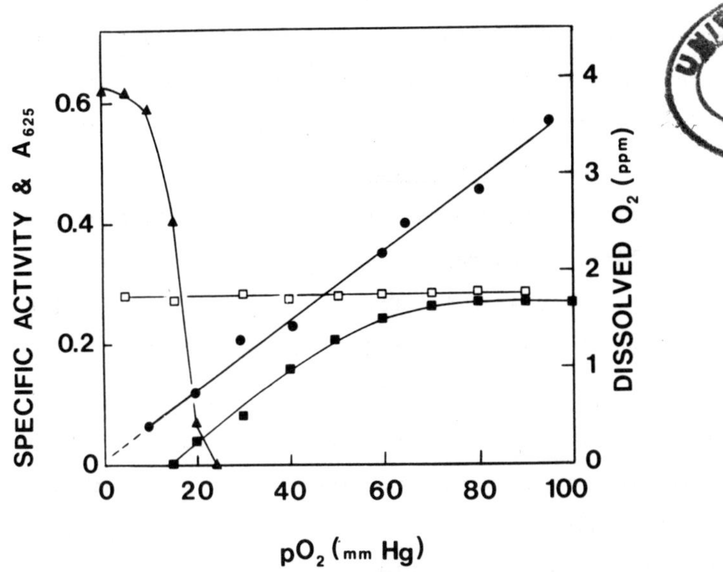

Fig. 2. Growth and NADH-nitrate reductase synthesis in relation to maintainance of fixed input pO_2. A fully induced inoculum (500 ml) from a culture in mid-exponential growth on CHSN medium was added to 6 liters CHSN medium in a small fermentor to give a starting density of 2 X 10^7 cells/ml. The input medium flow was maintained at a dilution of 0.25 hr^{-1}. The oxygen sensing probe was located 8.5 inches from the bottom of the fluid mass. The sparging gas was a mixture of 95 % nitrogen and 5% CO_2 or air and was delivered at an initial rate of 1400 cc/min to establish an inducing (1 ppm) or totally repressing (5 ppm) pO_2. The impeller rate was 300 rpm and the temperature was 65 C. Cell density with nitrate (□) and without (■), enzyme activity (▲) in units of millimoles/min, dissolved oxygen (●).

by previous growth in static culture in the presence of nitrate before subculture aerobically in the absence of nitrate. A typical aerobic culture, even if only partially induced, displayed hypoxic qualities due to the sudden decrease in available oxygen. In all such experiments, we noted a sharp decline in viability even in the presence of nitrate if there was insufficient oxygen to drive the biosynthesis of the transition to denitrification by the majority of the population. The above strain exhibited a meager fermentative capability with glucose, maltose, sucrose and does not apparently under the conditions of the experiment use the classical lactate fermentations to its advantage as do the Escherichia (15) and the Micrococcus (3).

In an effort to obtain more definitive information on the oxygen levels which afford repression of the nitrate reductase, we have examined the rate of its formation following induction in a sucrose limited chemostat with a dilution of 0.25 hr^{-1}. The population was subjected to a specific input pO_2 during induction and assays of enzyme activity were performed on samples retrieved at appropriate intervals during the experiment. The synthesis of nitrate reductase was totally repressed at dissolved oxygen concentrations of 1 ppm or greater (Fig. 2). Although the redox potential (rH) in millivolts of the fluid mass is not shown here for the sake of clarity in the figure, it was observed to remain steady at +200 to +300 mV until the input pO_2 was less than 20 mm Hg. It then fell in a steep transient which approached -180 to -200 mV at the point of detectable enzyme repression in our experiments. It is conceivable that it is the intracellular redox potentials, especially those which are more negative than -100 mV, that are actually repressing synthesis of the respiratory enzymes in bacteria.

The difference in the ability of an induced culture to achieve its maximum density in the presence of a prescribed pO_2 is shown in Fig. 2. The influence of the available nitrate on the growth of the culture is also evident here. A glucose or sucrose limited nitrate induced culture can achieve its maximum density independent of the pO_2, whereas an uninduced culture is oxygen limited at 1.5 ppm or less of dissolved oxygen (Fig. 2).

The thesis that functional enzyme is not formed and incorporated into the cell membrane under conditions of oxygen repression was supported by another kind of experimental approach. Electrophoresis of the enzyme from Triton dissociated protoplast membrane resulted in distinct differences in the amount of enzyme protein indirectly determined to be the nitrate reductase enzyme. A direct U.V. scan of 12.5% polyacrylamide gels at 250 nm revealed the position of the reduced methyl viologen-nitrate reductase protein among the membrane proteins separated (Fig. 3). The electrophoresis of proteins from Triton treated membrane from oxygen repressed cells indicated a significantly decreased level of the enzyme protein.

Although the macromolecular character of the membrane bound respiratory nitrate reductase is currently more elusive than that of the soluble assimilatory enzyme in algae or fungi (8,13), it may have one important mode of behavior in common and that is the effector role of the molybdoprotein itself in relation to its *in situ* catalytic activity. The oxidation state of the assimilatory enzyme is important in that it determines whether nitrate will effectively oxidize NADH, NADPH or $FADH_2$ (11,13). Substrate inhibition has been noted (14) and some form of allosteric behavior in relation to the existing level of terminal oxidant (NO_3^- or $\frac{1}{2}O_2$) seems reasonable. The nature of enzyme turnover in the membrane has not been examined directly (i.e., by use of isotope labeling) but we have considered the integration of induced nitrate reductase

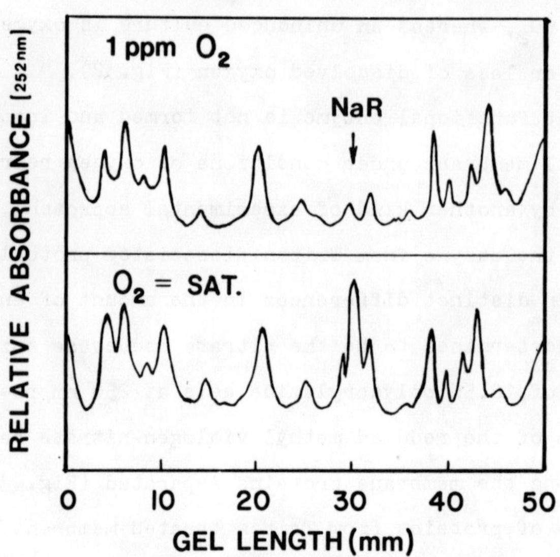

Fig. 3. Direct ultraviolet scan of Triton dissociated membrane proteins from cells grown under conditions of induction (top) or repression (bottom). The position of MVH-nitrate reductase was noted after conducting the conventional assay on material eluted from 2 mm serial slices of the gels.

to be essentially a matter of either replacement or dilution depending upon certain growth conditions. Our work indicates that the NADH-nitrate reductase can be repressed without active cell division and dilution by growth.

The question of whether or not the membrane bound NADH-nitrate reductase can be reversibly inactivated by oxygen _in vivo_ and _in vitro_ was examined by subjecting a preparation of the enzyme to a repressing oxygen tension (pO_2 = 100 mm Hg). The half life of the enzyme in intact membrane was noted to be about 10 min (Fig. 4) with virtually 100% inactivation of the enzyme in 30 min under the conditions of the experiment. Evidence that the enzyme is not degraded under these conditions of repression lies in the fact

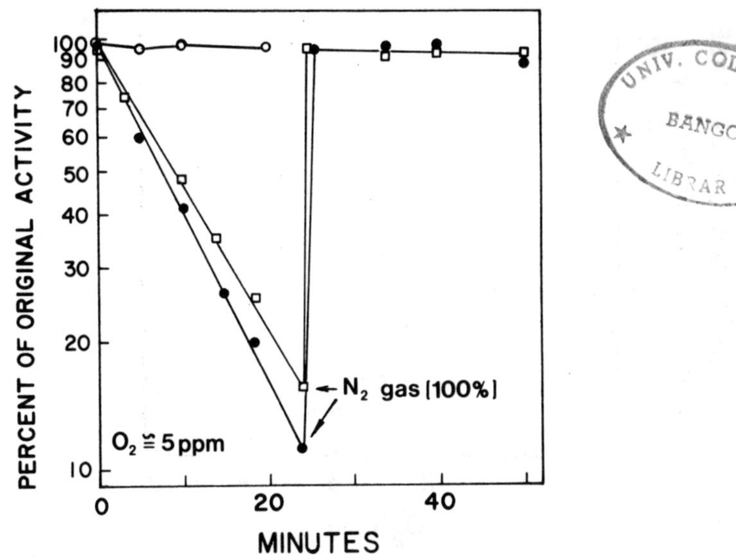

UNIV. COLL. N.W.
BANGOR
LIBRARY

Fig. 4. Inactivation and reactivation of the NADH-nitrate reductase in whole cells (●) and in isolated protoplast membranes (□) by inducing and repressing pO_2. The enzyme activity is plotted on a semilogarithmic scale. The activity in samples from the unaerated control mixture (○) is shown for the first 20 min. A 10 ml mixture of a washed, nitrate induced protoplast membrane fraction (10) containing 50 mM phosphate buffer, pH 7.0, 10 mM KNO_3 and 170 to 250 mg protein was kept saturated with air at 65 C by a fritted glass gas dispersion tube. At the indicated times samples were removed for assay of enzyme activity under aerobic conditions up to the addition of nitrogen and under anaerobic conditions each time thereafter.

that total restoration of activity resulted when the reaction mixture was gassed with a nitrogen:CO_2 (95:5%) mixture. It can be seen that the NADH-nitrate reductase in whole cells and isolated protoplast membranes behaves in a similar fashion to this treatment (Fig. 4).

The reversible inactivation with oxygen led us to attempt experiments which might provide some evidence for allosteric behavior of the enzyme. The input pO_2 (aeration) was maintained at

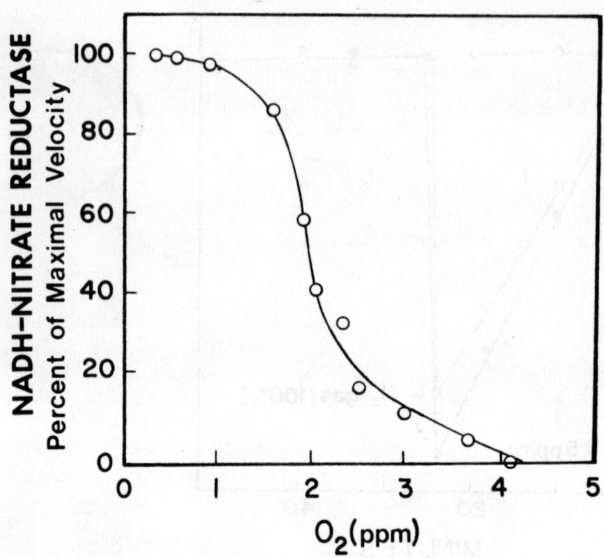

Fig. 5. Influence of oxygen partial pressure on NADH-nitrate reductase activity in a washed protoplast membrane fraction (10) incubated with fixed concentrations of NADH (600 µM) and KNO_3 (10 mM). The oxygen figures are relative values given in relation to the maximum oxygen (calibration level = 5 (± 0.25 ppm) in air saturated CHSN medium at 65 C. The figures carry corrections for solute and vapor pressure. The maximum specific activity of the enzyme preparations used for these experiments averaged 50 µmoles min^{-1}.

a steady level (meter fluctuations of less than 10% of the full measurement scale) for 10 min prior to initiation of the reaction. Sparging was halted, substrate was added quickly with a syringe and the velocity of the enzyme reaction was recorded in the presence of constant concentrations of either NADH or nitrate. The results show that the enzyme activity with NADH as substrate is maximal at oxygen concentrations of 1 ppm or less. It is also evident that the enzyme experiences a sharp transformation in the range of 1 to 2 ppm oxygen. This correlates closely with the range of oxygen concentrations which have been noted to influence synthesis. In these experiments, the oxygen level inactivating the enzyme is about twice repressing synthesis of the enzyme in a chemostat culture (see Fig. 2). This is open to a variety of interpretations but in the main it is in agreement with the hypothesis that oxygen has a definite modifying role in respect to the activity of the nitrate reductase enzyme. There is currently no direct evidence that this protein also functions as an effector in repressing the enzyme's synthesis.

How the oxygen tension or the intracellular oxidation-reduction potential ultimately achieves repression is unknown. The answer will doubtless require application of techniques which have proven useful in whole cell pH and rH measurements (9). The use of appropriate dyes and carefully derived cell samples will provide more direct evidence that the intracellular oxidation-reduction potential is in fact in the range noted to be effective in *in vivo* experiments.

It is quite possible that oxygen exerts its effect on another component in the membrane which in turn alters the NADH-nitrate reductase. Since a neutral detergent like Triton dissociates the structural mosaic sufficiently to disrupt the normal respiratory or denitrification sequence, it is evident that membrane integrity is crucial to the normal activity of the enzymes associated with it. The unique experiments of Souza and Esser (16) point out that the maximal and minimal growth temperatures of thermophiles seem to be regulated by the onset and conclusions of phase separations of the particular lipid components they synthesize.

How does the temperature of growth effect the activity of the membrane bound enzyme? The thermostability of the NADH-oxidase in the thermophile is closely tied to the growth temperature (17). An alkaline phosphatase and NADH-oxidase from cells grown at 55 C, 60 C, 65 C and 75 C were totally inactive at temperatures of 10 to 15 degrees higher than the growth temperature. We have noted this to be the case with the NADH-nitrate reductase also. Our measurements of dissolved oxygen (unpublished data) via the manganous ion oxidation method (18) revealed that the maximum oxygen in liquid CHSN medium at 90 C is 1.61 (±0.09) ppm. This means that the maximum available oxygen is insufficient for repression of synthesis or at the very least a controlling influence resembling that which is evident in thermophiles grown at 65 or 75 C.

If there is a broad range of temperature related adaptability, the organism seems to rely it and appears to manufacture the stability it needs. This may be the case with respect to adjustments to the concentration of the terminal oxidant available; although a 10 degree span in temperature in the range of 60 to 80 C has much less effect on the soluble oxygen than in the range of 20 to 50 C. Brock (2) has pointed our that there is no clearly established upper temperature limit for growth of thermophilic microorganisms. This is particularly germaine to the study of oxidative metabolism in these organisms because they are doubtless still evolving systems which exploit the various forms of soluble fixed oxygen as well as a variety of other electron acceptors.

If we accept the idea that nitrate predated free oxygen in evolution, then nitrate respiration has like the thermophiles been around for a long time. The preferential use of oxygen has yet to be completely elucidated. While there is comparatively little of it in solution at 100 C what there is is still capable of oxidation, as is steam. The work reported here reveals that oxygen has a dramatic effect on the use of nitrate by bacteria. Inordinately high levels of nitrate (100 mM) do not appear to produce similar effects on the use of oxygen. Further probes on the influence of oxygen are currently under investigation.

Acknowledgements

The authors are grateful for the excellent technical assistance of M.K. Young. This work was supported by Public Health Service Research Grant AI09405 from the National Institute of Allergy and Infectious Diseases.

References

1. Beard, J.P. and J.C. Connolly, *J. Bacteriol.* 122:59 (1975)
2. Brock, T.D., *Science* 158:1011 (1967)
3. Burke, K.A. and J. Lascelles, *J. Bacteriol.* 123:308 (1975)
4. Doelle, H.W., *F.E.B.S. Letters* 49:220 (1974)
5. Downey, R.J. and D.F. Kiszkiss and J.H. Nuner, *J. Bacteriol.* 98:1056 (1969)
6. Downey, R.J. and D.F. Kiszkiss, *Microbios* 2:145 (1969)
7. Downey, R.J. and W.J. Focht, *Microbios* 11A:61 (1975)
8. Garrett, R.H. and A. Nason, *J. Biol. Chem.* 244:2807 (1969)
9. Kashket, E.R., *Biochim. Biophys. Acta.* 193:212 (1969)
10. Kiszkiss, D.F. and R.J. Downey, *J. Bacteriol.* 109:803 (1972)
11. Kiszkiss, D.F. and R.J. Downey, *J. Bacteriol.* 109:811 (1972)

12. Lowry, O.H., N.J. Rosebrough, A.L. Farr and R.J. Randall *J. Biol. Chem.* 193:265 (1951)

13. Moreno, C.G., P.J. Aparicio, E. Palacian and M. Losada *F.E.B.S. Letters* 26:11 (1972)

14. Rigano, C. and G. Alliota, *Biochim. Biophys. Acta.* 384:37 (1975)

15. Showe, M.K. and J.A. DeMoss, *J. Bacteriol.* 95:1305 (1968)

16. Souza, K.A. and A.F. Esser, *Proc. Nat. Acad. Sci.* 71:4111 (1974)

17. Wisdom, C. and N.E. Welker, *J. Bacteriol.* 114:1336 (1973)

18. Winkler, L.W., *Ber. Deutsch. Chem. Gessellschaft* 21:2843 (1888)

THERMOPHILIC PHYCOCYANIN

ORANDA H. W. KAO, MERCEDES R. EDWARDS, ROBERT MAC COLL AND DONALD S. BERNS
Division of Laboratories and Research
New York State Department of Health
120 New Scotland Avenue
Albany, New York 12201 USA

INTRODUCTION

The questions central to the understanding of the adaptive mechanism of thermophilic organisms can best be addressed by selection of a model whose properties are likely to be a particularly sensitive function of temperature. Enzymes are useful for this type of analysis because, like other proteins, they possess a myriad of conformations as well as primary, secondary, tertiary, and quaternary structures, and their temperature adaptation might be reflected in subtle changes in enzymatic properties that are conformationally dependent. However, enzymes are generally present in very low concentrations in organisms, and their functions are difficult to determine, since in vivo condtions (such as protein concentration, pH and ionic strength) are not readily equated with in vitro conditions. We have chosen to characterize those properties of a protein system that would seem to have more promise as a model system.

The protein should be one present in large concentrations and as a demonstrated structural and functional element in the organism. Ideally it should be easily isolated from organisms that are rudimentary in evolutionary terms, so that the thermophily of some of these organisms will probably be, not the result of simple selection, but a property of one of the primitive forms of that class of organism. A good phylogenetic span of organisms of all environmental types should also be examined to characterize the natural variation in properties unrelated to thermophily. Characterization of the protein under conditions of varying concentration, pH, and ionic strength should be easy to accomplish with at least some tangible correlation to the protein function.

For all these reasons we have chosen as our model the nonenzymatic chromoprotein C-phycocyanin, which is found in procaryotic blue-green and eucaryotic red algae. Living blue-green algae resemble fossils which are about 3 X 10^9 years old (1), and these algae thrive in the widest possible span of environmental extremes, including thermal waters above 70°C (2, 3). C-Phycocyanin is an accessory pigment in photosynthesis and is attached to the outer surface of the thylakoid membranes. This protein may represent up to 25% of the total dry weight of the cell (4). The visible absorption of the protein--C-phycocyanin usually has a λ_{max} at 615-620 nm--results from linear tetrapyrrole groups which are covalently attached to the apoprotein (5-7).

EXPERIMENTAL

The thermophilic algae from which the C-phycocyanin was obtained in our laboratory are <u>Cyanidium caldarium</u> (8), two strains of <u>Synechococcus lividus</u>, labeled

Sy I and Sy III (9), and two strains of Mastigocladus laminosus (10). Additional algae discussed include a halophile, Coccochloris elabens (11), and the mesophiles Lyngbya sp. (12), Anabaena variabilis (12) and Phormidium luridum (12). The culture sources and conditions and the method of biliprotein purification are given in the references. C-Phycocyanin was generally extracted from disrupted cells into sodium phosphate buffer (pH 6.0, 0.1 ionic strength) and purified by ammonium sulfate fractionation to an A_{620}/A_{280} of at least 4.0. Antiserum was produced by injecting approximately 2-3 mg of the purified protein mixed with complete Freunds adjuvant into rabbits, followed by a second injection one month later. A week after the second injection the rabbits were bled. For Ouchterlony double-diffusion experiments, performed in 1% ionagar dissolved in sodium phosphate buffer (pH 6.0, 0.1 ionic strength), 100 μg of various C-phycocyanins were placed in the antigen wells; 0.10 ml of antiserum was used.

Analytic ultracentrifugation experiments were performed on a Model E ultracentrifuge, using either schlerien optics or the photoelectric scanner, as described elsewhere (9, 13). Results from sedimentation equilibrium studies were obtained by transmitting data from the scanner, via a data mover and magnetic tape, to a computer for analysis. These experiments were performed at 280 and 380 nm using a Model E equipped with an electronic speed control.

For electron microscopy of chloroplasts, isolated thylakoids, and phycobilisomes, a cell pellet weighing about 2 g (wet weight) was suspended in sodium acetate buffer (0.05 M, pH 5.5) and passed twice through a precooled French press at 20,000 psi. The material was then centrifuged at 48,000 g (Sorvall R-C2) for 10 min. The resulting pellet, containing some intact cells, cells at different stages of disruption, intact or fragmented chloroplasts, and isolated thylakoids, was fixed in 4% glutaraldehyde in the acetate buffer for 2 hr at room temperature. The material was then centrifuged at 48,000 g and left in the same buffer for 18 hr at 4°C. Again the material was centrifuged, and the resulting pellet was embedded in 1.5% agar prepared in the same buffer. The agar blocks were washed in the acetate buffer for 20 hr at 4°C and post-fixed in 1% OsO_4 in PO_4 (0.1 M, pH 6.0) for 2 hr at room temperature, then for 18 hr at 4°C. Dehydration and embedding were performed in Epon, and the ultrathin sections were examined in a Zeiss EM9 and a Siemens Elmiskop IA electron microscope.

C-Phycocyanin aggregates were studied with the electron microscope as follows: A cell pellet (1-2 g, wet weight) was suspended in 10 ml of phosphate buffer, pH 6.0, and passed twice through a French press at 20,000 psi. The suspension was spun at 12,000 g for 10 min to remove unbroken cells. The supernatant was further cleaned by repeated centrifugation at 27,000 g until no pellet was visible. The clear blue liquid was placed on sucrose gradients (2, 1, 0.75 and 0.5 M) prepared with the phosphate buffer. The bottom dark-blue or purple band which formed between the 1 and 2 M sucrose layers (density ± 1.4) was prepared for the electron microscope, either directly or after fixation in 2% glutaraldehyde solution in the phosphate buffer. A drop of dark-blue liquid (1 mg/ml) was placed on a carbon-Formvar-coated grid and negatively stained by 1% Na-PTA. After a washing in double-distilled H_2O, the grid was examined and photographed in a Siemens Elmiskop IA at an accelerating voltage of 80 kV.

PRIMARY STRUCTURE

The amino acid compositions of a variety of C-phycocyanins, compiled from the literature, are given in Table I. The general similarity among all those listed is evident, and the observed variations generally do not exhibit a consistent trend. It has previously been noted that the combined glutamic acid and glutamine content of thermophiles exceeds that of most mesophiles (9). There is no increase in hydrophobic amino acids for thermophilic C-phycocyanins.

All C-phycocyanins are composed of two polypeptide chains (16-22), which can be separated and sequenced. Partial N-terminal sequences of the two chains of both thermophiles and mesophiles are identical (10, 23-25); results from two mesophiles (10) and a thermophile (23) are shown in Table II. Furthermore, while the α and β chains are distinct, there are portions of homology between them (23, 25). Other sequencing findings (10) demonstrate that C-phycocyanin and another biliprotein, phycoerythrin, have regions of identical sequence. The peptide sequence of the region near the tetrapyrrole groups has also been reported (26).

IMMUNOCHEMISTRY

In addition to the close relationships among C-phycocyanins revealed by their amino acid compositions and partial sequencing data, immunochemical analysis by Ouchterlony double-diffusion experiments reveals that the C-phycocyanins studied are antigenically closely related (Figs. 1 and 2). Among proteins

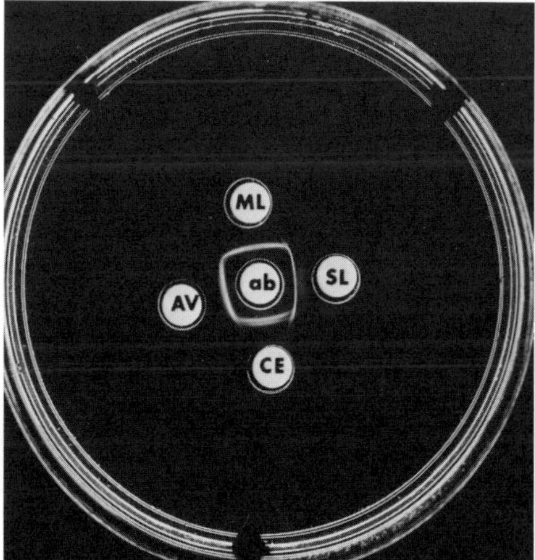

Fig. 1. Antigenic relationships shown by Ouchterlony double-diffusion experiment. Antiserum (ab) is antiphycocyanin from S. lividus (Sy III). The other wells contain C-phycocyanins from: SL, Sy III; CE, C. elabens; AV, A. variabilis; and ML, M. laminosus (strain NZ-DB2-m).

Table I. Amino acid compositions of C-phycocyanins (Number of residues per 30,000)*

Amino acid	Thermophilic			Halophilic	Mesophilic			
	Sy I (9)	Sy III (9)	*M. laminosus* (14)	*C. elabens* (11)	*Lyngbya* sp. (12)	*A. variabilis* (12)	*Synechococcus* sp. (15)	*Aphanocapsa* sp. (15)
Lys	10	11	11	10	11	11	11	10
His	3	1	2	2	2	2	1	2
Arg	20	15	16	17	16	16	17	15
Asp	27	30	32	30	33	31	36	29
Thr	14	16	17	15	17	16	16	18
Ser	17	16	15	14	20	18	21	23
Glu	30	30	27	29	21	27	19	20
Pro	11	10	9	8	9	9	9	8
Gly	21	22	19	21	23	23	23	20
Ala	41	46	45	32	42	41	51	43
Val	16	19	15	15	19	18	18	16
Met	6	7	3	8	8	3	3	5
Ile	15	17	19	10	15	15	15	14
Leu	26	25	25	22	23	24	26	22
Tyr	12	12	14	11	13	12	13	12
Phe	9	9	6	7	8	7	10	9
1/2Cys	6	2	3	2	2	2	1	2

*References given in parentheses.

Table II. Comparison of the N-terminal sequences of the α and β polypeptides of a thermophilic and two mesophilic algae*

Chain and species	Residue Number									
	1	2	3	4	5	6	7	8	9	10
α (*C. caldarium*)	Met	Lys	Thr	Pro	Ile	Thr	Glu	Ala	Ile	Ala
α (*P. luridum*)	Met	Lys	Thr	Pro	Leu	Thr	Glu	Ala	Val	Ala
α (*A. variabilis*)	Met	Lys	Thr	Pro	Ile	Thr	?	Ala	Ile	Ala
β (*C. caldarium*)	Met	Leu	Asn	Ala	Phe	Ala	Lys	Val	Val	Ala
β (*P. luridum*)	Met	Leu	Asp	Ala	Phe	Thr	Lys	Val	Val	
β (*A. variabilis*)	Thr	Leu	Asx	Val	Phe	Ser	Lys			

*References: *C. caldarium* (23); *P. luridum* and *A. variabilis* (10).

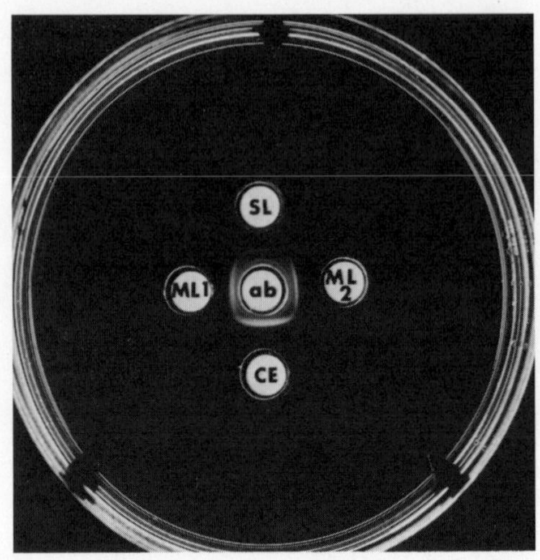

Fig. 2. Antigenic relationships shown by Ouchterlony double-diffusion experiment. Antiserum (ab) is antiphycocyanin from the halophile C. elabens. The other wells contain C-phycocyanins from: ML1, M. laminosus (strain I-30-m, clone 1); CE, C. elabens; ML2, M. laminosus (strain NZ-DB2-m); and SL, S. lividus (Sy I).

extracted from mesophiles, thermophiles, and halophiles some degree of partial identity is noted. When more than a single precipitation line was observed for the same antigen, various phycocyanin aggregates were shown to be present (27). All precipitation lines fluoresced red under ultraviolet light, as is characteristic of C-phycocyanin. The weakest reaction against antiserum to Sy III C-phycocyanin was observed with C-phycocyanin from C. elabens.

FUNCTION AND STRUCTURE

Biliproteins are accessory pigments in photosynthesis (28, 29); they harvest light and transfer the energy to the appropriate chlorophyll molecules. In addition, recent studies utilizing artificial bileaflet membranes suggest that C-phycocyanin and other biliproteins (30, 31) are capable of facilitating and directing electron flow across lipid bilayers. Certain photochemical properties of phycocyanin have also been observed (32, 33).

Electron microscope studies of blue-green and red algae have shown that the biliproteins, in order to carry out successfully their roles in photosynthesis, are assembled into large, discrete structures called phycobilisomes (34-36).

These are 350-450 Å in diameter and are found as well in thermophilic algae (37-40). An electron microscope examination of C. caldarium from which some C-phycocyanin has been extracted shows the presence of these sturctures (Fig 3).

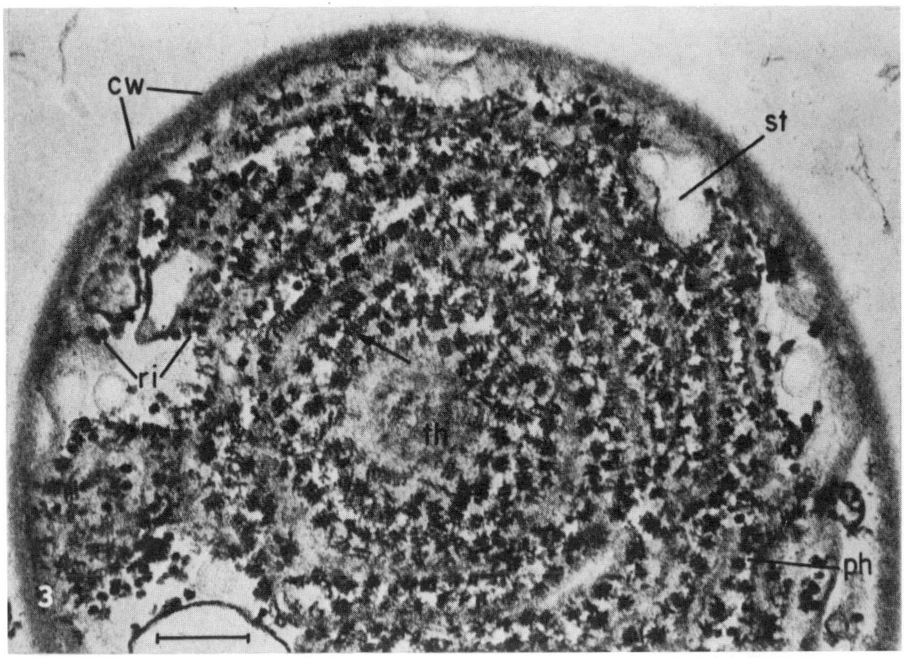

Fig. 3. Electron micrograph of a thin section grazing on a few thylakoid membranes (th) of a C. caldarium chloroplast. Phycobilisomes in longitudinal-tangential (rows of disclike structures; see arrow) and in oblique or transversal (see ph) views are shown. Vestiges of starch-granules (st), some ribosome-like particles (ri), and the cell wall (cw) are indicated. Bar = 200 nm.

Even after the thylakoid membranes are completely separated from the cells, some phycobilisomes remain attached to them, indicating a strong binding of the biliproteins to the membrane (Fig. 4). A general model for phycobilisome structure has been proposed, based on observations from the thermophile S. lividus (40). Elucidation of the three-dimensional structure of C-phycocyanin has been attempted by X-ray diffraction on protein from a thermophile, M. laminosus (41).

When C-phycocyanin is extracted into an aqueous medium, the phycobilisomes are usually disrupted. Experiments on these phycocyanin solutions reveal a variety of aggregates that may easily be manipulated as a unique probe into the nature of protein-protein interactions.

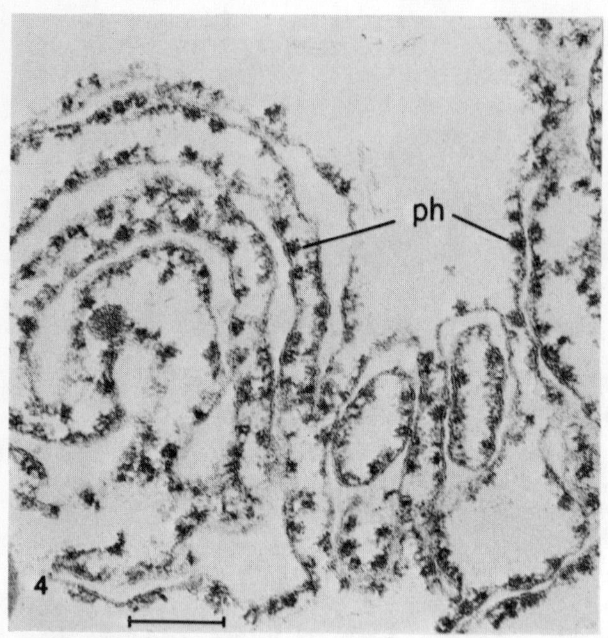

Fig. 4. Electron micrograph of thylakoids isolated from C. caldarium. A pellet of broken cells (passed through a French press) was fixed with glutaraldehyde-osmium, as explained in the text. Note phycobilisomes (ph) attached to the thylakoid membranes. Bar = 200 nm.

QUATERNARY STRUCTURE

The general aggregation properties of C-phycocyanin have been well established from a variety of mesophilic algal sources. The C-phycocyanin polymorphs are readily distinguished by their sedimentation velocity in the analytic ultracentrifuge; the most prevalent are 3S, 6S, 11S, and 17-19S. The 3S species is the monomer, composed of two polypeptide chains, with a molecular weight of 30,000 (42). The 11S aggregate is a hexamer (43-45). Reversible equilibria exist as a function of pH, protein concentration and solvent among the 3S, 6S, and 11S aggregates, and hydrophobic forces are important in the stabilization of the 11S species (12, 46-50). The larger aggregates (17S and greater) have been observed in extracts of mesophiles (51) and can likewise be identified in C-phycocyanin from thermophilic algae (Fig. 5). Properties of separated α and β chains have also been studied (15, 23, 52).

When the aggregation of C-phycocyanin from a halophile and some thermophiles was examined, valid comparisons to typical mesophile responses were thus quite feasible. C-Phycocyanin from S. lividus (Sy III) sedimented at 25°C contained comparatively elevated quantities of the smaller 6S aggregate and decreased amounts of the larger aggregates (53). However, when solutions were heated to 50°C, the distribution was nearly identical to that of mesophilic C-phycocyanin at 25°C. C-Phycocyanin from S. lividus is thermally denatured at 66°C, while mesophilic C-phycocyanin is denatured at 49-51°C (54, 55). Likewise, when C-phycocyanin from the halophile C. elabens was purified and studied, its aggregation in 0.1 ionic strength aqueous buffers was well below that of nonhalophiles (11), but when examined by sedimentation velocity or fluorescence

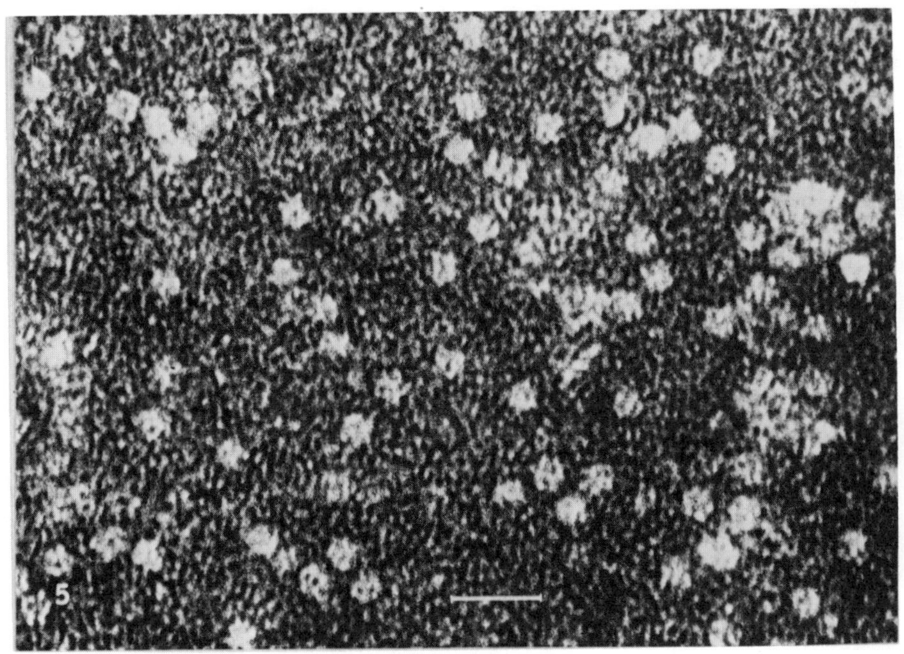

Fig. 5. Electron micrograph of negatively stained C-phycocyanin aggregates (17-21S). Bar = 30 nm.

polarization in 2.0 M NaCl, its aggregation resembled that of normal C-phycocyanin at 0.10 ionic strength (Fig. 6). In these two cases the environmental extreme appears to provide a necessary ingredient for successful protein assembly. Other intracellular factors may also be required.

Unexpectedly, C-phycocyanin from a second strain of thermophilic S. lividus (Sy I) exhibited the converse property. At 20°C, aggregation was augmented: the C-phycocyanin consisted of aggregates 17S and larger, and these aggregates were more stable to dissociation than the corresponding aggregates from mesophilic algae (9). Moreover, when fairly drastic solvent conditions were utilized to produce solutions consisting of 11S and 6S exclusively, the equilibrium distribution was altered so that more 11S was present than in the mesophilic cases (9). For mesophilic C-phycocyanin at pH 6.0 and 23°C, the equilibrium constant for the $6\ M \rightleftharpoons M_6$ reaction is 1×10^{30} (46), a ΔG of -41 kcal/mole; for Sy I the 11S aggregate is even more stable.

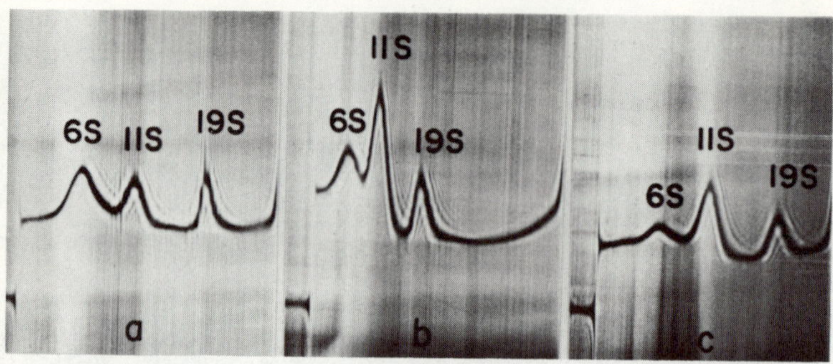

Fig. 6. Sedimentation velocity patterns of C-phycocyanin using schlieren optics. Sources and experimental conditions are: a, C. elabens in sodium phosphate buffer, pH 6.0, 0.1 ionic strength; b, C. elabens in 2.0 M NaCl, pH 6.0; and c, A. variabilis in sodium phosphate buffer, pH 6.0, 0.1 ionic strength. Sedimentation is from left to right at 59,780 rpm, 23°C, and 32 min after speed was reached.

Again, when C-phycocyanin from the thermophile C. caldarium was centrifuged in sedimentation velocity experiments (8), the larger aggregates appeared in increased quantities. An interesting aspect of this C-phycocyanin was an unusually high isoelectric point, 5.11, which was reduced to 4.69 after the large aggregates were dissociated. (C. caldarium is important, not only for its thermophily, but also because it is a taxonomically interesting eucaryote and an acidophile. A significant amount of study has been performed on its biliproteins [8, 23, 56].)

The effect of small, neutral aromatic molecules on C-phycocyanin is unusual. Molecules such as benzene, phenol, and naphthol increase its aggregation, converting 6S to 11S (47, 48). Since hydrophobic forces promote the stability of the 11S aggregate, we have proposed that the aromatic molecule can bind to nonpolar regions on the outer surface of the 6S particle, which are the protein-protein contact sites after aggregation occurs (47). We further postulate that increased hydrophobicity of the region upon aromatic binding, or increased van der Waals stabilization between 6S species, is responsible for the greater 11S stabilization. However, when thermophilic Sy III phycocyanin is treated with 0.10 molar phenol, this effect is reduced, even though the same concentration produces a dramatic aggregation in mesophilic phycocyanins. Work in progress with C-phycocyanin from two thermophilic strains of M. laminosus (57) has yielded results that confirm this lack of response to the aromatic molecules, since no 11S is produced at pH 7.0 for any of the aromatics tested. C-Phycocyanin from M. laminosus is very similar in its aggregation properties to that of S. lividus (Sy III), since the concentration of 11S is reduced relative to the 6S aggregate.

ALLOPHYCOCYANIN

A second biliprotein, allophycocyanin, has also been studied (14, 19, 58-64). Allophycocyanin from Sy I has a molecular weight by sedimentation equilibrium of 1.03×10^5 (9). Repeating this experiment with highly purified protein and the methods described in the experimental section confirmed this result. The ln concentration versus r^2 plot is linear (Fig. 7), suggesting a homogeneous protein. These results are in agreement with sedimentation equilibrium results on mesophilic allophycocyanin indicating a molecular weight of 9.6×10^4 (19) but are lower than results from gel filtration experiments (59, 63, 65). If allophycocyanin is composed of two polypeptide chains in a 1:1 ratio, each with a molecular weight of about 16,000-17,000, as has recently been reported (14, 19, 59, 62), the sedimentation equilibrium results indicate that the structure of this protein is $\alpha_3\beta_3$.

CONCLUSIONS

The amino acid compositions, partial N-terminal sequences, and immunochemistry all suggest strong similarities among C-phycocyanins derived from thermophilic, mesophilic, and halophilic blue-green algae and from a thermophilic eucaryote. However, the thermophilic C-phycocyanins are more stable at high temperatures and exhibit atypical aggregation compared with a large number of mesophiles. Aggregation of thermophilic C-phycocyanin is a complex phenomenon, since certain thermophiles have aggregates more stable than mesophiles (Sy I, C. caldarium), while others have much less 11S and larger aggregates (Sy III and M. laminosus, as well as a halophile, C. elabens). We call the first type "thermoresistant," and their adaptation seems simpler to achieve because a very stable aggregate may exist under usual dissociating conditions. The second type of behavior is more challenging. These C-phycocyanins must aggregate in vivo to form the very large phycobilisomes, but their ability to aggregate in vitro is suppressed (37, 38, 40, 9). It appears that extrinsic factors, such as high salt for the halophile and temperature and other cellular components for the thermophile, are required to activate these proteins to their functional structure.

The amino acid composition of both types of thermophiles is quite similar to that of mesophiles (Table I). The modifications that produce the diversity of aggregation must be fairly subtle. In a hemoglobin molecule a single amino acid substitution can result in a remarkable aggregation of the protein and produce sickle cell anemia (66). Presumably such a limited amino acid substitution in the contact area between subunits of C-phycocyanin could vastly alter the usual protein-protein attraction. Some support for this explanation comes from the lack of effect of aromatic molecules on the aggregation of thermophilic C-phycocyanin from Sy III and M. laminosus. Both this lack of effect and the low level

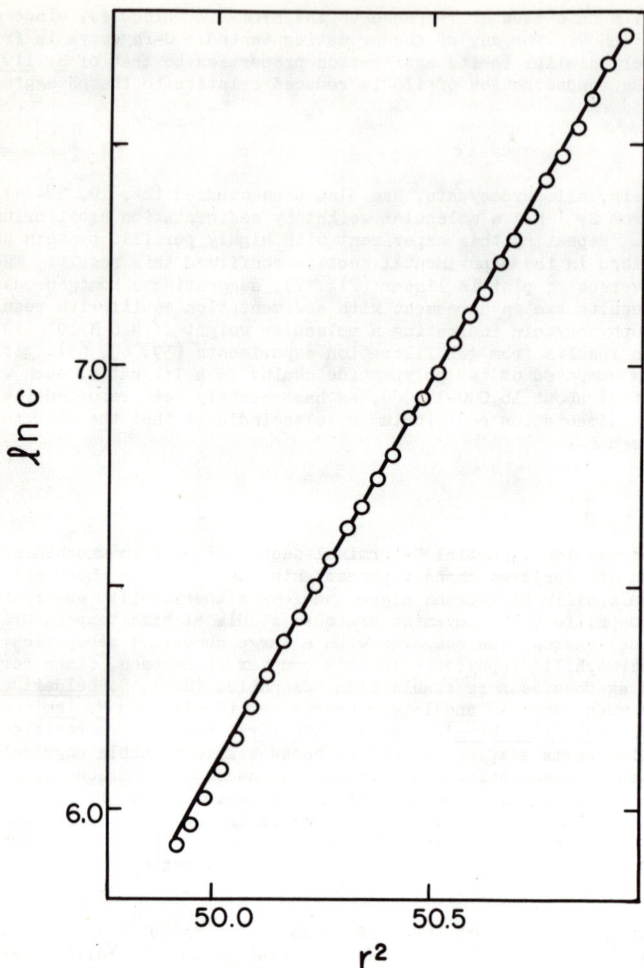

Fig. 7. Sedimentation equilibrium experiment with allophycocyanin from S. lividus (Sy I). Experiment was performed in a Model E ultracentrifuge, using the photoelectric scanner, in sodium phosphate buffer, pH 6.0, 0.1 ionic strength, at 20°C, 17,000 rpm, and using 280 nm light.

of aggregation can be explained by the same hypothesis. It is possible that amino acid substitutions in the contact region, where we have proposed that the aromatic molecules might bind in the mesophilic case (47), act to modify protein assembly and cause an environment wherein the neutral aromatics no longer are attracted. In the case of Sy III, the modification must be temperature-dependent, causing repulsion between adjacent subunits at temperatures below the optimal growth temperatures and reversing itself to promote assembly near 50°C. A study of the temperature dependence of the various structural parameters for each amino acid might help make a reasonable choice of a likely candidate.

ACKNOWLEDGMENT

This project was supported in part by NIH research grant AI08119-06, awarded by the Institute of Public Health Service, and by NSF research grant GB-24315.

REFERENCES

[1]J. W. Schopf, Biol. Rev. 45, 319 (1970).
[2]T. D. Brock, Science 158, 1012 (1967).
[3]R. W. Castenholtz, in The Biology of Blue-Green Algae (Eds. N. G. Carr and B. A. Whitton; University of California Press, Berkeley 1973) Vol. 9, pp. 379-414.
[4]J. Myers and W. A. Kratz, J. Gen. Physiol. 39, 11 (1955).
[5]W. J. Cole, D. J. Chapman and H. W. Siegelman, J. Am. Chem. Soc. 89, 3643 (1967).
[6]B. L. Schram and H. H. Kroes, Eur. J. Biochem. 19, 581 (1971).
[7]D. J. Chapman, in The Biology of Blue-Green Algae (Eds. N. G. Carr and B. A. Whitton; University of California Press, Berkeley 1973) Vol. 9, pp. 162-185.
[8]O. H. W. Kao, M. R. Edwards and D. S. Berns, Biochem. J. 147, 63 (1975).
[9]R. MacColl, M. R. Edwards, M. H. Mulks and D. S. Berns, Biochem. J. 141, 419 (1974).
[10]J. Harris and D. S. Berns, J. Mol. Evol. (In press).
[11]O. H. W. Kao, D. S. Berns and W. R. Town, Biochem. J. 131, 39 (1973).
[12]R. MacColl, D. S. Berns and N. L. Koven, Arch. Biochem. Biophys. 146, 477 (1971).
[13]R. MacColl, W. Habig and D. S. Berns, J. Biol. Chem. 248, 7080 (1973).
[14]J. Gysi and H. Zuber, FEBS Letters 48, 209 (1974).
[15]A. N. Glazer and S. Fang, J. Biol. Chem. 248, 663 (1973).
[16]P. O'Carra, Biochem. J. 119, 3p (1970).
[17]P. O'Carra and S. D. Killilea, Biochem. Biophys. Res. Commun. 45, 1192 (1971).
[18]A. Bennett and L. Bogorad, Biochemistry 10, 3625 (1971).
[19]A. N. Glazer and G. Cohen-Bazire, Proc. Natn. Acad. Sci., USA 68, 1398 (1971).
[20]P. A. Torjesen and K. Sletten, Biochim. Biophys. Acta 263, 258 (1972).
[21]A. Binder, K. Wilson and H. Zuber, FEBS Letters 20, 111 (1972).

[22]Y. Kobayashi, H. W. Siegelman and C. H. W. Hirs, Arch. Biochem. Biophys. 152, 187 (1972).
[23]R. F. Troxler, J. A. Foster, A. S. Brown and C. Franzblau, Biochemistry 14, 268 (1975).
[24]V. P. Williams, P. Freidenreich and A. N. Glazer, Biochem. Biophys. Res. Commun. 59, 462 (1974).
[25]G. Frank, H. Zuber and W. Lergier, Experientia 31, 23 (1975).
[26]P. G. H. Byfield and H. Zuber, FEBS Letters 28, 36 (1972).
[27]D. S. Berns, Plant Physiol. 42, 1569 (1967).
[28]R. Y. Stanier, in Evolution in the Microbial World (Soc. for Gen. Microbiol., Symposium No. 24; Cambridge University Press, Cambridge, England 1974), pp. 219-240.
[29]J. Myers, in Annual Reviews of Plant Physiology (Eds. L. Machlis, W. R. Briggs and R. B. Park; Annual Reviews Inc., Palo Alto 1971), pp. 289-312.
[30]A. Ilani and D. S. Berns, J. Membrane Biol. 8, 333 (1972).
[31]C-H. Chen and D. S. Berns, Proc. Natn. Acad. Sci., USA (In press).
[32]V. B. Evstigneev and O. D. Bekasova, in Proceedings of the First European Biophysics Congress (Baden 1971), Vol. IV, pp. 245-248.
[33]O. D. Bekasova and V. B. Evstigneev, Biofizika 18, 243 (1973).
[34]E. Gantt and S. F. Conti, Brookhaven Symp. Biol. 19, 393 (1966).
[35]R. B. Wildman and C. C. Bowen, J. Bacteriol. 117, 866 (1974).
[36]B. H. Gray and E. Gantt, Photochem. Photobiol. 21, 121 (1975).
[37]M. R. Edwards, D. S. Berns, W. C. Ghiorse and S. C. Holt, J. Phycol. 4, 283 (1968).
[38]S. C. Holt and M. R. Edwards, Canad. J. Microbiol. 18, 175 (1972).
[39]M. Lefort-Tran, G. Cohen-Bazire and M. Pouphile, J. Ultrastructure Res. 44, 199 (1973).
[40]M. R. Edwards and E. Gantt, J. Cell Biol. 50, 896 (1971).
[41]M. Dobler, S. D. Dover, K. Laves, A. Binder and H. Zuber, J. Mol. Biol. 71, 785 (1972).
[42]O. Kao, D. S. Berns and R. MacColl, Eur. J. Biochem. 19, 595 (1971).
[43]D. S. Berns and M. R. Edwards, Arch. Biochem. Biophys. 110, 511 (1965).
[44]D. S. Berns, Biochem. Biophys. Res. Commun. 38, 65 (1970).
[45]M. Kato, W. I. Lee, B. E. Eichinger and J. M. Schurr, Biopolymers 13, 2293 (1974).
[46]R. MacColl, J. J. Lee and D. S. Berns, Biochem. J. 122, 421 (1971).
[47]R. MacColl and D. S. Berns, Arch. Biochem. Biophys. 156, 161 (1973).
[48]C-H. Chen, R. MacColl and D. S. Berns, Arch. Biochem. Biophys. 165, 554 (1974).
[49]T. Saito, N. Iso and H. Mizuno, Bull. Chem. Soc. Japan 47, 1375 (1974).
[50]G. J. Neufeld and A. F. Riggs, Biochim. Biophys. Acta 181, 234 (1969).
[51]M. Kessel, R. MacColl, D. S. Berns and M. R. Edwards, Canad. J. Microbiol. 19, 831 (1973).
[52]A. N. Glazer, S. Fang and D. M. Brown, J. Biol. Chem. 248, 5679 (1973).
[53]D. S. Berns and E. Scott, Biochemistry 5, 1528 (1966).
[54]A. Hattori, H. L. Crespi and J. J. Katz, Biochemistry 4, 1213 (1965).
[55]D. S. Berns, Biochemistry 2, 1377 (1963).
[56]R. F. Troxler, Biochemistry 11, 4235 (1972).
[57]O. H. W. Kao, (Unpublished results).
[58]E. Gantt and C. A. Lipschultz, Biochim. Biophys. Acta 292, 858 (1973).
[59]A. S. Brown, J. A. Foster, P. V. Voynow, C. Franzblau and R. F. Troxler, Biochemistry (In press).

[60]N. Lazaroff, in The Biology of Blue-Green Algae (Eds. N. G. Carr and B. A. Whitton; University of California Press, Berkeley 1973), Vol. 9, pp. 279-319.
[61]A. N. Glazer and S. Fang, J. Biol. Chem. 248, 659 (1973).
[62]L. G. Erokchina and A. A. Krasnovskii, Mol. Biol. 8, 657 (1974).
[63]E. Gantt and C. A. Lipschultz, Biochemistry 13, 2960 (1974).
[64]O. D. Bekasova and V. B. Evstigneev, Izv. Akad. Nauk SSSR, Ser. Biol., 393 (1975).
[65]K. P. Koller and W. Wehrmeyer, Arch. Microbiol. 100, 253 (1974).
[66]M. Murayama, CRC Crit. Rev. Biochem. 1, 461 (1973).

FERREDOXIN FROM BACILLUS STEAROTHERMOPHILUS

R.CAMMACK, D.O.HALL, R.N.MULLINGER & K.K.RAO
King's College, Department of Plant Sciences
68, Half Moon Lane, London SE24 9JF, U.K.

Ferredoxins are low molecular weight electron transfer proteins found in all types of living organisms (1-3). They are used in many reactions ocurring at low redox potentials. The active centre of the ferredoxins consists of a chromophore containing iron and labile sulphur atoms.

Ferredoxins may be classified into three types, based on the number of iron atoms in the chromophore:

1. Two-iron ferredoxins with an active centre of two iron and two labile sulphur atoms. The soluble ferredoxins of plants and algae belong to this type. They have a molecular weight of about 10 500 , and they transfer single electrons at a redox potential of -420 mV.
2. Four-iron ferredoxins. This type of ferredoxin is a relative newcomer to the field - they have been isolated from a number of bacterial species, including Bacillus, Desulphovibrio and Rhodospirillum. Their molecular weights vary between 6000 and 8500. They are also one-electron transfer proteins.
3. Eight-iron ferredoxins, containing two 4Fe-4S clusters. Most of the ferredoxins isolated from anaerobic bacteria are of this type. They can transfer two electrons per molecule at a redox potential of about -400 mV.

The structures of the iron-sulphur clusters in the two-iron and eight-iron ferredoxins,which have been deduced from various physico-chemical measurements, are shown in Fig. 1. The active centre of the four-iron ferredoxins is probably of the latter type.

The biological activity of the ferredoxins is lost on incubation in aerobic solution at high temperatures, in parallel with a decrease in their visible optical absorption. The thermal stability of the ferredoxins of a particular type vary somewhat, depending on the species from which they are isolated.

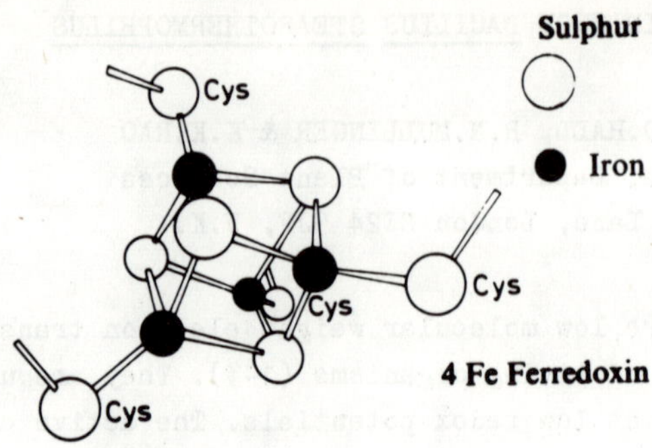

Fig. 1. Structures of four-iron and two-iron centres in ferredoxins.

Devanathan et al.(4) isolated eight-iron ferredoxins from two thermophilic clostridia, C. tartarivorum and C. thermosaccharolyticum, and found them to be more resistant to thermal and oxygen denaturation than ferredoxins from two mesophilic clostridia, C. acidi-urici and C. pasteurianum. The amino acid sequences of the thermostable ferredoxins (5,6) are homologous with those of other clostridial ferredoxins, but show some significant differences, notably the presence of histidine, which is not commonly found in bacterial ferredoxins. Perutz and Raidt (7) have compared the structures of these proteins, for which the X-ray crystal structure and amino acid sequences are known. They concluded that the increased stability of the thermostable ferredoxins was the result of an increased number of internal hydrogen bonds and, perhaps more important, salt bridges, between amino acid side-chains.

We report here the purification of a four-iron ferredoxin from the thermophile Bacillus stearothermophilus, and studies of its stability.

PURIFICATION OF FERREDOXIN

B. stearothermophilus cells grown at 60° C according to Sargeant et al.(8) were purchased from the Microbiological Research Establishment, Porton Down, Salisbury, Wilts., U.K. The cells were suspended in twice their volume of 20 mM Tris-Cl buffer, pH 8, containing 5 mM mercaptoethanol, and the suspension was homogenized by cycling through a Manton-Gaulin press. The cell debris was removed from the homogenate by centrifugation and the supernatant containing ferredoxin was stirred with DEAE-cellulose (Whatman DE 23). The ferredoxin adsorbed was eluted with 0.8 M NaCl, and was further purified by repeated chromatography on DE 23 using chloride buffers for elution, and finally by preparative electrophoresis. Details of the purification procedure and analysis have been reported by Mullinger et al.(9). The yield of ferredoxin was 8 mg/kg of cell paste.

PROPERTIES OF THE FERREDOXIN

Table 1 summarizes some of the properties of the purified protein. The optical absorption spectra of the oxidized and reduced forms (Fig. 2) are typical of bacterial ferredoxins containing 4Fe-4S centres (see 3).

TABLE 1. Properties of B. stearothermophilus ferredoxin (8)

Molecular weight	8200 - 8500
Redox potential (pH 8.0)	-280 mV
A390/A280 nm	0.61
No. of Fe and labile S per molecule	4

STABILITY OF THE FERREDOXIN

The ferredoxins from the non-thermophile Bacillus polymyxa are relatively unstable, and are prepared in the presence of thiols to prevent damage by oxygen (see 10). By contrast, ferredoxin from B. stearothermophilus is stable at room temperature under aerobic conditions, and appears to be more resisitant to denaturation at high temperatures than other ferredoxins. Fig. 3 shows a comparison with the eight-iron ferredoxin from C. pasteurianum, and the two-iron

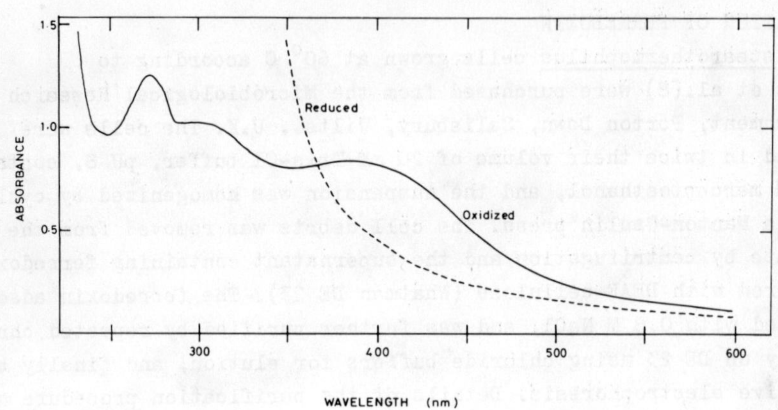

Fig. 2. Optical absorption spectra of B. stearothermophilus ferredoxin, oxidized, and reduced with dithionite.

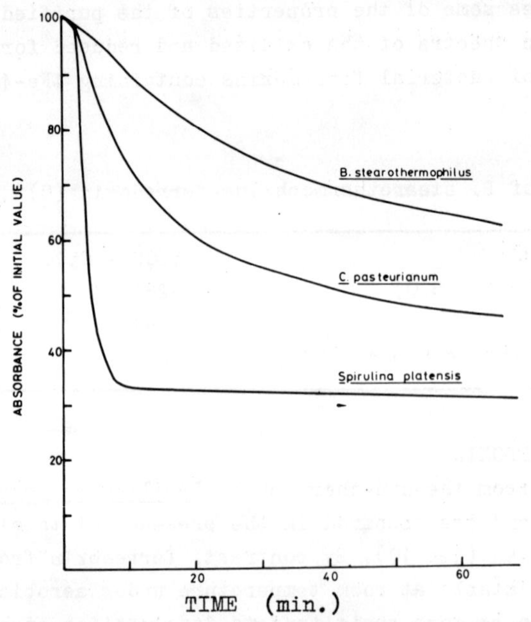

Fig. 3. Time-course of thermal denaturation of ferredoxins. 0.02 mM samples of ferredoxin were incubated in 0.15 M NaCl/ 0.01 M MOPS buffer, pH 7.2, at 65 °C under aerobic conditions. The absorbance of the samples was measured at 390 nm for B. stearothermophilus and C. pasteurianum ferredoxins, and 420 nm for S. maxima ferredoxin.

ferredoxin from Spirulina maxima, which is one of the most stable plant ferredoxins (11). The ferredoxin samples in this case were rather dilute; most ferredoxins are very stable to heat in concentrated solution under anaerobic conditions (12), but become less stable in dilute aerobic solution. The principal cause of decomposition under these conditions is probably the oxidation of the labile sulphur (13)

B. stearothermophilus ferredoxin is remarkably stable to reduction by sodium dithionite and reoxidation by oxygen (Fig. 4). This process normally leads to substantial decomposition of the ferredoxins (14).

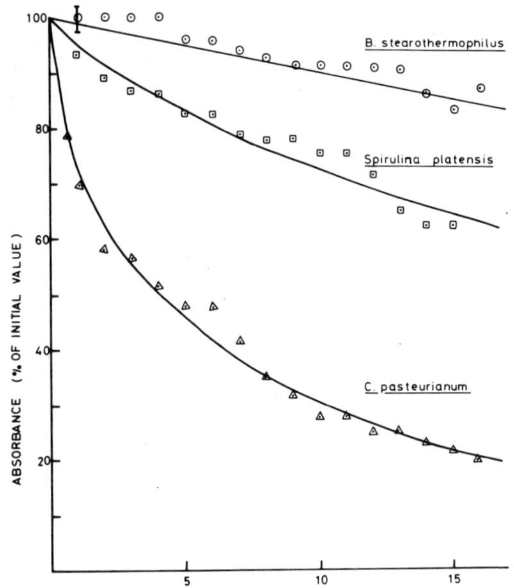

No. of oxidation-reduction cycles

Fig. 4. Decomposition of ferredoxins by oxidation and reduction. Samples of ferredoxin prepared as for Fig. 3 were successively reduced with 0.3 mM dithionite solution, buffered with Tris, and reoxidized by shaking with air, at room temperature. The absorbance of the samples was measured as for Fig. 3 after each oxidation. Absorbance values are corrected for dilution by dithionite solution.

The amino acid composition of B. stearothermophilus ferredoxin reveals no significant peculiarities when compared with other four-iron ferredoxins (Table 2). The molecule contains four cysteine residues, the minimum necessary to bind a 4Fe-4S cluster. It does not contain histidine, but has two residues of methionine. The explanation of the stability of this protein must await the determination of its sequence and structure.

TABLE 2. Amino acid composition of four-iron ferredoxins.

Amino acid	A	B	C	D	E	F
Lysine	4	1	2	3	4	6
Histidine	0	0	1	0	0	0
Arginine	0	1	0	0	1	1
Aspartic	15	11	10	8	15	9
Threonine	3	0	3	2	5	3
Serine	1	3	2	3	3	6
Glutamic	6	9	11	6	8	9
Proline	5	4	3	5	4	3
Glycine	5	1	6	3	6	8
Alanine	6	6	2	8	9	7
Cysteine	4	6	6	6	4	4
Valine	4	5	5	4	3	6
Methionine	2	2	0	0	0-1	1
Isoleucine	7	5	4	4	6	4
Leucine	2	1	2	1	4	5
Tyrosine	3	0	0	2	4	2
Phenylalanine	2	1	0	1	2	5
Tryptophan	0	0	0	0	0	0
TOTAL	69	56	57	56	78-79	79

Ferredoxin A: B. stearothermophilus (9)
B: Desulphovibrio gigas (15)
C: Desulphovibrio desulphuricans (16)
D: Spirochaeta aurantia (17)
E: Bacillus polymyxa I (18)
F: Bacillus polymyxa II (18)

ACKNOWLEDGEMENTS: We thank Mr K. Dawson for technical assistance. This work was supported by the U.K. Science Research Council.

REFERENCES

(1) Lovenberg, W. (1973) Iron-sulphur Proteins, vols 1 & 2, Academic Press, New York.

(2) Orme-Johnson, W.H. (1973) Ann. Rev. Biochem. 42, 159-204.

(3) Hall, D.O., Cammack, R. & Rao, K.K.(1974) in: Iron in Biochemistry and Medicine (Jacobs, A. & Worwood, M., eds), pp 279-334, Academic Press, London & New York.

(4) Devanathan, T., Akagi, J.M., Hersh, R.T. & Himes, R.H. (1969) J. Biol. Chem. 244, 2846-2853.

(5) Tanaka, M., Haniu, M., Matsueda, G., Yasunobu, K.T., Himes, R.H., Akagi, J.M., Barnes, E.M. & Devanathan, T. (1971) J. Biol. Chem. 246, 3953-3960.

(6) Tanaka, M., Haniu, M., Yasunobu, K.T., Himes, R.H. & Akagi, J.M. (1973) J. Biol. Chem. 248, 5215-5217.

(7) Perutz, M. F., & Raidt, H. (1975) Nature, Lond. 255, 256-259.

(8) Sargeant, K., East, D.N., Whitaker, A.R. & Elsworth, R. (1971) J. Gen. Microbiol. 65, iii.

(9) Mullinger, R.N., Cammack, R., Rao, K.K., Hall, D.O., Dickson, D.P.E., Johnson, C.E., Rush, J.D. & Simopoulos, A. (1975) Biochem. J., in press.

(10) Stombaugh, N.A., Burris, R.H. & Orme-Johnson, W.H. (1973) J. Biol. Chem. 248, 7951-7956.

(11) Hall, D.O., Rao, K.K. & Cammack, R. (1972) Biochem. Biophys. Res. Commun. 47, 798-802.

(12) McDonald, C.C., Phillips, W.D., Lovenberg, W. & Holm, R.H. (1973) Ann. N. Y. Acad. Sci. 222, 789-799.

(13) Petering, D., Fee, J.A. & Palmer, G. (1971) J. Biol. Chem. 246, 643-653.

(14) Mayhew, S.G., Petering, D., Palmer, G. & Foust, G.P. (1969) J. Biol. Chem. 244, 2830-2834.

(15) Travis, J., Newman, D.J., LeGall, J. & Peck, H.D. (1971) Biochem. Biophys Res. Commun. 45, 452-458.

(16) Zubieta, J.A., Mason, R., & Postgate, J.R. (1973) Biochem. J. 133, 851-854.

(17) Johnson, P.W. & Canale-Parola, E.(1973) Arch. Mik. 89, 341-353.

(18) Yoch, D.C. (1973) Arch. Biochem. Biophys. 158, 633-640.

STRUCTURAL AND METABOLIC BASIS OF THERMOPHILY

(Protein synthesis system, nucleic acids, membranes, temperature-adaptation, genetics etc.)

BIOCHEMICAL STUDIES ON AN EXTREME THERMOPHILE *Thermus thermophilus* : THERMAL STABILITIES OF CELL CONSTITUENTS AND A BACTERIOPHAGE

Tairo OSHIMA, Yoshiyuki SAKAKI, Nobuyuki WAKAYAMA,
Kimitsuna WATANABE, Ziro OHASHI* and Susumu NISHIMURA*
Mitsubishi-Kasei Institute of Life Sciences, Minamiooya,
Machida, Tokyo 194, Japan
and
* National Cancer Center Research Institute, Tsukiji, Tokyo, Japan

INTRODUCTION

Mechanism of thermophily has been the object of biochemical interest for many years, but has not been clearly answered at the molecular level. Most of the studies have been done using a moderate thermophile, *Bacillus stearothermophilus* which grows optimally at 60-65°C. Studies on more rigorously thermophilic bacterium in comparison with the moderate thermophile and mesophile will be one of the most fruitful way to the understanding of thermophily. Thus we have been studying the thermal stabilities of cellular components and bacteriophage of an extremely thermophilic bacterium, *Thermus thermophilus* (formerly named *Flavobacterium thermophilum*) HB8 which was isolated from a Japanese thermal spa[1,2].

Biopolymers and cell organella such as enzymes in glycolytic path[3-8], enzymes in protein synthesis[9], DNA[10], ribosomes[1,9], tRNA[11,12], membrane and membrane lipid[2,13], and bacteriophage[14] of the thermophile were studied. These cell components were resistant to heat without exception. We found that the mechanisms of thermal resistance of these biopolymers and organella could be classified into three types: (1) intrinsic stability, that is, a molecule is intrinsically resistant to heat without any specific protector and the stability is caused directly by the chemical structure coded in its gene, (2) stability supported by a ligand of small molecule, namely a cell component is resistant to heat only because of the presence of specific protector and, (3) heat induced stability, heat stability of a cell constituent is positively correlated to the growth temperature changing its chemical structure depending on the environmental temperature. Thermal stability of any biopolymer, cell organelle or bacteriophage of *T. thermophilus* hitherto studied could be interpreted by one of these or, in many cases, by the combination of two or three of these mechanisms. In this communication, we will report some typical examples of these three mechanisms on thermal stability.

Intrinsic stability; thermostability of enzymes

A number of investigations[3-8] on enzymes from *T. thermophilus* have shown that the enzyme proteins are, in many cases, stable to heat without the presence of specific cofactors, as reported for many enzymes from other thermophilic sources[15]. For instance, glyceraldehyde-3-phosphate dehydrogenase of *T. thermophilus* was found to possess remarkable heat stability as shown in Fig. 1[16]. No significant difference was observed between heat stability of the crystalline preparation (Fig. 1) and that of the enzyme in a crude extract, indicating the absence of specific protector for the thermophile glyceraldehyde-3-phosphate dehydrogenase in the cell. The apoenzyme prepared by treating the holoenzyme with charcoal was stable to heat. Thus unusual thermostability of the thermophilic enzyme arises from protein structure itself. Molecular mechanism of the thermostability is still speculative.

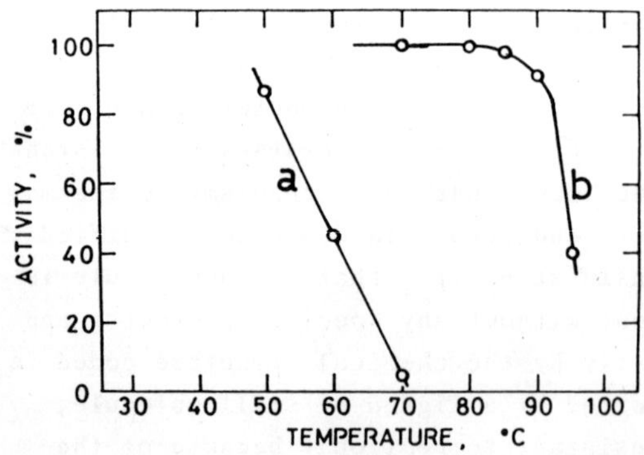

Fig. 1. Heat stabilities of glyceraldehyde-3-phosphate dehydrogenase from mesophilic source (pig muscle) (a) and *T. thermophilus* (b). Solution of crystalline enzyme was heated for 5 min at various temperatures and the remaining activity was assayed at 30°C.

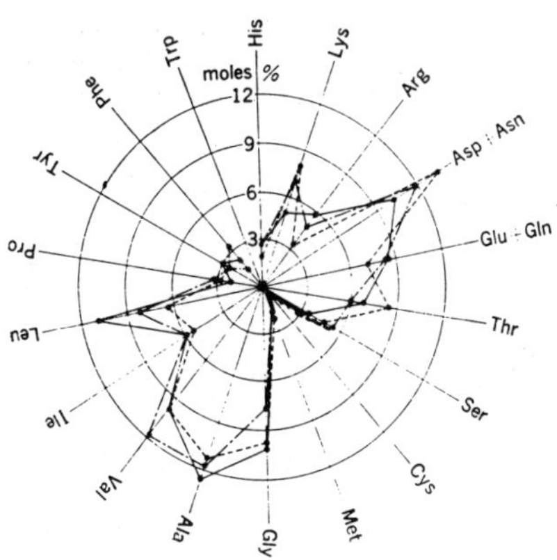

Fig. 2. Comparison of amino acid compositions of glyceraldehyde-3-phosphate dehydrogenase from *E.coli* (----), *B.stearothermophilus* (—·—·), and *T.thermophilus* (——).

The comparative studies on the thermophilic enzyme with the mesophilic counterpart revealed similarities in catalytic properties such as pH-activity profile, behavior to SH reagents, subunit structure and specificity, and dissimilarity in stability to heat and denaturating agents such as organic solvents, detergents and urea, suggesting that the structure around the active site of the thermophilic enzyme is similar to that of the mesophilic dehydrogenase, and the active site structure has been preserved during the evolutionary process to thermophile while the other parts of the molecule are modified to make the enzyme thermostable. It was also shown that there is no special features such as covalent cross linkage or the presence of unusual amino acid residues in the thermophile enzyme protein. Studies on the stability of the enzyme perturbed by denaturating agents suggested that the thermostability arises from subtle difference in the architecture of the molecule[16,17]. This assumption was also supported by the numerical simulation of the denaturation[17] and by comparison of amino acid composition of the thermophile dehydrogenase with those of a moderate thermophile and mesophile enzymes. In Fig. 2, molar contents of amino acid of the enzymes from various sources are illustrated. Although contents of some amino acids such as Arg, Ala and Leu are positively, and some such as Asp and Lys are negatively correlated to the denaturation temperature, generally speaking there is no remarkable difference in amino acid composition. Similarly no significant difference was found in comparing average hydrophobicity defined by Bigelow[18], number of amino acid residues which may participate in hydrogen

Table 1. Comparison of average hydrophobicity and number of amino acid residues which may participate in hydrogen bonding or electrostatic interactions of glyceraldehyde-3-phosphate dehydrogenase from various sources.

	H_ϕ per residue	Lys + His + Arg + Asp + Glu + Ser + Thr + Cys + Met + Thr, per subunit	Glu + Asp per subunit	Lys + His + Arg per subunit
T. thermophilus	4.32 KJ	152	57	44
B. stearothermophilus	4.45	165	65	46
E. coli	4.20	173	67	44

bonding, and number of residues which may be positively or negatively charged and thus participate in electrostatic interactions as given in Table 1. Data in Table 1, do not deny the contributions of hydrophobic, hydrogen bonded or electrostatic forces to the thermostability, but suggest that the increment of these bonding in the thermophile enzyme, if any, must be quantitatively small.

The structural difference in relation to the intrinsic stability of the thermophilic enzyme could be considered so small that the exact three dimensional structure of the thermophilic and mesophilic enzymes would be necessary to interpret the difference in stability at the molecular level. Search for explanation of the heat lability of mesophilic enzymes ironically would help the understanding of the abnormal stability of enzymes from thermophilic sources as well.

Stability supported by a protector; stability of a phage

Some enzymes have been reported to contain factors which increase their thermal stability. For instance, some enzymes from moderate thermophiles such as thermolysin[19], alkaline protease from thermophilic Streptomyces[20], α-amylases from *B.stearothermophilus*[21,22] and from an unspecified thermophile, strain V-2[23], were stabilized by the presence of calcium ion. α-Amylase from V-2 was shown to be more heat labile upon removal of tightly bound calcium ion than calcium free α-amylase from a mesophilic organism.

Here we would like to present another example in which the presence of metal ion or other small molecule plays a crucial role for the thermostability. The example is thermophilic bacteriophage (ϕ)YS40

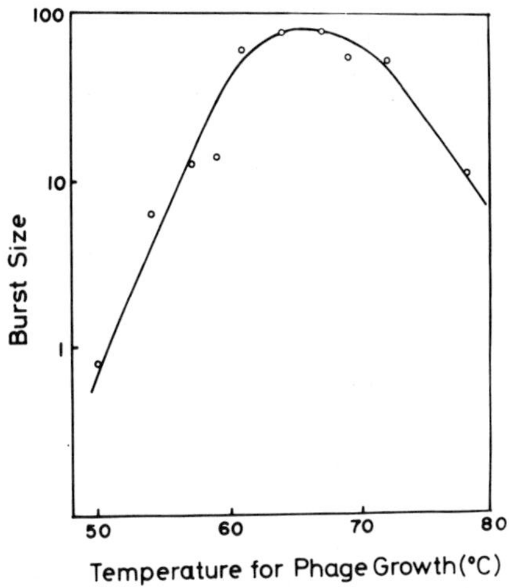

Fig. 3. Growth of φYS40 at various temperatures. φYS40 was grown in HB8 broth[1,2] at the indicated temperature for 5 hr. Burst size means the number of progeny phages per infected cell.

A bacteriophage (φYS40) infectious to *T.thermophilus* HB8 was isolated and characterized[14]. This is the first phage of extreme thermophiles. Phage YS40 grows over the temperature range of 56 to 78°C, and the optimum growth temperature is 65°C (Fig. 3). The phage has a hexagonal head, a tail, a base plate and tail fibers. The phage was easily inactivated by heating at 80°C in a buffer containing 10mM Tris (pH 7.5) and 10 mM $MgCl_2$ (The half life is about 4 min). Also, the phage was sensitive to high concentrations of salt such as NaCl or CsCl (Fig. 4). Electronmicroscopic observation suggested that the head structure was destroyed by these treatments.

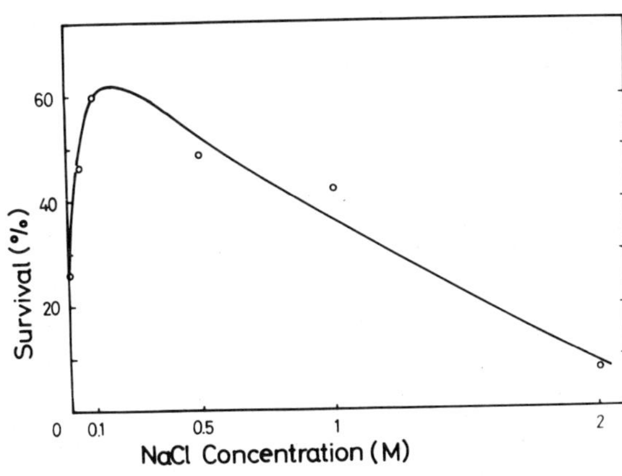

Fig. 4. Salt sensitivity of φYS40. The phage was suspended in 10mM Tris-HCl buffer, pH 7.5, containing 10mM $MgCl_2$ and NaCl at the indicated concentration, and heated at 70°C for 30 min.

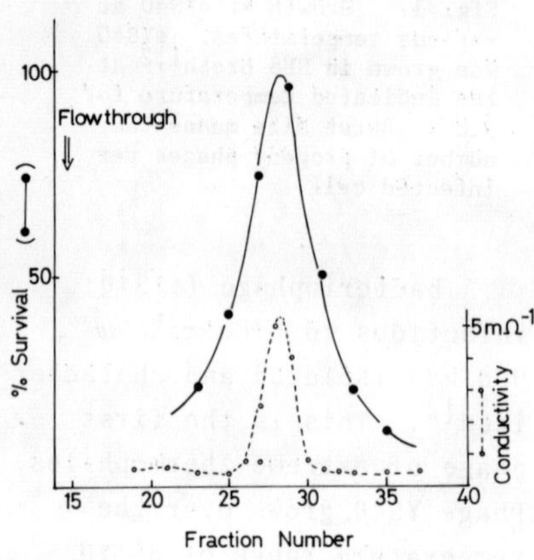

Fig. 5. Elution profile of the stabilizing factor in the hot spring water. The water of hot spring was concentrated 100 fold by evaporation and applied on a Sephadex G-15 column (2×55 cm). The factor was eluted with 10mM Tris (pH 7.5) containing 10mM $MgCl_2$. ϕYS40 was added to each fraction and heated at 80°C for 30 min.

The thermostability of the phage was restored by the addition of casamino acid (1%), yeast extract (0.4%), polypeptone (0.8%), NaCl (0.1M) or spermidine (1mM). The phage is also thermostable in the water of the hot spring from which this phage was isolated. These results suggested that a certain factor(s) is required to stabilize the phage. Such a stabilizing factor contained in the hot spring water was characterized. The effective factor was eluted at the same position as salts on a column of Sephadex G-15 (Fig. 5). The factor lost its effectiveness after treatment with chelating agents such as EDTA or Chelex 100. The results suggested that the effective factor contained in the hot spring water is a metal ion(s) (other than magnesium or calcium which could not protect the phage from heat inactivation) which is easily chelated by EDTA or Chelex 100. Identification of the effective metal ion is now in progress.

Heat induced stability; thermal properties of biomembranes

The most remarkable example of the third feature of adaptation to high temperature, heat induced stability, is the cell membrane. Chemical analysis of lipid components of the cell was carried out by M.Oshima and collaborators[13,24]. They showed that the main component of the lipid fraction is a novel glycolipid with iso and ante-iso branched fatty acids. The dependence of the fatty acid composition on culture temperature was also studied. The ratios of iso $C_{17:0}$/ iso $C_{15:0}$ and ante-iso $C_{17:0}$/ante-iso $C_{15:0}$ were much greater at higher growth temperature. The results strongly suggest that the thermal properties of the membrane may vary depending upon the environmental temperature. This view is supported by microviscosity measurements of the membrane.

So called "microviscosity" η of the hydrophobic region of the membrane fraction were measured in the following way[25].

Cells were harvested at their logarithmic phase and treated with lysozyme-CyDTA in isotonic sucrose solution. Membrane ghosts were obtained by transferring the cells into hypotonic solution. After washing, the membranes were further purified by sucrose density gradient.

Ghost membranes were suspended in 0.05M phosphate buffer (pH 7.0). Perylene in acetone was mixed with the membrane suspension and the mixture was rigorously stirred. Fluorescence anisotropy r of perylene molecules was measured as a function of temperature. A fluorometer for the measurement of the anisotropy was deviced in our laboratory (detail will be published elsewhere).

Fluorescence anisotropy r is defined as

$$r = \frac{I_{\parallel} - I_{\perp}}{I_{\parallel} + 2I_{\perp}}$$

where $I_{\parallel}$ and $I_{\perp}$ are the fluorescence intensities observed with the analyzing polarizer parallel and perpendicular to the polarized excitation beam. Microviscosity is obtained by the following equation

$$\eta = \frac{KT\tau}{V} / [(ro/r) - 1]$$

where ro is anisotropy at $T \rightarrow 0$, τ is the fluorescence life time obtained by single photon counting method, K Boltzmann constant, V is the effective volume of the probe. The value of V was used from the literature[26]. The temperature dependence of microviscosity was obtained for each sample of different growth temperature.

A typical example of the Arrhenius plot of microviscosity is shown in Fig. 6, where the cells were grown at 75°C. The figure is characterized by three slopes. It is known with synthetic phospholipid dispersion that the fusion activation energy ΔE of solid phase is smaller than that of liquid phase. Thus our experimental results suggest that three phases exist there; liquid crystalline at

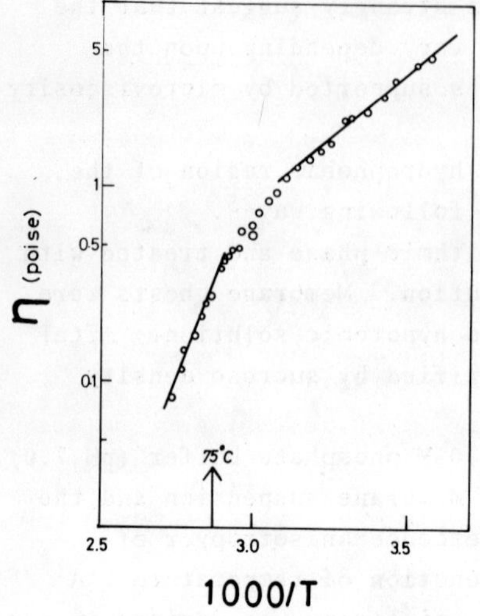

Fig. 6. Temperature dependence of the microviscosity of the membrane dispersion. The cells were grown at 75°C.

T > 77°C, solid at T < 48°C and intermediate phase at 77°C < T < 48°C. Phase separation[27] (coexistence of both liquid and solid phase) seems to take place over the intermediate region. Similar results were obtained with cells grown at other temperatures. The results are summarized in Table 2. It is noted that the growth temperature is close to the higher inflexion temperature with each sample. In other words, the membrane is adapted to the environmental temperature so that it can be in an appropriate physical state at the specific temperature.

tRNA : A model for thermostable biopolymers

tRNAs from *T.thermophilus* were more stable to heat than those from *B.stearothermophilus* and *E.coli*[1,3]. The thermostable tRNA is equivalent

Table 2. Inflexion points of log η − 1/T relations.

Growth temperature	Inflexion I	Inflexion II
50°C	27°C	53°C
55	44	60
60	43	69
75	48	77
88	52	80

to that of *E.coli* in its function in protein biosynthesis[3,9]. The thermophile tRNA would be a good model compound for elucidation of the mechanism of thermal stability, because (i) the molecule is small enough to easily determine its primary sequence, (ii) when the primary structure is determined, its three dimensional structure can be speculated based on the recent knowledge obtained by X-ray analysis[28] on yeast $tRNA^{Phe}$, and (iii) it will be easy to identify the structural difference which relates to the thermostability of the thermophile tRNA since only a small part of the molecule may be modified to make tRNA thermostable without the loss of its biological functions. Structures around the sites interacting with mRNA, rRNA, ribosomal proteins, elongation factors, and aminoacyl-tRNA synthetase should be preserved to retain its role in protein synthesis during adaptation to high temperature.

Formylmethionine specific tRNA ($tRNA_f^{Met}$) was extracted from *T.thermophilus* cells, purified and its nucleotide sequence was studied. By analyzing the nucleotide sequences of fragments obtained by RNase T_1 and RNase A digestions, the primary structure was proposed as illustrated in Fig. 7, in which a clover-leaf model of the thermophile $tRNA_f^{Met}$ is shown. Six bases were different from the structure of *E.coli* $tRNA_f^{Met}$; 2'-O-methyl-guanosine at position 19 instead of guanosine in *E.coli* tRNA, cytosine at 51 instead of uridine, guanosine at 52 instead of C, 5-methyl-2-thiouridine (m^5s^2U) at 55 instead of ribothymidine (T), 1-methyladenosine at 59 instead of A, and C at 64 instead of G.

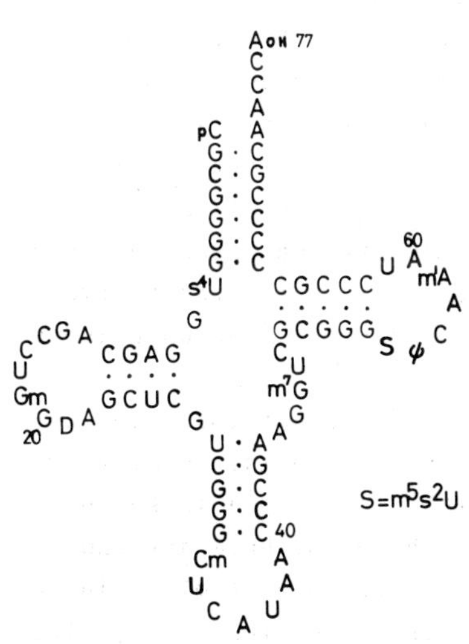

Fig. 7. Proposed primary sequence of *T.thermophilus* $tRNA_f^{Met}$.

The G-C pair content in base-paired regions of the thermophile $tRNA_f^{Met}$ is estimated to be 90% and the high G-C pair content may, at

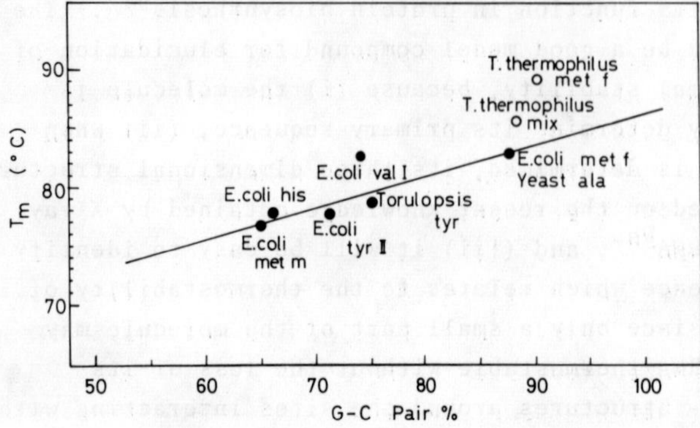

Fig. 8. Relation between G-C pair content and melting temperature of tRNA.

least partly, be responsible for the high melting temperature of the tRNA (89°C). It was found that circular dichroic (CD) difference spectra of tRNA upon heat denaturation correlated with G-C pair content[11]. Based on this observation, the CD difference spectrum study suggested that tRNA from *T.thermophilus* contains more G-C pairs than those from *E.coli* and average G-C pair content in the thermophile tRNA is 85-90%. As an average, three to four A-U pairs per molecule would be replaced by G-C pairs in tRNA from *T.thermophilus*, thus increment of three or four hydrogen bonds would be expected inside of the thermophile tRNA molecule. We have found that the high melting temperature of tRNAs from *T.thermophilus* could not be explained fully by the increment of G-C pairs as shown in Fig. 8. A linear relation was observed between G-C pair content and melting temperature of tRNA from *E.coli*, but $tRNA_{mixture}$ and $tRNA_f^{Met}$ from the extreme thermophile did not fall on the straight line in Fig. 8. There must be some other mechanism(s) responsible for the thermostability than the G-C pair content. One possible candidate is the presence of modified base(s).

Among the minor components, m^1A and m^5s^2U are not found in *E.coli* tRNA. It is, however, unlikely that m^1A has a crucial role in the thermostability of the thermophile tRNA, since m^1A is known to be present widely in eukaryotic tRNAs which do not have any unusual thermal stability. On the other hand, the presence of m^5s^2U in the primary sequence of the thermophile tRNA is unique feature[12]. The view that m^5s^2U may play an important role in the thermostability was supported by measurements of CD bands in near-ultraviolet region

(300-400 nm). The CD bands in this region are correlated to the conformations of such minor components as s^4U or m^5s^2U. A positive band at around 310 nm was assigned to m^5s^2U, because this band did not disappear when s^4U base was chemically converted to U. As temperature of the thermophile $tRNA_f^{Met}$ solution was raised, the main band centered at 265 nm gradually decreased suggesting partial loss of the secondary structure of the tRNA, but band at 310 nm scarcely changed until up to 86°C, and then it rapidly decreased as shown in Fig. 9. The profile of the change in the band at 310 nm upon heating was close to the melting profile measuring the change in absorbancy at 260 nm. This observation suggests a close relation between heat denaturation of the tRNA and the conformation of m^5s^2U base.

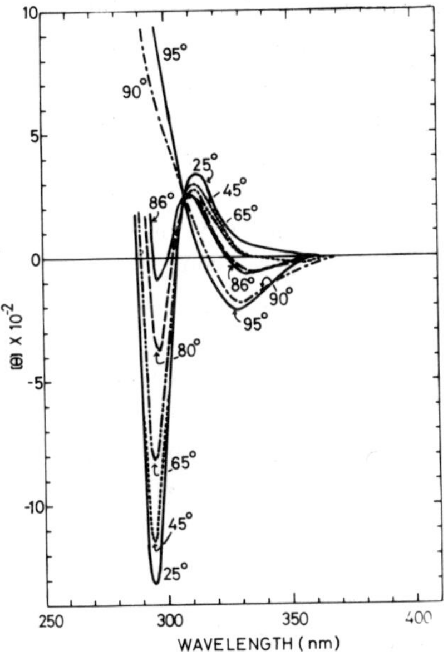

Fig. 9. Circular dichroic bands of *T.thermophilus* tRNA in near ultraviolet region at various temperatures. The spectra were recorded on a JASCO J-20 Spectropolarimeter connecting a circulating thermostatic bath.

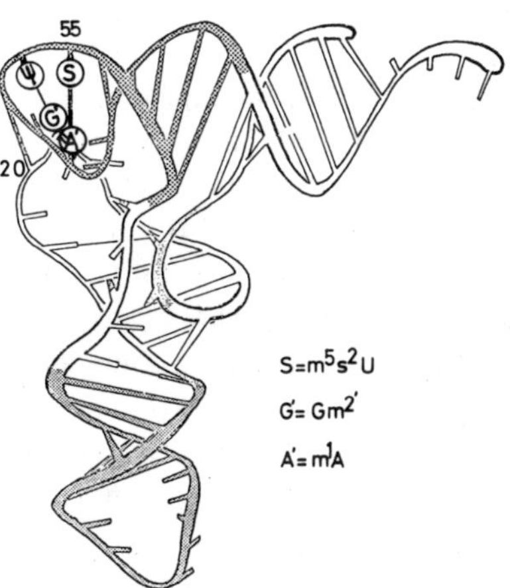

Fig. 10. A proposed three dimensional structure of *T.thermophilus* $tRNA_f^{Met}$. Four nucleotide residues were superimposed on the structure reported[28] for yeast $tRNA^{Phe}$.

Table 3. Effect of magnesium ion on melting temperature of tRNA. After tRNA was treated with sodium CyDTA to remove tightly bound Mg, melting temperature was measured in 10mM Tris buffer (pH 7.5) containing the indicated concentration of Mg^{++}.

tRNA	Mg^{++} concentration(M) 0	10^{-4}	10^{-2}	G-C pair content(%)
E.coli $tRNA_m^{Met}$	53	76	77	65
E.coli $tRNA_f^{Met}$	55	81	83	87.5
T.thermophilus $tRNA_f^{Met}$	54	87	89	90

2-Thiouridine derivatives are known to strengthen the stacking interaction with the neighboring bases[29,30]. In three dimensional structure assumed for the thermophile $tRNA_f^{Met}$ based on the data reported[28] for yeast $tRNA^{Phe}$ (shown in Fig. 10), m^5s^2U at position 55 which is base-paired with m^1A-59, stacks on the neighboring base G_m-19 thus TψC loop is connected with DHU loop. It is conceivable that the stacking force between m^5s^2U and G_m is stronger than that between T and G in mesophile tRNA and thus TψC loop is connected more tightly to DHU loop in the thermophile tRNA resulting in high stability of maintenance of the folded structure of the tRNA.

Another mechanism which is also responsible for the thermostability of the thermophile tRNA is the binding of magnesium ion. As reported for mesophile tRNA, the stability of tRNA from the thermophile was also greatly affected by the concentration of magnesium. As shown in Table 3, the remarkable thermostability of tRNA from the thermophile was observed only in the presence of magnesium indicating the crucial role of magnesium in the stability of tRNA molecule.

Finally we found that the thermal stability of tRNA from *T.thermophilus* is dependent on the growth temperature. Unfractionated tRNA was isolated from cells grown at various temperatures ranging from 50° to 80°C. The melting temperature was correlated positively with the growth temperature as shown in Fig. 11. Nucleotide analysis revealed that contents of only two components, m^5s^2U and T, varied upon changing the growth temperature. The sum of the mole fraction of m^5s^2U plus that of T was constant. A linear relation was observed

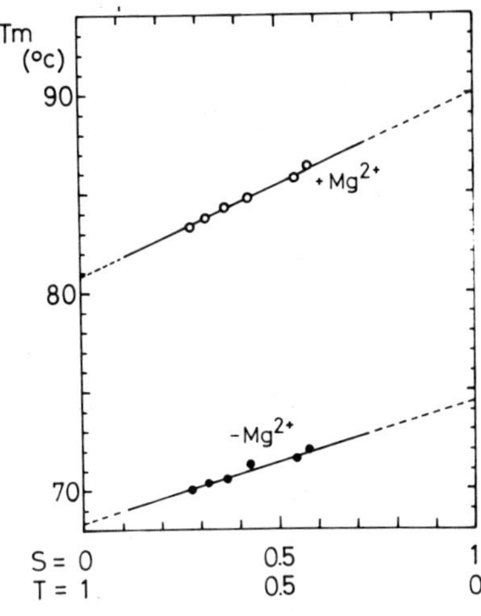

Fig. 11. The relationship between mole fraction of 5-methyl-2-thiouridine residue and melting temperature of tRNA. S = 5-methyl-2-thiouridine, T = ribothymidine.

between mole fraction of m^5s^2U and the melting temperature of tRNA preparation as shown in Fig. 11. This observation also strongly supports the idea that the presence of m^5s^2U is directly related to the thermostability of the thermophile tRNA. The results stated here indicate that the modification of T to m^5s^2U (thiolation) is dependent on the environmental temperature resulting in the change of melting temperature of the tRNA.

The thermostability of *T.thermophilus* tRNA is explained by the combination of three mechanisms for the thermal resistance; (i) intrinsic stability arised by the increment of G-C pair content and by tight stacking between m^5s^2U-55 and G_m-19, (ii) protection by magnesium ion, and (iii) temperature induced stability by thiolation of T base to m^5s^2U base depending upon the environmental temeprature.

REFERENCES

1. Oshima, T. & Imahori, K. (1971) *J. Gen. Appl. Microbiol. 17*, 513-517.
2. Oshima, T. & Imahori, K. (1974) *Intern. J. Syst. Bacteriol. 24*, 102-112.
3. Oshima, T. (1972) in *Molecular Evolution : Prebiological and Biological* (Rohlfing, D.L. and Oparin, A.I. eds) P.399-423, Plenum Publishing Co., New York.

4. Yoshizaki, F., Oshima, T. & Imahori, K. (1971) *J. Biochem. 69,* 1083-1089
5. Yoshida, M., Oshima, T. & Imahori, K. (1971) *Biochem. Biophys. Res. Commun. 43,* 36-39.
6. Yoshida, M., Oshima, T. & Imahori, K. (1973) *J. Biochem. 74,* 1183-1191.
7. Yoshida, M. (1972) *Biochemistry 11,* 1087-1093.
8. Yoshida, M. & Oshima, T. (1971) *Biochem. Biophys. Res. Commun. 45,* 495-500.
9. Ohno-Iwashita, Y., Oshima, T. & Imahori, K. (1975) *Z. Allg. Mikrobiol. 15,* 131-134.
10. Oshima, T. & Imahori, K. (1974) *J. Biochem. 75,* 179-183.
11. Watanabe, K., Seno, T., Nishimura, S., Oshima, T. & Imahori, K. (1973) *Polymer J. 4,* 539-552.
12. Watanabe, K., Oshima, T., Saneyoshi, M. & Nishimura, S. (1974) *FEBS Letters 43,* 59-63.
13. Oshima, M. & Yamakawa, T. (1974) *Biochemistry 13,* 1140-1146.
14. Sakaki, Y. & Oshima, T. (1975) *J. Virology 15,* 1449-1453.
15. Singleton, Jr., R. & Amelunxen, R.E. (1973) *Bacteriol. Rev. 37,* 320-342.
16. Fujita, S.C., Oshima, T. & Imahori, K., submitted
17. Fujita, S.C. & Imahori, K., in *Peptides, Polypeptides & Proteins* (Blout, E.R., Bovey, F.A., Goodman, M. & Lotan, N. eds) P.217-229, John Wiley & Sons, London and New York.
18. Bigelow, C.C. (1967) *J. Theoret. Biol. 16,* 187-211.
19. Endo, S. (1962) *Hakko Kogaku Zasshi 40,* 346-353.
20. Mizusawa, K., Ichishima, E. & Yoshida F. (1964) *Agr. Biol. Chem. 28,* 884-895.
21. Pfueller, S.L. & Elliott, W.H. (1969) *J. Biol. Chem. 244,* 48-54.
22. Ogasahara, K., Imanishi, A. & Isemura T. (1970) *J. Biochem. 67,* 77-82.
23. Hasegawa, A., Miwa, N., Oshima, T. & Imahori, K. *J. Biochem.* in press
24. Oshima, M. & Miyagawa, A. (1974) *Lipids 9,* 476-480.
25. Shinitzky, M., Dianoux, A.C., Gitler C. & Weber, G. (1971) *Biochemistry 10,* 2106-2113.
26. Cogan, U., Shinitzky, M., Weber, G. & Nishida, T. (1973) *Biochemistry 12,* 521-528.

27. Linden, C.D., Wright K.L., McConnell, H.M. & Fox, C.F. (1973) *Proc. Nat. Acad. Sci. USA 70,* 2271-2275.

28. Kim, S.H., Suddath, F.L., Quigley, G.J., McPherson, A., Sussman, J.L., Wang, A.H.J., Seeman N.C. & Rich, A. (1974) *Science 185,* 435-440.

29. Mazumdar, S.K., Saenger, W. & Scheit, K.H. (1974) *J. Mol. Biol. 85,* 213-229.

30. Bähr, W., Faerber, P. & Scheit, K.H. (1973) *Eur. J. Biochem. 33,* 535-544.

THE EFFECT OF POLYAMINES ON THE THERMOSTABILITY OF A CELL FREE PROTEIN SYNTHESIZING SYSTEM OF AN EXTREME THERMOPHILE

Y. OHNO-IWASHITA, T. OSHIMA* and K. IMAHORI
Department of Agricultural Chemistry, Faculty of Agriculture, the University of Tokyo, Tokyo 113, Japan
and
*Mitsubishi-Kasei Institute of Life Sciences, Machida, Tokyo 194, Japan

INTRODUCTION

The enzymes of thermophilic organisms, because of their high thermostability, have been used widely not only for the study of the molecular mechanism of thermostability but also for the elucidation of structure and function relationship of the enzymes. In our laboratory we have purified and studied several enzymes from an extreme thermophile, *Thermus thermophilus* HB8, which was isolated from Japanese hot spring[1]. This organism is a gram-negative, non-sporulating rod shaped bacterium with its maximum growing temperature at 85°C and the minimum at 47°C.

Generally, all the enzymes so far tested were thermostable enough to account for the maximum growing temperature of the organism[2,3]. The enzymes involved in the protein synthesizing system were also thermostable in the sense that they were not irreversibly inactivated at high temperature[4]. However, the cell free protein synthesizing system of this organism showed only very low activity at high temperature. By our studies in detail on the system it was found that some special polyamines as spermine were obligatory for the stimulation of protein synthesis at high temperature[5]. In this communication we will report a detailed mechanism of polyamine actions.

MATERIALS AND METHODS

Bacterial strains and preparation of subcellular components

The thermophiles used in this study were *T.thermophilus* HB8 (=ATCC 27634)[1] and another thermophile, V-2[6]. The cells were harvested at the early exponential phase. Ribosomes and supernatant fractions

(S-100) of these bacteria were prepared by the method of Nirenberg and Matthaei[7].

Ternary complex of poly(U), phenylalanyl-tRNA and a ribosome

Formation of a ternary complex was detected by the method of Nirenberg and Leder[8]. A mixture (0.1ml) containing 45μg of L-[^{14}C] phenylalanyl-tRNA obtained from *T.thermophilus* HB8 (5050 cpm, 20.4 pmoles of phenylalanine), 3 A_{260} units of ribosomes and 12μg of poly(U) was incubated at 45°C or 65°C. At intervals a 20μl aliquot was filtered through a Millipore filter.

Polymerization assay

The standard reaction mixture (0.1ml) consisted of 50mM Tris-HCl, pH 7.8, 100mM NH_4Cl, 6mM 2-mercaptoethanol, Mg^{++} and polyamines at the indicated concentrations, 3.2mM ATP, 0.4mM GTP, 54-81μg of the thermophile tRNA, 50-100μg of fraction S-100, 2 nmoles of L-[^{3}H] phenylalanine (specific activity 0.5Ci/mmole), 2 nmoles each of 19 other amino acids, 12μg of poly(U), and 1 A_{260} unit of ribosomes.

Two different assays were applied —— with preincubation and without preincubation. In the experiments without preincubation, the reaction was started with the addition of ribosomes. In the other experiments, the preincubation was carried out in the mixture (0.1ml) consisted of 50mM Tris-HCl, pH 7.8, 100mM NH_4Cl, 6mM 2-mercaptoethanol, 18-22mM Mg^{++} or 6mM Mg^{++} plus 1.5mM spermine, 1 A_{260} unit of ribosomes, 10-100μg of poly(U) and 45μg of L-[^{14}C] phenylalanyl-tRNA. This mixture was incubated at the indicated temperatures and the reaction was started by adding the other components of the reaction mixture described above.

When L-[^{14}C]phenylalanyl-tRNA (22pmoles) was used as a starting material, ATP, tRNA and L-[^{3}H]phenylalanine were omitted from the reaction mixture.

The extent of polymerization was estimated by measuring the radioactivity in trichloroacetic acid-insoluble materials.

RESULTS AND DISCUSSION

Effect of polyamines on phenylalanine incorporation

Poly(U)-dependent polyphenylalanine synthesizing activity of the cell free system of *T.thermophilus* HB8 was examined on the temperature range of 20°C-85°C. Above 50°-55°C the activity was observed very low in the absence of spermine. However, the synthesis was enhanced reasonably high by an addition of spermine. Figure 1 indicates the time course of polymerization reaction at 65°C, both in the presence and absence of spermine. This effect of spermine cannot be to prevent the irreversible inactivation of the system. As indicated in Figure 1, when spermine was added to the system after the 10 min. heating at 65°C, the reaction started at almost the same rate as in the mixture supplemented with spermine.

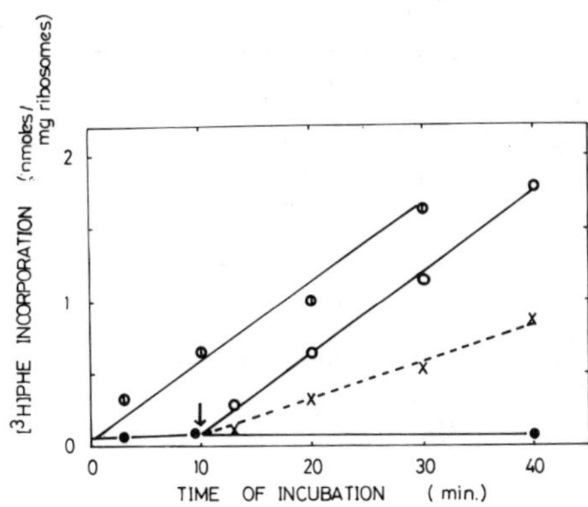

Fig. 1. Time course of L-[^{3}H]phenylalanine incorporation at 65°C. The reaction was carried out in the presence of 18 mM Mg^{++}(●—●), or 18mM Mg^{++} plus 2mM spermine(⦶—⦶). In some experiments the reaction was started without spermine and at 10 min. 2mM spermine(o—o), or 2mM spermine plus 20mM putrescine (×--×) was added. The arrow indicates the point of addition of the polyamines.

Such a stimulation effect was tested with several available polyamines listed in Table I. Among them, three amines, spermine, spermidine and thermine, a new polyamine found in this organism[9], were found to be effective. Other polyamines including putrescine or Mg^{++} alone were ineffective.

Next, we examined the exact step of protein synthesis reaction which spermine would work on to stimulate the reaction. Amino acyl-tRNA synthetase was active at high temperature regardless of the absence or presence of spermine : the addition of spermine did not show any stimulation. Truly, as shown in Figure 2, when polymerization reaction was carried out using phenylalanyl-tRNA the reaction did not proceed unless spermine was added. Thus it became evident spermine stimulates some of the steps after amino acyl-tRNA formation.

Table I. Effects of various polyamines on phenylalanine incorporation

Name	Structure	Assay conditions (mM)	(Optimum) (mM)	Effect
spermine	$NH_2(CH_2)_3NH(CH_2)_4NH(CH_2)_3NH_2$	1 - 15	(2 - 3)	+++
thermine	$NH_2(CH_2)_3NH(CH_2)_3NH(CH_2)_3NH_2$	0.5 - 4	(1.5)	++
spermidine	$NH_2(CH_2)_4NH(CH_2)_3NH_2$	2 - 30	(10)	+
putrescine	$NH_2(CH_2)_4NH_2$	5 - 40		–
1,3-propanediamine	$NH_2(CH_2)_3NH_2$	20		–
1,2-ethylenediamine	$NH_2(CH_2)_2NH_2$	20		–
1,8-octamethylene-diamine	$NH_2(CH_2)_8NH_2$	8 , 17		–
1,12-dodecamethylene-diamine	$NH_2(CH_2)_{12}NH_2$	4 , 13		–
N,N'-dimethylethylene-diamine	$CH_3NH(CH_2)_2NHCH_3$	20		–
N,N-dimethylpropane-diamine	$NH_2(CH_2)_3N(CH_3)_2$	20		–

The reaction was measured after incubation for 10 min. at 65°C in the presence of polyamine. The symbol (-) indicates that no stimulation was observed.

Effect of cold preincubation

In the course of our trial to determine the step of spermine stimulation we found a very interesting phenomenon as described in the following.

Phenylalanyl-tRNA, poly(U) and a ribosome were first incubated in a cold for a longer time (for example at 10°C for 30 min.). Other necessary components being added, the mixture was brought up to 65°C to start the reaction. Polymerization reaction proceeded normally even without the addition of spermine. Omission of any component from the cold preincubation mixture reduced the polymerization rate at high temperature down very low.

This suggests that the formation of a ternary complex of poly(U), phenylalanyl-tRNA and a ribosome in the cold was crucial for the polymerization reaction at high temperature. The existence of such a ternary complex was indicated by the following experiment which is shown in Figure 3. When phenylalanyl-tRNA, poly(U) and a ribosome were incubated together in the cold and passed through a CPG-10 column,

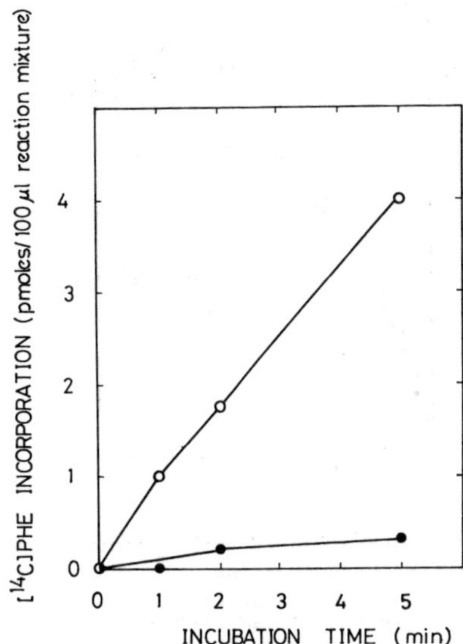

Fig. 2. Effect of spermine on phenylalanine incorporation when L-[^{14}C]phenylalanyl-tRNA was used instead of phenylalanine and tRNA. The incorporation was assayed at 65°C in the presence of 18mM Mg^{++} (●—●), or 3mM spermine plus 10mM Mg^{++} (o—o).

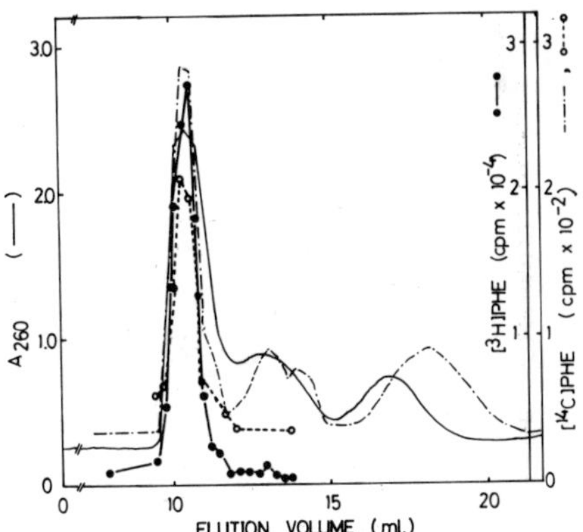

Fig. 3. Isolation of the ternary complex by a CPG-10 (Controlled-Pore Glass, Electro-nucleonics, Inc.) column chromatography. A reaction mixture (50μl) containing 20mM Mg^{++}, 30μg of poly(U), 135μg of L-[^{14}C]phenylalanyl-tRNA (6500 cpm, 66pmoles of phenylalanine), and 5 A_{260} units of ribosomes was incubated for 30 min. at 10°C and was applied to the CPG-10 column (7.5 × 600 mm) equilibrated with the buffer containing 50mM Tris-HCl, pH 7.8, 20mM Mg^{++}, 100mM NH_4Cl, 6mM 2-mercaptoethanol. ——, absorbance at 260 nm; —·—, radioactivity of each fraction (per 50μl). Each effluent fraction (50μl) was supplemented with the other necessary components including L-[^{3}H]phenylalanine and 5 × 10^{-5}M ATA, and [^{3}H] (●—●) or [^{14}C] (o--o) phenylalanine incorporation into hot TCA insoluble material was assayed at 65°C for 5 min..

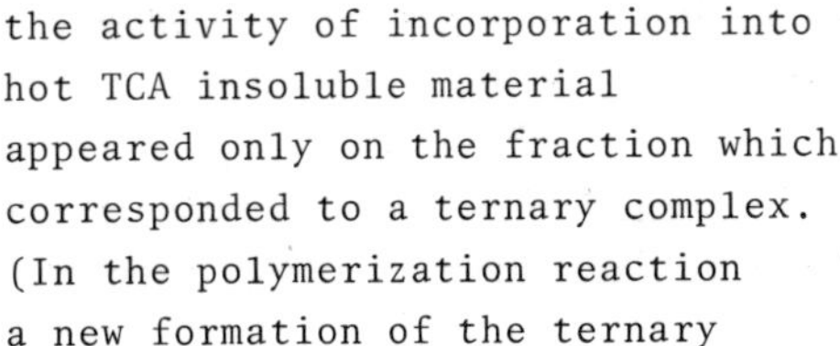

the activity of incorporation into hot TCA insoluble material appeared only on the fraction which corresponded to a ternary complex. (In the polymerization reaction a new formation of the ternary complex was prevented by the additon of aurin tricarboxylic acid (ATA))[10]. Such a ternary complex which is potent to initiate the polymerization reaction at high temperature is called "active ternary complex" by the reasons explained later. It will also be

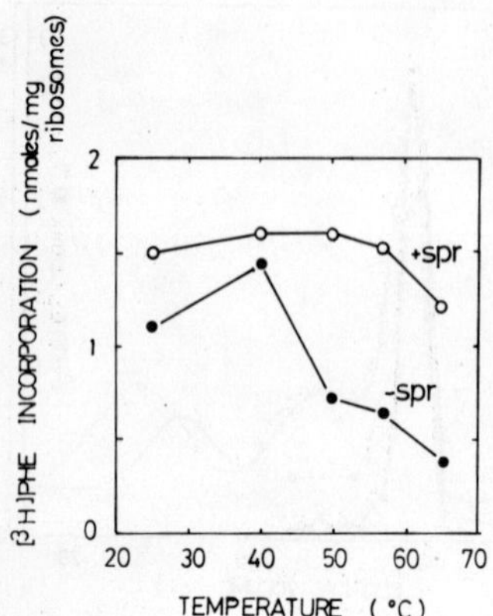

Fig. 4. Active ternary complex formation at various temperatures. The reaction mixture was preincubated for 1 min. at the indicated temperature in the presence of 6mM Mg^{++} plus 1.5mM spermine (o—o), or of 20mM Mg^{++} (●—●). And then the other necessary components were supplemented and polymerization was assayed in the presence of 6mM Mg^{++} plus 1.5mM spermine at 65°C for 3 min.

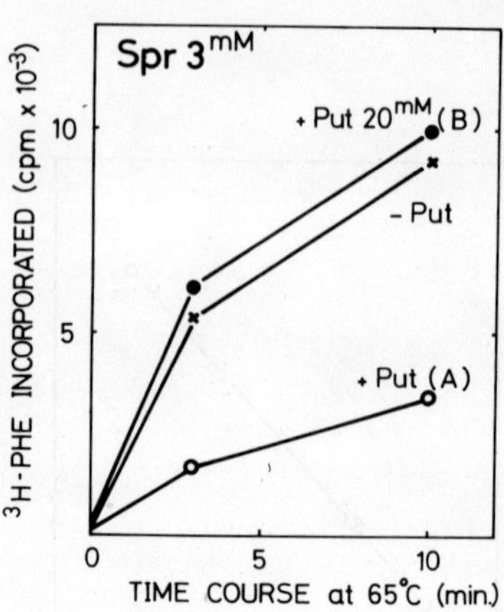

Fig. 5. The suppression of the spermine effect by putrescine. The reaction mixture containing ribosomes, poly (U) and phenylalanyl-tRNA was preincubated at 65°C for 2 min. in the presence of 3mM spermine (×—×, ●—●), or of 3mM spermine plus 20mM putrescine (o—o). After the other necessary components were supplemented L-[^{3}H]phenylalanine incorporation was measured at 65°C in the presence of 3mM spermine (×—×) or 3mM spermine plus 20mM putrescine (●—●, o—o).

indicated that the amount of active complex depends on the time of cold incubation and the concentration of Mg^{++}, and that this amount relates to the activity of polyphenylalanine synthesis at high temperature.

Thus the essential role of the cold preincubation would exist in the formation of active ternary complex of phenylalanyl-tRNA, poly(U) and a ribosome.

The effect of spermine on the active ternary complex formation

Then we examined if the stimulation effect of spermine at high temperature can be explained also by enhancement of such active ternary complex formation.

After the reaction mixture containing L-[^{14}C]phenylalanyl-tRNA, poly(U) and a ribosome was incubated for one minute at each temperature with or without spermine, aurin tricarboxylic acid (5×10^{-5}M) was added to the reaction mixture in order to prevent further formation of ternary complex. The mixture was supplemented with other necessary components, brought up to 65°C and polymerization activity was assayed. In such experiment the polymerization activity would reflect the number of active ternary complex formed during the preincubation. The results are shown in Figure 4. Now it is clear that in the presence of spermine almost the same amount of active complex was formed at high temperature as well as at low temperature. In the absence of spermine active complex was formed only at low temperature. Probably, at high temperature the formation of active complex was prohibited in the absence of spermine, and thus very low polymerization activity was resulted. The cold preincubation or the addition of spermine would be effective to remove the prohibiting effect.

We have conducted further experiments to prove that spermine works in the step of active complex formation, which are shown in Figure 5. Putrescine, as shown in Table I, did not demonstrate any stimulation effect. However, when putrescine was added to the reaction mixture with spermine prior to the active ternary complex formation, putrescine suppressed the stimulatory effect of spermine proportionally to its amount. On the contrary if putrescine was added after active complex formation it did not show any inhibitory effect. Thus it is deduced that the effect of spermine exists in the step of active complex formation.

As shown above, the effect of cold preincubation and that of spermine addition run in parallel. In fact, when the addition of spermine at 10 min. in Figure 1 was replaced by cold preincubation (at 10°C for 3 min.) the reaction proceeded with a comparable rate as the addition of spermine.

The effect of cold preincubation and spermine addition can be compared more clearly in the following results. Figure 6 represents the polymerization rate as Arrhenius plot. Once active ternary complex was formed by cold preincubation the reaction rate at each temperature gave a straight line up to 65°C, however, without cold preincubation a break point was observed at around 55°C. In the presence of spermine the reaction rate at each temperature gave a

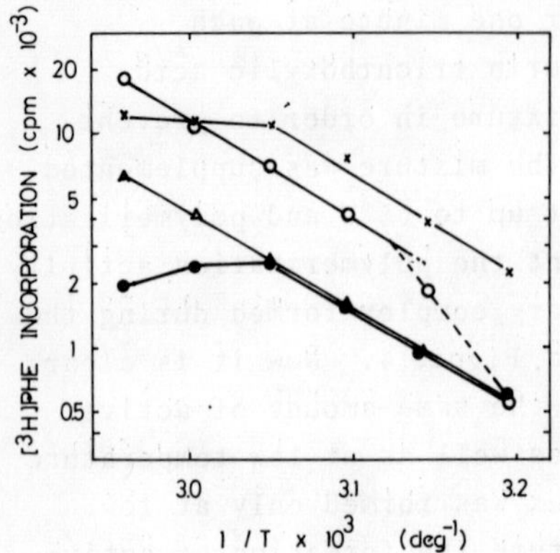

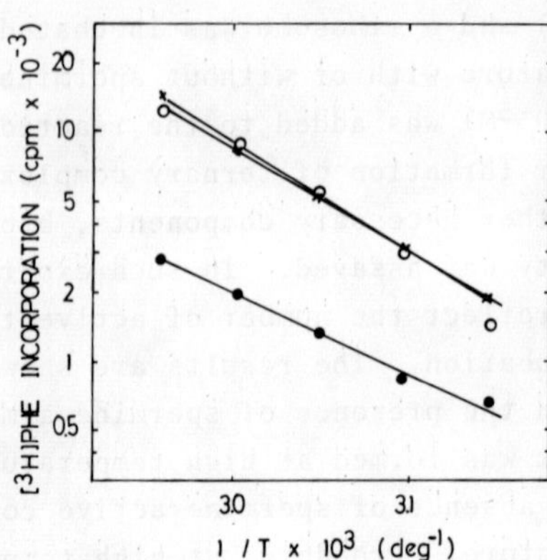

Fig. 6. The Arrhenius plot of L-[^{3}H] phenylalanine incorporation reaction. Δ—Δ, with preincubation of poly(U), phenylalanyl-tRNA and ribosomes at 10°C for 30 min. in the presence of 20mM Mg^{++}; ●—●, o—o, ×—×, without preincubation. L-[^{3}H]phenylalanine incorporation was measured in the presence of 20mM Mg^{++} (Δ—Δ, ●—●), or of 6mM Mg^{++} plus 1.5 mM spermine with (×—×) or without (o—o) 20mM putrescine. In this experiment the concentration of NH_4Cl was 75mM.

Fig. 7. The Arrhenius plot of L-[^{3}H] phenylalanine incorporation reaction. After the formation of the complex between poly(U) and a ribosome in the presence of 20mM Mg^{++} at 10°C for 30 min., the other necessary components including 5 × 10^{-5}M ATA were supplemented and phenylalanine incorporation was measured in the presence of 20mM Mg^{++} (●-●), or 5mM Mg^{++} plus 2mM spermine (o-o), or 8mM Mg^{++} plus 2mM spermine plus 20mM putrescine (×-×).

straight line up to 65°C regardless of the preincubation. On the contrary the addition of both spermine and putrescine prior to the active complex formation resulted in an appearance of a break point at around 55°C. Thus it became clear that both cold preincubation and spermine addition are effective to rescue the reaction system from leveling off at high temperature.

It is conceivable that such a rescuing effect of spermine or cold preincubation affects the step of the association of a ribosome with poly(U) (Step I of spermine effect). As shown in Figure 7, once a complex of a ribosome and poly(U) was formed by cold preincubation or spermine reaction, Arrhenius plot of polymerization gave a straight line. The addition of putrescine did not result a break of the line in the plot.

Another effect of spermine

It can be seen from both Figure 6 and 7 that the addition of spermine stimulated the reaction rate throughout the whole temperature range.

It should be noted that in Figure 7 Step I was accomplished first by cold preincubation and then polymerization reaction was conducted with or without spermine. Thus, the stimulation observed in Figure 7 cannot be assigned to Step I effect. Furthermore the addition of putrescine with spermine did not inhibit the reaction but rather stimulated it. This effect of spermine (Step II of spermine effect) should be different from the one previously mentioned (Step I of spermine effect) in the sense that it was not repressed by putrescine as well as that it is effective throughout the whole temperature range. The parallel shift observed in Arrhenius plot may suggest the difference in the amount of the initiation complex, although it is so far ambiguous.

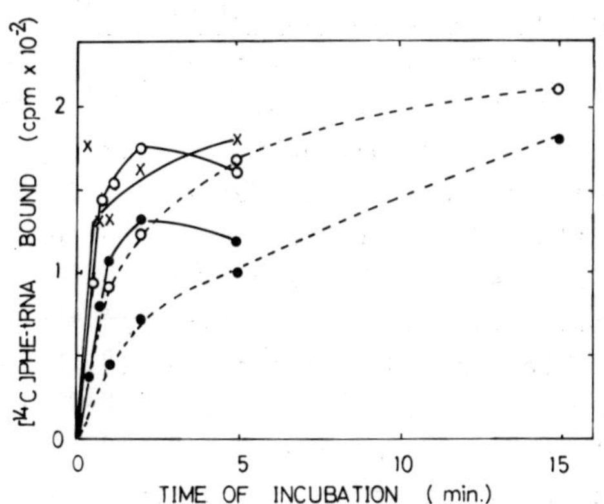

Fig. 8. Ternary complex formation at 65°C and 45°C. Details in the procedure was presented in Materials and Methods. ●—●, 20mM Mg^{++}; o—o, 6mM Mg^{++} plus 1mM spermine; ×—×, 6mM Mg^{++} plus 1mM spermine plus 20mM putrescine. Dotted line shows that the reaction was carried out at 45°C and solid line at 65°C.

Figure 8 shows poly(U)-dependent binding of phenylalanyl-tRNA to the ribosomes assayed by Millipore filter method. It is shown that the binding is stimulated by spermine at 45°C as well as at 65°C. The addition of putrescine moreover does not inhibit the binding. The results suggest that the spermine effect observed in the phenylalanyl tRNA binding experiments reflects stimulation of the Step II.

Conversion of inactive ternary complex to the active one

Ternary complex observed in the above experiment, however, does not necessarily mean that it is fully active. Figure 8 indicates that almost the same amount of ternary complex was likely formed at 65°C in the absence of spermine. Such complex, however, cannot initiate enough the polymerization reaction, as previously indicated in Figure 4. In this sense such complex is called "inactive complex" in contrast to the "active complex" previously defined. Then it may be deduced that one of the spermine effects would be to convert "inactive complex" to "active complex".

The condition of the 10 min. incubation at 65°C applied in Figure 1 corresponds to that of inactive complex formation in Figure 8. The fact that addition of spermine initiated the polymerization reaction should mean nothing but that inactive complex was converted to the active one. That the addition of putrescine together with spermine repress the reaction suggests that the conversion of inactive complex to the active one is related to the Step I of spermine effect.

Now, the effects of spermine can be classified into two major categories. The first one is the effect which normalizes the reaction at high temperature, can be replaced by cold preincubation and

Table II. Stimulation effect of spermine on polymerization reaction using various combinations of ribosomes and fraction S-100 from two thermophiles.

S100	RIB	-Spr			+Spr 0.5mM			+Spr 1.5mM		
		Mg^{++} opt.	Activity		Mg^{++} opt.	Activity		Mg^{++} opt.	Activity	
		mM	nmoles/mg ribosomes	(%)	mM	nmoles/mg ribosomes	(%)	mM	nmoles/mg ribosomes	(%)
V-2	V-2	18-22	0.323	(100)	10	0.287	(89)	5-8	0.384	(119)
HB8	V-2	18-22	0.117	(100)	10	0.205	(175)	7	0.155	(132)
V-2	HB8	22	0.407	(100)	12	0.583	(143)	5	1.115	(274)
HB8	HB8	20-25	0.240	(100)	10	0.482	(201)	5	1.527	(637)

Polymerization reaction was measured at 57°C in the presence of 75mM NH_4Cl.

suppressed by putrescine. Thus this effect is suggested to affect the step of active ternary complex formation —— probably the step of association of poly(U) with ribosomes (Step I). The second effect is the stimulation of polyphenylalanine synthesis observed throughout the temperature range of 40°-65°C, effective even after the cold preincubation and is not suppressed by putrescine.

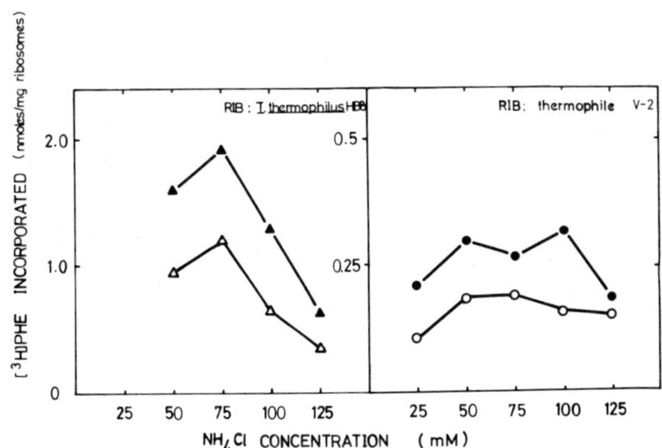

Fig. 9. The suppression of spermine effect by NH_4Cl or putrescine tested with subcellular components from two thermophiles. Polymerization reaction was assayed at 57°C with various combinations of ribosomes and S-100 fraction from *T.thermophilus* HB8 and a thermophile, V-2. The combinations are represented as follows:

S-100 \ ribosomes	HB8	V-2
HB8	▲-▲	o-o
V-2	Δ-Δ	●-●

Site of action of spermine

In order to identify the component of protein synthesizing system which spermine would affect directly, we conducted following experiment.

Another moderate thermophile, V-2, was isolated from a Japanese hot spring. The cell free system of this organism is quite active at 60°C without the supplement of spermine. Several combinations of ribosomes and S-100 fractions from the two organisms were tested in Table II, and Figure 9. It was revealed that both stimulation effect of spermine and the suppression of it by putrescine could be observed only when *T.thermophilus* ribosomes were used. Thus inability of polymerization without spermine could be assigned to the ribosomes.

It is easier to explain the results of the experiments discussed above by postulating ribosomes can have two forms in equilibrium —— active form and inactive form. The equilibrium would shift to

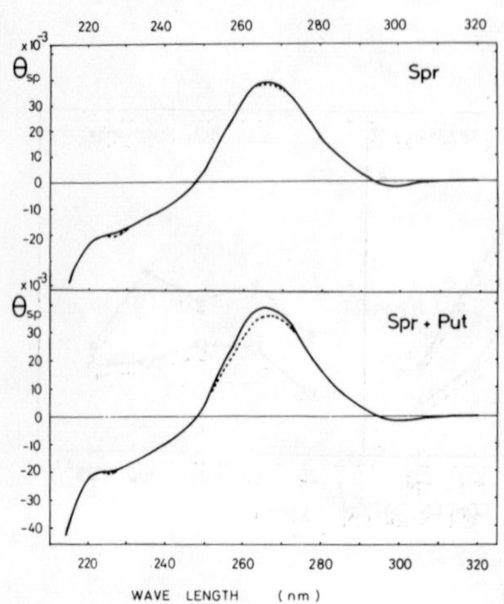

Fig. 10. Circular dichroic spectra of *T.thermophilus* HB8 ribosomes. The spectra of the ribosomes were recorded in 0.05M Tris-HCl buffer, pH 7.8, containing 75mM NH_4Cl, 6mM 2-mercaptoethanol, 5mM Mg^{++}, and 1.5mM spermine with (below) or without (above) 20mM putrescine. Solid line shows the spectrum at 45°C and dotted line at 62°C.

the active form in the cold, but at high temperature inactive form becomes prevalent. In the presence of spermine the equilibrium would shift to the active one but the addition of putrescine would revert this equilibrium shift. Once the "ternary complex" is formed from the active form ribosomes, it cannot be reverted to the inactive form by high temperature or by the addition of putrescine.

Truly, these ribosomes showed conformational change, detected by circular dichroism, at around 55°C, when spermine was omitted or when both spermine and putrescine were added. However, in the presence of spermine no significant conformation change was observed (Figure 10).

Native polyamines extracted from *T.thermophilus* cells

We have revealed a very important function of spermine in the cell free system from *T.thermophilus* HB8. It is generally accepted, however, that spermine does not occur in microorganisms[11]. Thus we have tried to extract polyamines from the present organism.

It was found that this becterium had mainly two kinds of polyamines —— spermine and thermine[9]. As shown in Table I, these polyamines were the most effective on *in vitro* polymerization reaction. Besides, the content of spermine in this bacterium was high at an early log-phase of its growth and decreased sharply as it went into the late log-phase.

These observations suggest that these polyamines, especially spermine, also play important roles on the protein synthesis *in vivo*. Furthermore, thermine was detected in all extreme thermophiles so far tested, and spermine was in moderate and extreme thermophiles, and found scarcely in mesophiles.

Thus it is conceivalbe that spermine and thermine are obligatory for thermophile growth and play a significant role *in vivo*.

REFERENCES

1. Oshima, T. & Imahori, K. (1974) *Intern. J. System. Bacteriol. 24*, 102-112.
2. Yoshida, M., Oshima, T. & Imahori, K. (1971) *Biochem. Biophys. Res. Commun. 43*, 36-39.
3. Yoshizaki, F., Oshima, T. & Imahori, K. (1971) *J. Biochem. 69*, 1083-1089.
4. Ohno-Iwashita, Y., Oshima, T. & Imahori, K. (1975) *Z. Alleg. Mikrobiol. 15*, 131-134.
5. Ohno-Iwashita, Y., Oshima, T. & Imahori, K. *Arch. Biochem. Biophys.* , in press.
6. Hasegawa, A., Miwa, N., Oshima, T. & Imahori, K. submitted to *J. Biochem.*
7. Nirenberg, M.W. & Matthaei, J.H. (1961) *Proc. Natl. Acad. Sci. U.S.A. 47*, 1588-1602.
8. Nirenberg, M.W. & Leder, P. (1964) *Science 145*, 1399-1407.
9. Oshima, T. (1975) *Biochem. Biophys. Res. Commun. 63*, 1093-1098.
10. Grollman, A.P. & Stewart, M.L. (1968) *Proc.Natl. Acad. Sci. U.S.A. 61*, 719-725.
11. Cohen, S.S. (1972) *Advan. Enzyme Regul. 10*, 207-223.

that tested, and [illegible] was [illegible] and [illegible] and found [illegible] respectively.

Thus it is conceivable that [illegible] and their [illegible] are obligatory for the [illegible] and play a [illegible] role [illegible].

REFERENCES

1. [illegible] (1974) [illegible]
2. [illegible] & [illegible] (1977) [illegible]
3. [illegible] & [illegible] (1979) [illegible] 1089-1094.
4. [illegible] & [illegible] (1977) [illegible]
5. [illegible] [illegible]
6. [illegible] submitted to [illegible]
7. [illegible] (1981) [illegible] 1488-1494.
8. [illegible] 1399-1405.
9. [illegible] (1982) [illegible]

PURIFICATION, CHARACTERIZATION AND MECHANISM OF ACTION OF SEVERAL AMINOACYL-tRNA SYNTHETASES FROM BACILLUS STEAROTHERMOPHILUS

H. Grosjean, J. Charlier, C. Darte, G. Dirheimer*, R. Giege*, S. de Henau, G. Keith*, R. Parfait and C. Takada.

Laboratoire de Chimie Biologique, Université Libre de Bruxelles, 65 rue des Chevaux, B-1640 Rhode-St Genèse.

The 20 aminoacyl-tRNAsynthetases, one specific for each amino acid, catalyse the following reaction:

$$aa_1 + ATP + tRNA_1 \xrightleftharpoons{\text{aa-tRNAsynthetase, } Mg^{++}} aa_1\text{-}tRNA_1 + AMP + PP_i$$

In vivo these reactions are extremely specific and constitute a key step in the overall protein biosynthesis process.

The aminoacyl-tRNAsynthetases from Bacillus stearothermophilus were purified and studied as a mean of identifying some parameters of the recognition mechanism between the enzymes and their cognate tRNAs. The use of thermostable enzymes from B. stearothermophilus would lead us to study the interactions with the tRNA under conditions where their tertiary and secondary structure might be affected without too much incidence on the enzyme activity. In this work we report mostly on the effect of an organic solvent such as dimethylsulfoxide, which was shown to modify the tertiary stucture of the tRNAs (1, 2) without any significant loss of activity of the aminoacyl-tRNAsynthetases from B. stearothermophilus up to 40% Me_2SO (3). One might also expect to find some particularities in the cross reactions between the synthetases from a thermophilic organism and tRNAs from Escherichia coli and yeast. Indeed several authors have demonstrated that in heterologous in vitro systems, where the synthetase and the tRNA are from different origin, like E. coli and yeast (4,5 and references therein), the specificity of the interactions is not so absolute as in the homologous systems and several misacylations occur. Our experiments of misacylations between the aminoacyl-tRNAsynthetases from B. stearothermophilus and pure tRNAs from E. coli and yeast give us some information about the essential parameters for the discrimination of the tRNA by the synthetase.

* I.B.M.C. Strasbourg (France)

1. Purification and general characteristics of the aminoacyl-tRNAsynthetases from Bacillus stearothermophilus.

Our classical purification procedure applied to about 100 g Bacillus stearothermophilus NCA 1518 cells includes lysis by a French pressure cell, ultracentrifugation at 10,000 g and 105,000 g for 2 hours, fractionation by $(NH_4)_2SO_4$, adsorption and elution from Cγ alumina gel, chromatography on DEAE-cellulose followed by hydroxylapatite chromatography of the several fractions after DEAE-cellulose (details have already been described 6, 7, 8). The activity for the 20 aminoacyl-tRNAsynthetases was followed in all the purification steps. We obtained homogeneous preparations for 6 synthetases specific for ile, val, met, arg, gly and lys and 3 fractions for pro, thr and cys, free of other synthetases. The enzymes specific for his and trp were contaminated with each other. The synthetases specific for asp, ser, ala and also for phe, leu, tyr and glu were not completely pure and contaminated with each other. The complete purification of the leucyl-tRNAsynthetase was reported elsewhere (6).

Table 1 summarizes the great resistance to heat inactivation of the enzymes from B. stearothermophilus compared to some of those from E. coli and also the molecular weights and subunit structure as determined by sucrose gradient centrifugation and SDS gel electrophoresis respectively. The melting temperatures (T_m corresponding to the temperature at which 50 % of the initial activity remains after heating the enzyme in a waterbath with increasing temperature of 1 °C per minute, cooling to 0 °C and measuring the residual activity at 40 °C) referred to the intrinsic stability of the proteins in absence of the substrates. The presence of saturating amounts of substrates, amino acid, ATP and tRNA generally enhances the T_m by 5 to 10 °C. The optimal temperatures for the aminoacylation reaction are less than the T_m's, as for example the leucyl-tRNAsynthetase, which exhibits an optimal temperature of 60 °C (6), the arginyl-tRNAsynthetase, which has an optimal temperature of 58 °C. The isoleucyl-tRNAsynthetase has one of 45 °C and the valyl-tRNAsynthetase one of 52 °C (3, 10).

The aminoacyl-tRNAsynthetases from B. stearothermophilus show a large diversity in molecular weights ranging from 52,000 to 221,000 with most of them about 100,000 daltons. The repartition is already very similar to that of E. coli and yeast (5, 11).

Table 1 : Characterization of the aminoacyl-tRNAsynthetases from B.stearothermophilus compared to E.coli

synthetase specific for:	B.stearothermophilus Tm (°C)	MW x10^{-3} (subunits)	E.coli Tm (°C)	MW x10^{-3}
Ala	66	173		
Arg	75	62 (α_2 or $\alpha\beta$)**		75 (α)
Asn	65	114		
Asp	66	129		
Cys	79	61		
Glu	68	42		102 ($\alpha\beta$)
Gln		126		69 (α)
Gly	69	120 (α_2)		227 ($\alpha_2\beta_2$)
His	52	102		85 (α_2)
Ile	67	110 (α)	55	114 (α)
Leu	73	95 (α)	54	105 (α)
Lys	62	130 (α_2)		104 (α_2)
Met	71	153 (α_2)		173 (α_2)
Phe	68	221		180 (α_4)
Pro	68	125		94 (α_2)
Ser	67	92 (α_2)		95 (α_2)
Thr	73	120		117
Trp	67	76 (α_2^*)		74 (α_2)
Tyr	66	90 (α_2^*)		95 (α_2)
Val	73	99 (α)		110 (α)

The melting temperatures were measured as described in the text. The molecular weights for the synthetases from B. stearothermophilus were determined by sucrose gradient centrifugation and the subunits by SDS gel electroforesis. The molecular weights from E. coli enzymes are from Söll and Schimmel (5) and the values * from Koch et al. (11) ; ** the Arginyl-tRNAsynthetase from B. stearothermophilus was studied by J-M. Godeau (submitted to Eur. J. Biochem (1975)

A more systematic comparative study of the isoleucyl-, leucyl- and arginyl-tRNAsynthetases from Bacillus stearothermophilus and E.coli reveal that the identity of properties and mechanism of action are very striking except for the difference in thermostability. This comparison includes: 1. the kinetic constants for the substrates amino acid, ATP and cognate tRNA (6,7,8).

2. the activation energy for the activation and transfer of the amino acid onto tRNA (7,12).

3. the stability of the intermediary complexes aminoacyl-AMP-enzyme (13) and enzyme-tRNA (7).

4. the effect of Mg^{2+} (14) and the sensibility towards SH-reagents (15).

The great similarity between all the aminoacyl-tRNAsynthetases from Bacillus stearothermophilus and E.coli could also be deduced from the results of cross-reactions at 37°C between enzymes and tRNAs from the two sources. In all cases they are completely exchangeable which is not the case between Bacillus stearothermophilus and yeast or E.coli and yeast. Moreover when a synthetase from yeast failed to aminoacylate a tRNA from E.coli the equivalent tRNA from Bacillus stearothermophilus is never acylated too and vice versa tRNAs from yeast which are not recognized by the E.coli synthetases behave in the same manner with the Bacillus stearothermophilus enzymes. Similarly other proteins from Bacillus stearothermophilus involved in the protein biosynthesis as the initiation and elongation factors and the ribosomal proteins are generally exchangeable with the equivalent from E.coli without any appreciable loss of activity of the whole E.coli system. Evolutionary speaking, the structure and function of the proteins and nucleic acids involved in protein synthesis in E.coli and Bacillus stearothermophilus must be very close.

2. tRNAs from Bacillus stearothermophilus

Two tRNAs specific for phenylalanine and valine have been purified from Bacillus stearothermophilus strain NCA 1518 and their sequences determined with the collaboration of Dr Keith in the laboratory of professor Dirheimer (Strasbourg). Figure 1 represents these two nucleotide sequences folded in the clover leaf configuration of $tRNA^{phe}$ and $tRNA^{val}$. Comparison with the corresponding tRNAs from yeast and E.coli shows that at least 50 to 70% of the bases are identical. We have not found any particularity in the base composition and sequence of the two nucleic acids from Bacillus stearothermophilus that could be related to thermophily. Oshima et al (in this revieuw) have shown the existence of a 5-methylthiouracyl in position 55 and also a high G-C content in F-met-tRNA from Thermus thermophilus which they could related to the greater thermostability of this tRNA. The difference with our result is perhaps reliable to the organisms studied: Bacillus stearothermophilus being a moderate thermophile with an optimal growth rate at 65°C and Thermus thermophilus being an obligate thermophile which grows at higher temperatures up to 88°C.

3. Specificity of the aminoacyl-tRNAsynthetase - tRNA interactions

Arca et al. (16 - 17) first reported that, in vitro, a partially purified preparation of isoleucyl-tRNAsynthetase from Bacillus stearothermophilus, devoided of valyl-, seryl- and threonyl-tRNAsynthetase activity acquires above 70 °C the ability to catalyse the formation of valyl-, seryl- and threonyl-tRNA besides the normal isoleucyl-tRNA. Apparently above 70 °C, the isoleucyl-tRNAsynthetase starts to attach "wrong" aminoacids onto $tRNA^{ile}$. From the results, the authors suggested that this kind of errors might be connected with the upper limit of life, consistently found at above 70 °C for Bacillus stearothermophilus (17).

<u>Figure 1</u> : The primary sequences of 2 transfer RNAs from Bacillus stearothermophilus (NCA 1518), specific for phenylalanine and valine.
For details, see TAKADA, GROSJEAN, DIRHEIMER and KEITH, FEBS letters 1975 (submitted)

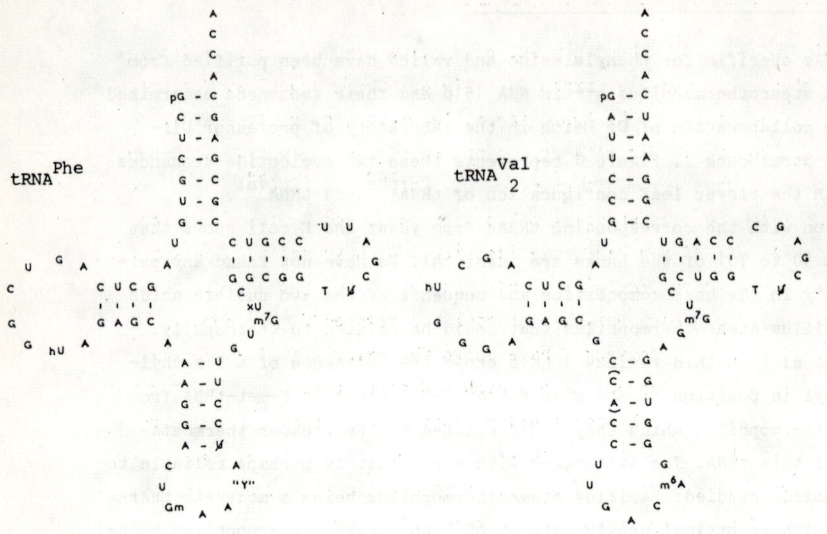

<u>Figure2</u> : Effect of heat on the loading of different aminoacids on unfractionated tRNA from B. stearoth. catalysed by Ile-RS and Val-RS from B. stearoth. △,▲ referred to the charging of valine, ○ ● to isoleucine and □,■ to threonine. The incubation time was 5 min. at each temperature; for other details, see ref7.

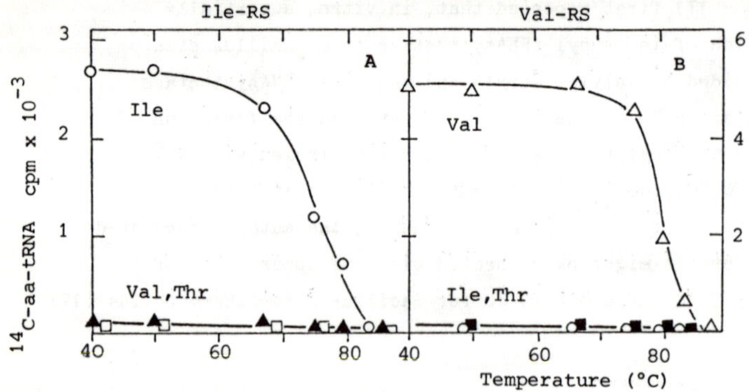

Using pure preparations of isoleucyl-tRNA from B. stearothermophilus (strain NCA 1518), we have never been able to reproduce the results of Arca et al. We found that only isoleucine was significantly attached to tRNA from B. stearothermophilus or from E.coli in the temperature range from 30 to 80°C (fig 2A). Similar results were obtained with the pure valyl-tRNAsynthetase from B. stearothermophilus in the presence of valine, threonine and isoleucine (fig 2B).

We concluded that, in vitro, high temperature did not lead to esterification of a wrong amino acid onto tRNA by the isoleucyl- and valyl-tRNA-synthetase from B. stearothermophilus.

More recently, increasing literature on mischarging of tRNA's catalysed by several aminoacyl-tRNAsynthetases from E.coli and yeast appeared (4,18,19,20 and references therein). In this case, the reaction results from a misrecognition of the tRNA by the aminoacyl-tRNAsynthetase. These reactions however, are slow compared to the normal acylations of the cognate tRNA ; they have been detected especially in heterologous systems where the enzyme and the tRNA come from different sources and under particular assay conditions. Important misacylation was also detected in homologous systems, when pure tRNA species and pure enzymes were used, even under normal assay conditions ; however significant mischarging was found more easily in the presence of an organic solvent.

Fig 3A and 3B show the effect of heat and of dimethylsulfoxide (Me_2SO) on the yield of aminoacylation of unfractionated tRNA or purified $tRNA^{Phe}$ from B. stearothermophilus by the pure isoleucyl-tRNAsynthetase from the same origin. In the range of temperature between 35 and 75°C, only very small amounts of isoleucine have been enzymatically attached to the noncognate pure $tRNA^{Phe}$ (corresponding to less than 5% of its maximum acceptor capacity for phenylalanine). Under the same conditions, full charging of $tRNA^{Ile}$, occurs up to 65°C (fig 3A). If the reaction is performed at 30 °C in presence of increasing concentrations of Me_2SO, $tRNA^{Phe}$ from B. stearothermophilus becomes misacylated with isoleucine and reaches about 78% of its full acceptor capacity in presence of 35% Me_2SO. With unfractionated tRNA from B. stearothermophilus (containing

Figure 3 : Comparison of the effect of heat (panel A) and dimethylsulfoxide (panel B,C) on the loading of isoleucine on unfractionated tRNA (-O-) and on purified tRNA-Phe (-●-) from B. stearothermophilus in the presence of an excess of ile-RS from B. stearoth. In fig. 3C, the experiments are the same as in 3B, except that 0.04 mg/ml of pure B.stearoth. tRNA-Phe was mixed with 0.8 mg/ml of the unfractionated tRNA(O); also 0.8 mg/ml of periodate oxidized unfractionated tRNA was added to 0.04 mg/ml of pure tRNA-Phe (●).For other details,see ref. 3

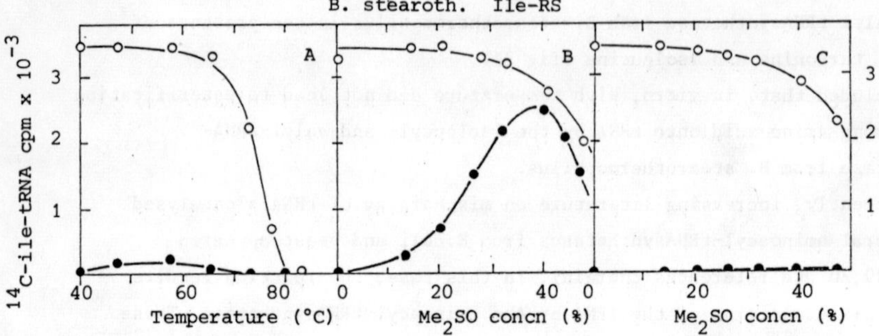

Figure 4: Effect of dimethylsulfoxyde on the aminoacylation extent of various pure tRNA species from E. coli and yeast by pure Ile-RS or pure Val-RS from B. stearothermophilus (heterologous systems). × E.coli tRNA-ile; ▼ yeast tRNA-Met-F; ○ E.coli tRNA-Phe; △ E·coli tRNA-Met-F; ● yeast tRNA-Phe; ▲ E.coli tRNA-val-1; ▽ yeast tRNA-val-1; + yeast tRNA-Ala; □ E.coli tRNA-Glu-2. For detailled experimental conditions, see reference 3.

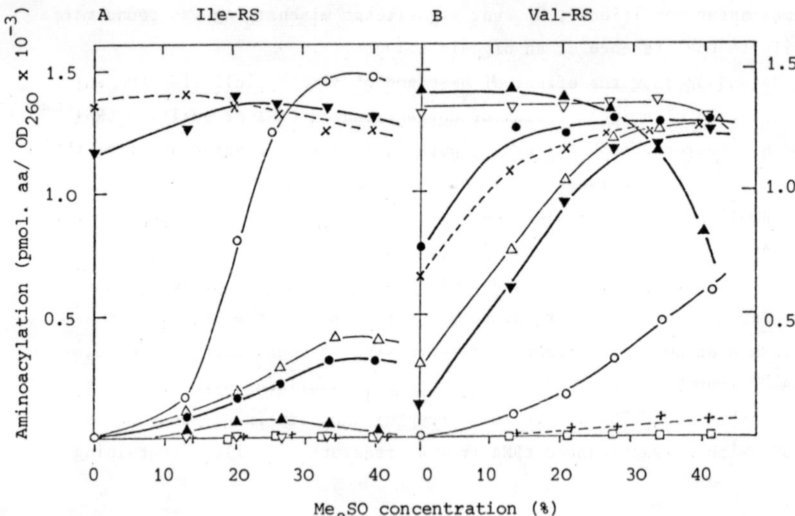

both $tRNA^{Ile}$ and $tRNA^{Phe}$), no such extra aminoacylation occurs in the same range of organic solvent concentration. This results most probably from a strong competition of the cognate $tRNA^{Ile}$ with the noncognate $tRNA^{Phe}$ for the same binding site on the isoleucyl-tRNAsynthetase. In agreement with this interpretation the addition of unfractionated tRNA (oxidized with periodate in order to avoid the normal acylation of $tRNA^{Ile}$), completely suppress the mischarging of B. stearothermophilus $tRNA^{Phe}$. Similarly, the addition of pure $tRNA^{Phe}$ in a mixture containing unfractionated tRNA did not enhance the attachement of isoleucine to tRNA to a higher level than the acylation of the cognate $tRNA^{Ile}$.(fig.3C) Similar results were obtained with other synthetases from B. stearothermophilus (3), E.coli (1,21) and yeast (4); they demonstrate that under certain conditions the aminoacyl-tRNAsynthetases aminoacylate significantly non cognate tRNAs from the same origin. Beside the important competition effect that must ex ist inside the cell (22) our results also demonstrate that the temperature cannot induce misacylations in vitro as the organic solvent does . It seems therefore improbable that the upper limit of life of 70°C for B. stearothermophilus might be connected primarily with a sudden increase of amino - acylation errors.

One approach to determine the specific nucleotides involved in the recognition of a tRNA molecule by an aminoacyl-tRNAsynthetase is to compare the sequences of several tRNAs aminoacylated by a single synthetase and look for common nucleotides at specific sites. In this way Dudock and coworkers (19,23) propose the involvement of some specific nucleotides located in the dihydrouridine arm and at the fourth position from the 3'end of the $tRNA^{Phe}$ molecule for the recognition with the yeast phenylalanyl-tRNAsynthetase.

For the valyl-tRNAsynthetases from E.coli and yeast, Ebel and coworkers conclude similarly, but point out that some other bases located in the anticodon loop and in the extra arm are important (24,25). It might be that different synthetases may recognize different bases located at different sites in the tRNA molecules. In order to check this hypothesis we studied several synthetases from B. stearothermophilus in collaboration with R.Giege from Ebel's laboratory. Figures 4_A and 4_B illustrate the results obtained respectively with the isoleucyl- and valyl-tRNAsynthetases from B. stearothermophilus in combination with 24 pure tRNAs from E.coli and yeast for which the sequences were known. The heterologous but cognate tRNAs (specific for isoleucine and valine) are fully charged by the corresponding synthetase from B. stearothermophilus and increasing the concentration of Me_2SO up to 35 - 40% has almost no effect on the yield of this reaction. Several noncognate tRNAs from E.coli and yeast are misacylated by the isoleucyl- and valyl-tRNA-synthetases even without organic solvent ; however increasing the concentration of Me_2SO lead to almost complete charging of these tRNAs with the wrong amino acid (at 100% of their full acceptor capacity). Several noncognate tRNAs which are not misacylated in absence of Me_2SO, become charged with the wrong amino acid to an appreciable extend at 35 to 40% of the organic solvent while most of the other noncognate tRNAs remain completely inactive. Table 2 sumarizes these results with those obtained under identical experimental conditions for three other pure aminoacyl-tRNAsynthetases specific for serine, arginine and glutamic acid. Clearly, increasing Me_2SO concentrations lead to a less efficient discrimination of the different tRNAs by the synthetase. As first suggested by Yarus (1) this property may result from the effect of the organic solvent on

TABLE 2 : Summary of cross reactions between several aminoacyl-tRNA synthetases from B. stearothermophilus and different pure tRNA from E. coli and yeast (see fig. 4).

	VAL-RS	ILE-RS	SER-RS	ARG-RS	GLU-RS
Most easily aminoacylated tRNA's.	Val-1 , coli Val-2A,Bcoli Val-1 ,yeast Ile , coli Phe ,yeast Met-F , coli Met-M , coli Met-F ,yeast	Ile , coli Met-M , coli Met-F ,yeast Phe , coli	ser-1 , coli Ser-3 , coli Ser 1+2,yeast Leu-1 , coli Leu-2 , coli	Arg-1 , coli Arg-3 ;yeast Arg-2 ,yeast Ser-3 , coli Gly-3 , coli	Glu-2 , coli Met-F , coli Met-F ,yeast Leu-2 , coli Asp ,yeast
Poorly aminoacylated tRNA's	Phe , coli Tyr , coli Leu-1,2 coli Gly ,yeast Ala ,yeast	Phe ,yeast Met-F , coli Val-1,2,coli	Arg-3 ,yeast Val-1 , coli Tyr , coli		Phe , coli Val,coli,yeast Met-M , coli Leu-1 , coli Tyr,coli,yeast Ile , coli Arg-2,3,yeast Ala ,yeast Ser-1,3,coli Gly ,yeast
tRNA's not aminoacylated at all.	His , coli Ser,coli,yeast Gly-3 , coli Ala ,yeast Asp ,yeast Arg,coli,yeast Glu-2 , coli	His , coli Ser,coli,yeast Gly,coli,yeast Ala ,yeast Asp ,yeast Arg,coli,yeast Glu-2 ,coli Val ,yeast Tyr , coli Leu-1,2,coli	His , coli Tyr ,yeast Gly,coli,yeast Ala ,yeast Asp ,yeast Arg-1 ,coli Glu-2 , coli Val ,yeast Ile , coli Phe,coli,yeast Met,coli,yeast	His , coli Ser ,yeast Ser-1 , coli Ala , yeast Asp ,yeast Leu-1,2coli Glu , coli Val,coli,yeast Ile,coli,yeast Phe,coli,yeast Met,coli,yeast Gly ,yeast Tyr , coli	His , coli Ser ,yeast Gly-3 , coli Ala ,yeast

the conformation of the tRNA molcule, but also on the conformation of the synthetase or both. It seems possible therefore that a special rigid structure of the nucleic acid and/or of the synthetase plays an essential role in the specificity of their interaction. However, despite the fact that the enzymes from B. stearothermophilus become less specific in the presence of organic solvent, they remain quite selective for only certain tRNA species from E. coli and yeast. This property might result from the existence in the different tRNA families, of some common structural features in their tertiary, but also in their primary and secondary structures.

Figure 5

Sequence comparison of the most easily aminoacylated tRNA species by different aminoacyl-tRNA synthetases from B.stearothermophilus. (see table 2).
● = differences; only nucleotides common to all tRNA aminoacylated by a given synthetase are encircle; pu = purine; py = pyrimidine; ✱ = modified nucleosides in some tRNA species.

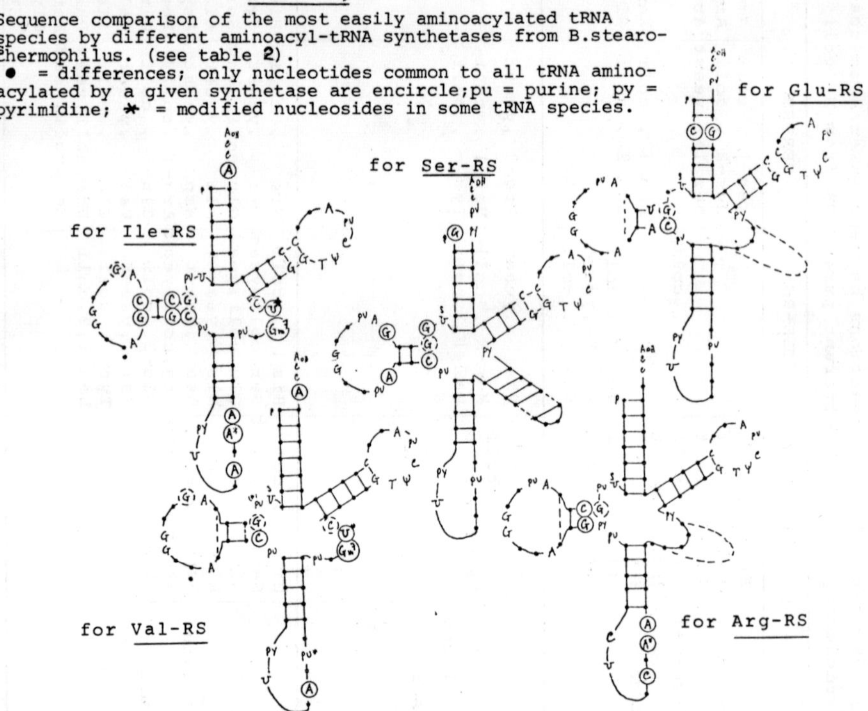

Figure 5 summarizes the results obtained by comparison of the cloverleaf structures of all tRNA's that have been most easily misacylated by the synthetases from B. stearothermophilus, specific for isoleucine, valine, serine, arginine and glutamic acid. For each group of tRNA's and beside the common nucleotides that are found anyway in all tRNA's sequenced until now, only a few specific nucleotides are common. These common nucleotides are localized mainly in four different regions of the tRNA molecule : at the end of the -CCA arm, in the dihydrouracyl arm, in the anticodon loop and in the extra loop. Assuming that all the tRNA molecules have almost the same conformation as that proposed for yeast $tRNA^{Phe}$ from X-ray crystallography (26,27), it is obvious (Fig 6) that all these common nucleotides are localized on the same side of the nucleic acid which might therefore constitute the contact side for the synthetase. For each synthetase however, the common nucleotides are different and also localized at different places of the tRNA molecules. A similar hypothesis concerning the recognition between tRNA's and synthetases in yeast and E.coli systems, was also proposed by Rich and Schimmel (28 and paper in preparation). These authors supported their conclusions by cross-linking experiments between tRNA's and synthetases, by hydrogen-tritium exchange in tRNA engaged in a complex with the enzyme as well as by the existence of tRNA mischarging.

In the particular case of valyl-tRNAsynthetase from B. stearothermophilus, a detailled analysis of the kinetic parameters (K_m and v_{max}) of the misacylation reactions was performed (3). The K_m for all the tRNA species which have been found to cross react easily with the synthetase, was almost the same ; however, considerable difference exists in the v_{max} of the several aminoacyl-acylation reactions. We believe that the common nucleotides we found in all tRNA's, recognized by this particular synthetase (Fig 5), constitute the possible condidates for the primary discriminatory sites which determine the K_m of the aminoacylation reaction. Considering that the v_{max} of all the misacylation reactions are very low compared to the v_{max} for the correct acylation of tRNA, it is quite possible that other specific nucleotides, or some other structural features, such as a specific local conformation (or mobility) of the tRNA molecule are also important in determining the rate of the initial enzyme-tRNA complex and actually the rate and the specificity of the whole aminoacylation process.

Figure 6 : Localisation on the threedimensional structure of the nucleotides common to all tRNAs acylated by a given synthetase (see figure 5).

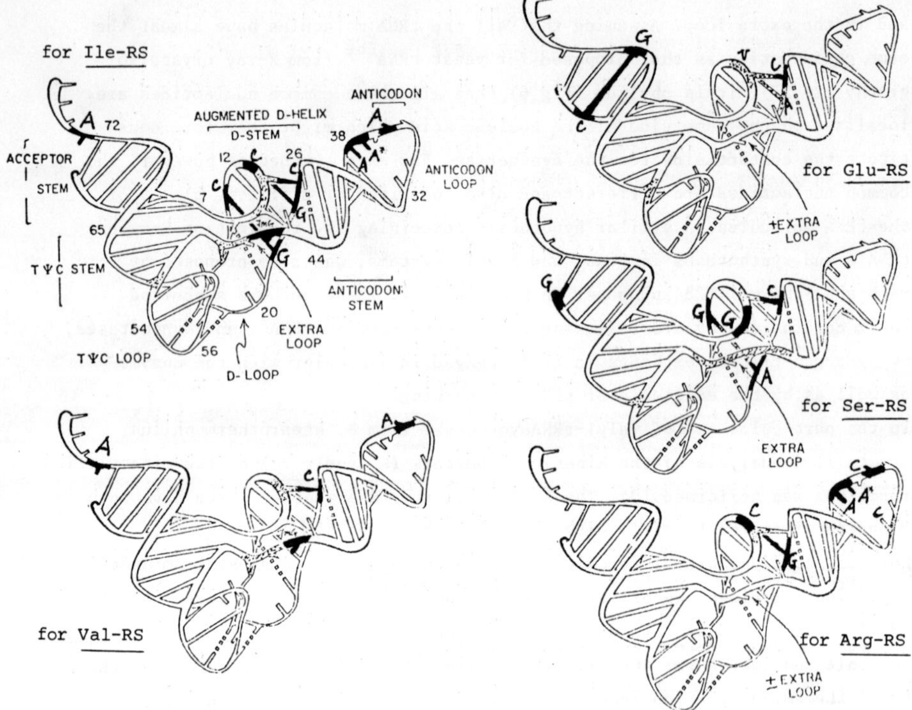

References

1. M.Yarus Biochemistry 11, (1972) 2050 - 2060
2. H.Prinz, A.Maelicke and F.Cramer Biochemistry 13, (1974) 1322 - 1326
3. R.Giege, D.Kern, J-P.Ebel, H.Grosjean, S.deHenau and H.Chantrenne Eur.J.Biochem. 45, (1974) 351 - 362
4. J-P.Ebel, R.Giege, J.Bonnet, D.Kern, N.Befort, C.Bollack, F.Fasiolo, J.Gangloff and G.Dirheimer Biochimie 55, (1973) 547 - 557
5. D.Söll and P.R.Schimmel in The Enzymes 10, (1974) 489 - 538
6. J.VanHumbeeck and P.Lurquin Eur.J.Biochem. 10, (1969) 213 - 218
7. J.Charlier and H.Grosjean Eur.J.Biochem. 25, (1972) 163 - 174
8. R.Parfait and H.Grosjean Eur.J.Biochem. 30, (1972) 242 - 249
9. R.Parfait FEBS letters 29, (1973) 323 - 325
10. S.Wilkinson and J.Knowles Biochem.J. 139, (1974) 391- 398
11. G.Koch, Y.Boulanger and B.Hartley Nature NB 249; (1974) 316 - 320
12. M.Yarus and P.Berg J.Mol.Biol. 42, (1969) 171
13. J.Charlier, H.Grosjean, P.Lurquin, J.VanHumbeeck and J.Werenne FEBS letters 4, (1969) 239 - 242
14. H.Grosjean and J.Charlier FEBS letters 18, (1971) 342 - 346
15. J.VanHumbeeck PHdthesis (1969) Université Libre de Bruxelles
16. M.Arca, L.Frontali and G.Tecce Biochim.Biophys.Acta 108, (1965) 326 - 328
17. M.Arca, L.Frontali, O.Sapora and G.Tecce Biochim.Biophys.Acta 145, (1967) 284 - 291
18. K.B.Jacobson Prog.Nucl.Acid.Research 11, (1971) 461 - 487
19. B.Roe, M.Sirover and B.Dudock Biochemistry 12, (1973) 4146 - 4154
20. M.Yarus and M.Mertes J.Biol.Chem. 248, (1973) 6755 - 6758
21. M.Mertes, M.A.Peters, W.Mahoney and M.Yarus J.Mol.Biol. 71, (1972) 671-685
22. M.Yarus Nature NB 239, (1972) 106 - 108

23. B.Roe and B.Dudock Biochem.Biophys.Res.Commun. 49, (1972) 399-406
24. R.Giege, D.Kern and J.P.Ebel Biochimie 54, (1972) 1245-1255
25. D.Kern, R.Giege and J.P.Ebel Eur.J.Biochem. 31, (1972) 148-155
26. S.H.Kim, F.L.Suddath, G.J.Quigley, A.Mc Pherson, J.L.Sussman, A.H.J.Wang, N.C.Seeman and A.Rich Science 185 (1974) 435-440
27. J.D.Robertus, J.E.Ladner, J.T.Finch, D.Rhodes, R.S.Brown, B.F.C.Clark and A.Klug Nature 250 (1974) 546-551
28. A.Rich Biochimie 56 (1974) 1441-1449

Acknowledgments

The authors are particularly indebted to Dr. Nishimura (Tokyo), Dr. Zachau (Munich), Dr. Weimann (from Boehringer Mannheim GmbH), Dr. Favre (Paris) and Drs Bonnet and Kuntzel (Strasbourg) for kind gifts of pure tRNA species. This work was supported by the Ministère de la politique et de la programmation scientifique and by the Fonds de la recherche fondamentale collective (Belgium).

THERMOPHILICITY AND THERMOSTABILITY OF RNA POLYMERASES FROM TWO STRAINS OF CALDARIELLA ACIDOPHILA, AN EXTREMELY THERMO-ACIDOPHILIC BACTERIUM

M. G. CACACE, M. DE ROSA and A. GAMBACORTA

Laboratory of Molecular Embryology and Laboratory for the Chemistry of the Molecules of Biological Interest, C. N. R., via Toiano, 2, 80072 Arco Felice, Naples, Italy

INTRODUCTION

Transcription in highly thermophilic microorganisms requires stringent conditions both for the structural stability of RNA polymerase and the conformational state of the DNA template.

It has generally been observed that the (G+C) content of DNA from various thermophiles is not directely correlated to their temperature of growth (1, 2). On the other hand, RNA polymerases extracted from two other thermophilic bacteria, *Thermus aquaticus* (3) and *Bacillus stearothermophilus* (4), exhibit optimal activity *in vitro* at temperatures which are several degrees lower than that of growth.

We report here some peculiar enzymological characteristics of the DNA-dependent RNA polymerases from two strains of an acidophilic microorganism, *Caldariella acidophila*, growing at 75 ^{o}C and 87 ^{o}C respectively.

GENERAL PROPERTIES OF CALDARIELLA ACIDOPHILA

C. acidophila naturally occurs in acidic hot springs located in the Flegrean area (Agnano, Naples) and has been found to live at temperatures ranging from 74 °C to 89 °C and at pH from 1.4 to 2.6 (5). Two of the strains isolated, MT-3 and MT-4, which had respectively the lowest and highest temperature of growth, were cultured in liquid media and used for RNA polymerase preparation. The DNA of the strains MT-3 and MT-4 had a (G+C) content of 42% and 39% respectively as determined by CsCl gradient centrifugation. The microorganism is very resistant to thermal stress, in fact MT-3 remained viable after 40-60 min at 90 °C and strain MT-4 survived at 100 °C for 15 min. The internal pH is close to neutrality rising upon lysis from 3.5 to 6.3 .

PURIFICATION AND BASIC PROPERTIES OF THE ENZYMES

The purification procedure of RNA polymerase from C. acidophila involves: DNase treatment of the crude extract, ammonium sulphate precipitation, DEAE cellulose chromatography and high and low salt glycerol gradient centrifugations. At the last stage of purification the enzyme resulted 180-fold purified over the DNase treated fraction (6).

RNA polymerase from MT-3 strain was active at high ionic strenght as shown in Fig. 1. Identical behaviour was observed for the MT-4 enzyme. The values of optimal salt concentration

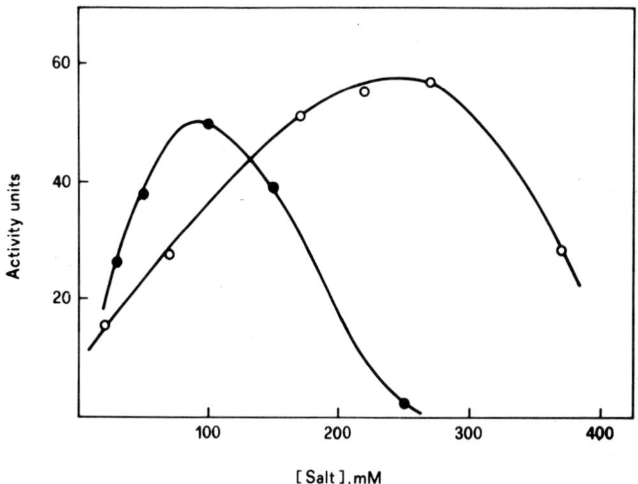

Fig. 1. Effect of ionic strenght on the RNA polymerase activity from strain MT-3. The assay mixture contained in a final volume of 0.1 ml: 0.1 mM each GTP, CTP and ATP, 14.8 μM unlabelled UTP, 5.2 μM [^{14}C(U)] UTP (367 mCi/mmole), 0.2 mg/ml BSA, 1 mM dithiothreitol, 30 mM borate buffer pH 8.0 at 20 °C. Salt concentration was adjusted to the final values indicated in the figure. The mixture was incubated at 70 °C for 10 min and the reaction stopped by TCA precipitation. The precipitate was collected on GF/C Whatman glass fiber filters and counted in 4% Liquifluor-toluene scintillation fluid. 10 μl aliquots of DEAE cellulose peak fraction were used for each assay. -●-●-, $(NH_4)_2SO_4$; -o-o-, KCl.

for RNA synthesis, 0.27 M for KCl and 0.1 M for $(NH_4)_2SO_4$, are among the highest reported for bacterial RNA polymerases.

The effect of the two activating cations, Mg^{++} and Mn^{++} on

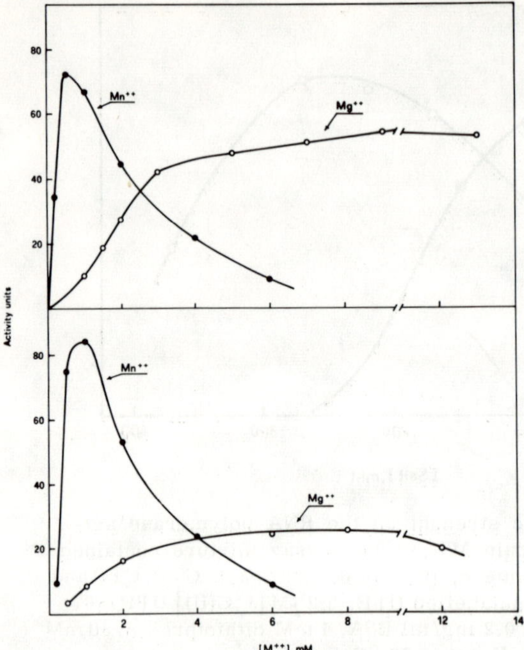

Fig. 2. Effect of Mg^{++} and Mn^{++} on the RNA polymerase activities from strains MT-3 and MT-4. Assays were as reported in fig. 1, at 0.18 M KCl final conc. ; metal ion conc. was adjusted to the final values given in figure. 10 μl aliquots of low salt glycerol gradient fraction were used for each assay. Upper curves: MT-4 enzyme, T:80 °C; lower curves: MT-3 enzyme, T: 70 °C.

both RNA polymerases is shown in Fig. 2. Interestingly, Mn^{++} has a higher activating capacity than Mg^{++} for both enzymes and is optimally effective at a concentration of 0.5-1 mM. No definite peak of activation was found in the presence of Mg^{++}. The Mn^{++}/Mg^{++} activation ratio was 3.3 and 1.3 for MT-3 and MT-4 enzymes respectively.

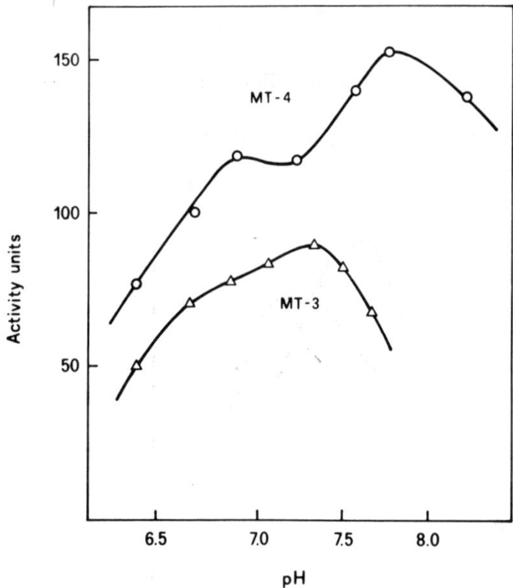

Fig. 3. Dependence on pH of the RNA polymerase activities from strains MT-3 and MT-4. Assays were as reported in fig. 2 except that Tris-HCl buffer was used at 80 mM final conc.; pH was measured at the assay temperatures using model mixtures having the same composition. 10 μl of low salt glycerol gradient fraction were used per assay.

The pH dependence of enzymic activities is shown in Fig. 3.

In the presence of Tris-HCl buffer MT-3 RNA polymerase has an optimum at pH 7.35, whereas the MT-4 enzyme is maximally active at pH 7.8. The inflection present in the curve relative to the MT-4 polymerase has also been observed for the MT-3 enzyme at temperatures close to 70 °C.

THERMOPHILICITY AND THERMOSTABILITY OF THE ENZYMES

The effect of temperature on the rate of RNA synthesis <u>in</u>

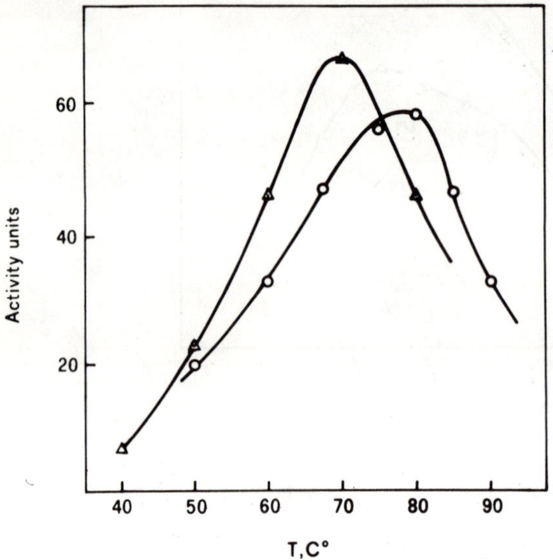

Fig. 4. Dependence on temperature of the RNA polymerase activities from strains MT-3 and MT-4. Assays were as reported in fig. 2. 10 μl of low salt glycerol gradient fraction were used per assay. -o-o- , MT-4 enzyme; -△-△-, MT-3 enzyme.

vitro for both RNA polymerases is shown in Fig. 4. The optimal temperatures for MT-3 and MT-4 enzymes were 70 °C and 80 °C respectively. These values reflect the difference in the optimal growth temperatures of the two strains and are in both cases only 5 °C-7 °C lower than these.

The thermostability of the enzyme from the most thermophilic strain has been studied in the conditions of the assay.

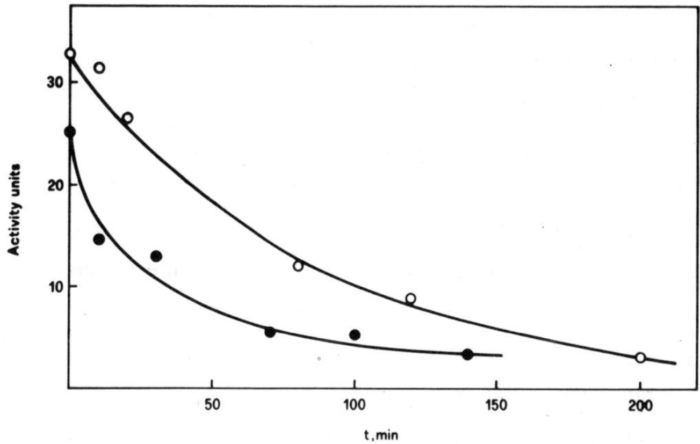

Fig. 5. Thermal inactivation of RNA polymerase from strain MT-4 at 90 °C. A complete assay mixture was divided into 0.05 ml aliquots and incubated at 90 °C in sealed glass ampoules. At the time indicated nucleotides and DNA, where absent, were added and the activity assayed at 80 °C for 10 min; 10 µl of low salt glycerol fraction were used per assay and incubated in the absence (-●-●-) or in the presence (-o-o-) of DNA (0.2 mg/ml final concentration)

Fig. 5 shows the kinetic of inactivation of RNA polymerase from MT-4 in the presence and in the absence of DNA. The presence of the template strongly increases the thermostability of the enzyme which in this case presents a $t_{1/2}$ of inactivation of 60 min.

DISCUSSION

The efforts for understanding the mechanism of RNA synthesis have focused the interest toward RNA polymerases exhibiting special characteristics such as resistance to temperature or to specific inhibitors. It has already been shown that some steps of the RNA synthesis process (e. g. initiation) are critically controlled by temperature (7). Moreover, the isolation of RNA polymerases from highly thermophilic bacteria enables investigations on transcription to be performed over an unexplored temperature range where cooperative transitions of most templates occur.

Remold-O'Donnell and Zillig (4) have isolated the enzyme from B. stearothermophilus and found that it had physical and enzymological properties very close to the E. coli RNA polymerase. The optimal activity for the enzyme from the most thermophilic strain investigated by these authors was found at 50 °C.

Air and Harris (3) studied the thermophilicity and subunit composition of the RNA polymerase from T. aquaticus; the enzyme was optimally active at 50 °C-60 °C and had a subunit composition which resembled very much that of the E. coli enzyme. Unfortunately the temperature range for optimal activity of both enzymes was lower than the range of T_m for natural DNA, even of low (G+C) content.

RNA polymerases from both strains of C. acidophila exhibit peculiar characteristics of a) optimal ionic strenght and metal ion requirements necessary for the in vitro transcription and b) resistance to denaturation at temperatures ranging from 70 °C to 90 °C. The exceptionally high ionic strenght at which the enzymes are optimally trancribing DNA indicates that the enzyme-DNA-RNA complex is extremely stable . The two cations Mg^{++} and Mn^{++} differ significantly in activating both RNA polymerases and the sharp peak of activation by Mn^{++} contrasts with the general preference for Mg^{++} of all other bacterial RNA polymerases. It is known that ionic strenght up to 0.5 M and divalent cations stabilize DNA and shift its T_m to higher values (8). This effect of stabilization of the template is probably correlated to the unusual requirements of both RNA polymerases and could play an important role for efficient transcription in thermophiles.

The small difference between the optimal temperatures of growth and that of enzymic activity indicates that the integrity of the two enzymes has been preserved in the course of purification. Less favourable results have been obtained for B. stearothermophilus and T. aquaticus RNA polymerases in that the optimal temperature for the in vitro assay was found to be considerably lower than that of growth.

C. acidophila RNA polymerase from strain MT-4 when preincubated at 90 °C in the absence of DNA exhibits high thermostability with a $t_{1/2}$ of 20 min. The subunit composition of the enzyme, as determined by SDS-acrylamide gel electrophoresis, shows that, apart from the smaller subunits, it contains two high molecular weight chains, β' and β, of 127,000 and 120,000 daltons respectively (6). These values are the smallest found for the mesophilic and thermophilic RNA polymerases so far investigated (9). The intrinsic stability of the molecule to thermal denaturation could be a consequence of a particularly stable quaternary structure of the enzyme itself. The reduction in molecular weight of the larger subunits probably confer, at least in part, that unusual resistance to the molecule. The effect of stabilization by DNA, whose presence results into a 3-fold increase of the $t_{1/2}$ of inactivation at 90 °C, indicates that C. acidophila RNA polymerase binds with a high affinity to the DNA template and that this interaction causes profound changes in its conformation.

REFERENCES

1) Marmur, J., Biochim. Biophys. Acta, 38, 342-353 (1960)
2) Mangiantini, M. T., Orlando, P., Schiesser, A. and Tecce G., La Ricerca Scientifica, 4, 97-101 (1964)

3) Air, M. G. and Harris, J. I., FEBS Letters, 38, 277-281 (1974).

4) Remold-O'Donnell, E. and Zillig, W., Eur. J. Biochem., 7, 318-323 (1969).

5) De Rosa, M., Gambacorta, A. and Bu'Lock, J. D., J. Gen. Microbiol., 86, 156-164 (1975)

6) Cacace, M. G., De Rosa, M. and Gambacorta, A. (1975), manuscript submitted for publication.

7) Kerrich-Santo, R. E. and Hartmann, R. G., Eur. J. Biochem., 43, 521-532 (1974).

8) Dove, W. F. and Davidson, N., J. Mol. Biol., 5, 467-478 (1962).

9) Darnall, D. W. and Klotz, I. M., Arch. Biochem. Biophys., 166, 651-681 (1975).

THERMOPHILIC AND MESOPHILIC ENZYMES FROM B. CALDOTENAX AND B. STEAROTHERMOPHILUS: PROPERTIES, RELATIONSHIPS AND FORMATION

G. FRANK, H.-U. HABERSTICH, H.P. SCHAER, J.D. TRATSCHIN and H. ZUBER
Institut für Molekularbiologie und Biophysik
Eidgenössische Technische Hochschule
8049 Zürich, Switzerland

INTRODUCTION

The isolation of several thermostable enzymes from thermophilic microorganisms demonstrates that one of the primary contributing factors towards the survival of these organisms is the inherent thermostability of these enzymes.[1,2,3] An important goal in the study of structure-function relationships in thermophilic enzymes is the understanding of the structural details upon which not only the thermostability, but also the special catalytic and regulatory properties of these enzymes rest. This may be accomplished by comparison of the structures of thermophilic and mesophilic enzymes. However, even the study of enzymes from closely related microorganisms leaves open the question whether it is possible to elucidate these structural details simply by comparing the amino acid sequences and the 3-dimensional structure of these enzymes since a) thermophilic and mesophilic enzymes from closely related organisms may still show extensive species differences and b) only a few amino acid residues difficult to localize may contribute to thermostability and the special "thermophilic" properties.

B. caldotenax and some strains of B. stearothermophilus adapted to and cultivated at high or low temperatures produce thermostable (thermophilic) or thermolabile (mesophilic) enzymes[4,5]. This adaptive system seems to be appropriate a) for comparative studies (structure and properties) of these enzymes and b) for an investigation of the metabolic adaptation process leading to different enzymes. It would be particularly interesting to see whether a close relationship in the amino acid sequences exists between thermophilic and mesophilic enzymes, produced by the same bacterium. If there are close relationships it might be worthwhile to compare these sequences to find out "thermophilic details" in these sequences. Furthermore homologies or differences in the amino acid sequences between the thermophilic and mesophilic enzymes from B. caldotenax or B. stearothermophilus should point to the processes in the protein synthesizing system leading to different enzymes at different temperatures. Finally comparison of the primary structures of the enzymes from B. caldotenax and B. stearothermophilus should demonstrate the degree of phylogenetic relationship between these bacteria.

In the present paper some results are described concerning 1) the process of thermoadaptation in B. caldotenax leading to thermophilic or mesophilic enzymes. 2) comparative studies on thermophilic and mesophilic lactate dehydrogenase from B. caldotenax and B. stearothermophilus.

1) SYNTHESIS OF THERMOPHILIC AND MESOPHILIC ENZYMES IN TEMPERATURE ADAPTED B. CALDOTENAX

Production of thermophilic and mesophilic enzymes synthesized at high and low temperatures, respectively, was studied with the extremely thermophilic B. caldotenax, isolated by Heinen[6] from hot springs (at 85°)[4]. Within the temperature range from 30° to 70° the organism adapts to the thermophilic or mesophilic variant depending on the temperature.

For example (cf. Fig. 3-8[4]), a 37° preculture from B. caldotenax, adapted from the thermophilic form (70°) was further cultivated between 30° and 70° (in different samples) at 5° intervals (30°, 37°,40°,45°,50°,55°,60°,65°,70°). Whereas growth begins within 2-3 hours at 30° to 50°, there is an extended lag at temperatures higher than 55°. The metabolic changes that obviously take place should be reflected in the changing enzyme-pattern. Investigation of the lag period indicated, for example, that between 37° and 50° thermolabile glucose-6-phosphate isomerase (M-type) is produced whereas above 60° the thermostable enzyme is formed and that at 55° probably both are found. Extracts and pure enzymes were also compared as well in respect to thermostability. The other enzymes examined from the glycolytic pathway (glucokinase, lactate dehydrogenase, glyceraldehyde-3-phosphate dehydrogenase) and from the citrate cycle (isocitrate dehydrogenase) behaved similarly.

In an analysis of the lag-period, following the temperature change from 37° to 70°(during which the mesophilic cells become smaller), the following results were obtained with respect of the enzymes tested:

1] The mesophilic enzyme activity disappears after 20 min at 70° evidently as a result of denaturation. Inhibition of protein synthesis (the mesophilic enzymes continue to be synthesized) by chloramphenicol (5 or 30 min. before temperature change) increases the apparent denaturation rate considerably.

2] One hour before the beginning of growth of the thermophilic cells (after about 6-9 hours) measurable synthesis of the corresponding thermostable (thermophilic) enzymes begins. Growth of the thermophilic variant starts when the specific enzyme activity has already reached the maximum value. During the period of adaptation in the lag phase some enzymes (of the glycolysis or citrate cycle) are not present or not accessible. This raises the question as to the state of the resting cell. Not all enzymes can be absent; for example - as we will see later - the important protein synthesizing system (and perhaps the protein degrading enzymes (API)) can in some phases not be inactivated.

The results of the reverse experiments, the adaptation process of 70° cells to 37° cells are also of interest. In this case the lag phase is, as expected, longer at lower temperatures. Analysis of this lag-period yielded the following results:

1] In the 70° cells, during the lag (4 1/2 hours), about 80% of the thermophilic enzymes (for example glucose-6-phosphate isomerase) are still present, i.e. slow growth of the thermophilic variant is possible. After about 6 hours the activity of the thermophilic enzyme vanishes. This is shown by thedecrease of the activity of the thermostable enzyme, whereas the total activity is increased after about 3 1/2 hours. This means that the synthesis of the mesophilic enzyme begins after 3 1/2 hours, about 1 hour before the start of the mesophilic growth.

The process of adaptation from high to low temperatures appears to be different from the reverse process. Thermostable enzymes cannot be removed by denaturation (at 37°). It is obvious, however, that they must be replaced by mesophilic enzymes in order to make an optimal metabolism at low temperatures possible. Functional requirements (catalytic activity adjusted to low temperatures), too, effect a displacement of the thermophilic enzymes. This shows indirectly that the function (and lifetime) of the enzyme molecule are just as important as the right thermostability. Otherwise the mesophilic bacterium could accomplish its metabolism with thermophilic enzymes. It seems that after synthesis of the thermophilic enzyme has ceased, these enzymes are removed in the growing mesophilic culture by dilution.

The thermal adaptation of B. caldotenax is a metabolic process, whereby apparently only two forms of the enzymes are formed. Both forms have the same substrate specificity, but are different in specific activity, thermostability and other properties. It is reasonable to assume that each form is represented by a single type of polypeptide chain, as is shown by the sequence analysis in the case of lactate dehydrogenase.

This fact and the other findings mentioned pose the question concerning the mechanism of thermal adaptation. It seems reasonable, and inhibition during the lag following transition at 37° to 70° with chloramphenicol and actinomycin D indicates, that protein synthesis is involved in the process of adaptation.

Some general questions arise[4]:

1] Do the two enzyme-variants derive from two different structural genes, repressed or derepressed as a function of the temperature?

2] Is only one gene present and does the enzyme adaptation take place in a special phase of the protein synthesis?

3] or are the finished enzymes modified by enzymatically catalyzed reactions?

To answer these questions the present investigations are advancing in two directions:

1] Study of the metabolic state of the bacterial cell in thermal adaptation during the lag period.

2] Investigation of the protein synthesizing system involved in the thermal adaptation. Where and how does the change from "thermophilic" to "mesophilic" protein synthesis take place? How does the "temperature-information" enter the system?

At 70° the metabolic state of the formerly mesophilic cell of B. caldotenax during the lag period seems to be different from that of the normal vegetative and reproducing cell. The mesophilic cells heated to 70° are smaller and the proteins are difficult to extract. Although the lag period lasts as long as 7-10 hours (temperature change from 37° to 70°), the transition from the small cells of the lag to the vegetative (dividing) thermophilic cells (from 8% to 80% of the total cell population, which remains constant) takes only 30 min.

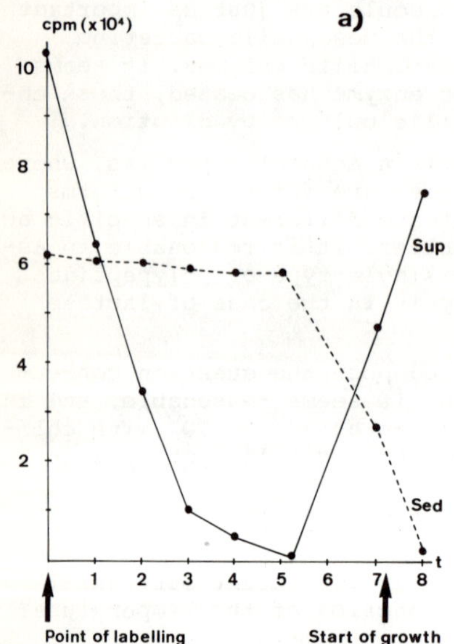

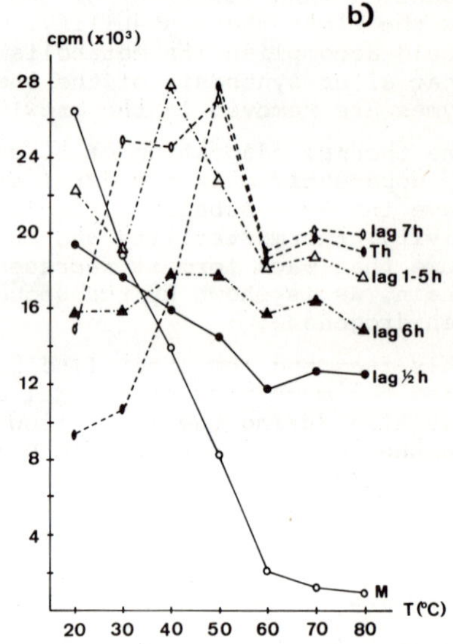

Fig. 1a: Radioactivity (^{14}C-amino acids) incorporated in the soluble (●—●) and insoluble (●----●) protein fractions of B. caldotenax during the lag period. Arrow at start of growth indicates the beginning of growth of the thermophilic variant.

Fig. 1b: Temperature optimum for incorporation of ^{14}C-amino acids into the proteins of whole bacterial cells of B. caldotenax during the lag period. M = mesophilic cells, Th = thermophilic cells.

In order to study the fate of the denatured mesophilic enzymes throughout the lag period, the mesophilic proteins were labelled with ^{14}C by incubating the mesophilic cells for 10 min. with a ^{14}C-chlorella hydrolysate before the temperature change (37° to 70°). During the lag period, the label was found only in soluble and insoluble protein fractions and not in free amino acids or peptides. This means that free amino acids splitted off from denatured polypeptide chains are rapidly metabolized or secreted into the culture medium.

Radioactivity was determined, for comparison, in the soluble extract and in the insoluble sediment of the cell during the lag period (Fig. 1a). In the soluble extract the radioactivity decreases steadily with

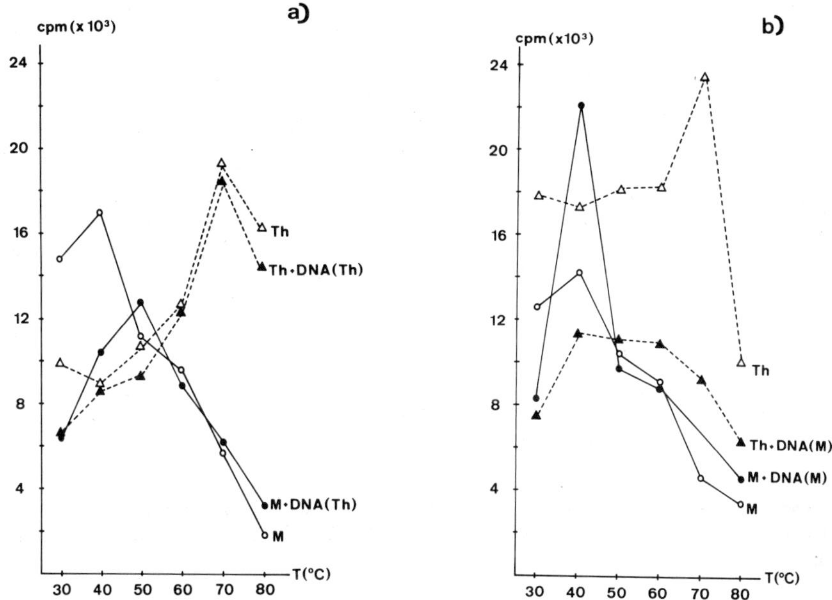

Fig. 2a and 2b: Optimum temperature for protein synthesis in vitro for the thermophilic (70°) and mesophilic (40°) system of B. caldotenax.

Fig. 2a: Shift of the temperature optimum of the mesophilic protein synthesizing system after addition of "thermophilic" DNA (M+DNA (Th). Control: thermophilic system plus "thermophilic" DNA (Th+DNA (Th)).

Fig. 2b: Shift of the temperature optimum of the thermophilic protein synthesizing system after addition of "mesophilic" DNA (Th+DNA(M). Control: mesophilic system plus "mesophilic" DNA (M+DNA (M)).

the lengthening lag, presumably as a result of the degradation of proteins to amino acids which disappear into the medium. After about 5-8 hours, two hours before the thermophilic (logarithmic) growth, a rapid increase of radioactivity in the soluble fraction occurs, although the cell has been in the unlabelled (cold) medium. Simultaneously, the hitherto constant radioactivity of the insoluble sediment decreases. This could mean the de novo synthesis of soluble thermophilic proteins (enzymes) with ^{14}C-marked amino acids taken up and incorporated before the temperature change and stored during the lag in the insoluble fraction.

The incorporation of amino acids into the protein of the whole bacterial cells was studied during the lag-period (37° to 70°) at different temperatures (Fig. 1b). Mesophilic cells had an optimum temperature for incorporation of amino acids of 20°, during the lag period the optimum was shifted: After 5 1/2 hours to 40°, after 7 hours to the optimum of the thermophilic cells, i.e. 50°. The incorporation rate in the lag period up to 5 hrs after temperature change was 2.5%, after 6 hours 12% and after 7 hours 25% ("thermophilic rate") of the mesophilic incorporation rate. The optimum temperature for protein synthesis in vitro is 40° for the mesophilic system, and 70° for the thermophilic system (Fig. 2a and 2b). It is reasonable to assume that the enzymes involved in protein synthesis change their temperature optimum. Addition of DNA from thermophilic cells to the mesophilic system displaced the "mesophilic" optimum to 50° (Fig. 2a). Conversely the "mesophilic" DNA displaces the optimum temperature for incorporation to lower temperatures (Fig. 2b). This direct influence of theDNA-preparations could mean that the information for the alternative synthesis of thermophilic or mesophilic enzymes in the protein synthesizing system comes from DNA. Since, by definition, the DNA of the thermophilic and mesophilic variant of B. caldotenax should be identical (an identical C+G content of about 67% was found for both variants), the state of the DNA for gene expression of the different enzyme variants should be different at high or low temperatures.

2) COMPARISON OF THE PROPERTIES AND STRUCTURES OF THERMOPHILIC AND MESOPHILIC LACTATE DEHYDROGENASE FROM B. CALDOTENAX AND B. STEAROTHERMOPHILUS

a) Properties

For comparative studies L-lactate dehydrogenase (LDH) was isolated from thermophilic cells (70° or 55° cells) as well as from mesophilic cells (37° cells) of both B. caldotenax and B. stearothermophilus (adapted through spores). Mesophilic LDH from B. caldotenax was very labile and could not be isolated in pure form yet. No activity of D-lactate dehydrogenase was found in these bacteria. In the two thermophilic bacteria LDH was produced at the end of the log-phase, in the mesophilic types at the beginning to the middle of the log-phase.

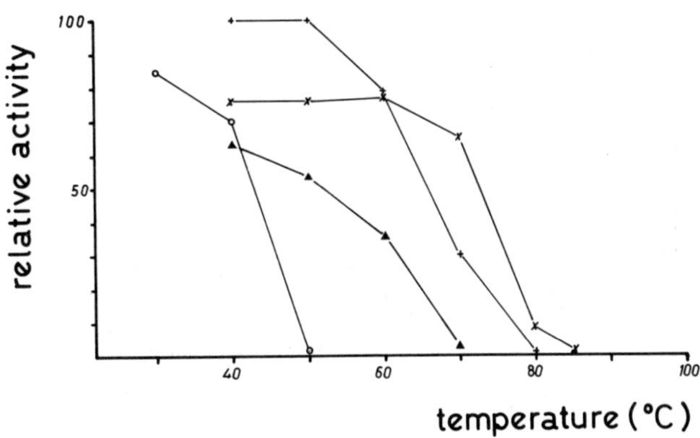

Fig. 3: Thermostability of thermophilic and mesophilic bacterial LDHs (heating 30 min.)
B. stearothermophilus grown at 55° (x —— x —— x)
B. stearothermophilus grown at 37° (o —— o —— o)
B. caldotenax grown at 70° (+ —— + —— +)
B. caldotenax grown at 37° (▲ —— ▲ —— ▲)

Isolation was accomplished mainly by affinity chromatography after ammonium sulfate fractionation. Differences in affinity were evident: The thermophilic LDH are bound on AMP-Sepharose as well as on oxamate-Sepharose, whereas under the same conditions the mesophilic LDH from B. stearothermophilus is adsorbed only on oxamate-Sepharose. The mesophilic LDH from B. caldotenax has no affinity to both types of affinity columns under the conditions used. No explanation for these differences is possible at present.

As can be expected significant differences between the thermophilic and mesophilic LDHs in their respective thermostabilities can be observed (Fig. 3). Thermophilic LDH from B. caldotenax is stable for 30 min. up to 50°. In contrast thermophilic LDH from B. stearothermophilus shows a small decrease of activity (about 20%) already at a temperature above 30°; perhaps a conformational change to a less active but more thermostable (up to 60°) molecule takes place. In contrast the mesophilic LDHs (B. stearothermophilus, B. caldotenax, pig muscle) became denatured above 30°. Considerable problems of the stability of the mesophilic LDH exist: Both bacterial mesophilic LDHs must be stabilized during isolation by β-mercaptoethanol, the thermophilic species are stable without thiols.

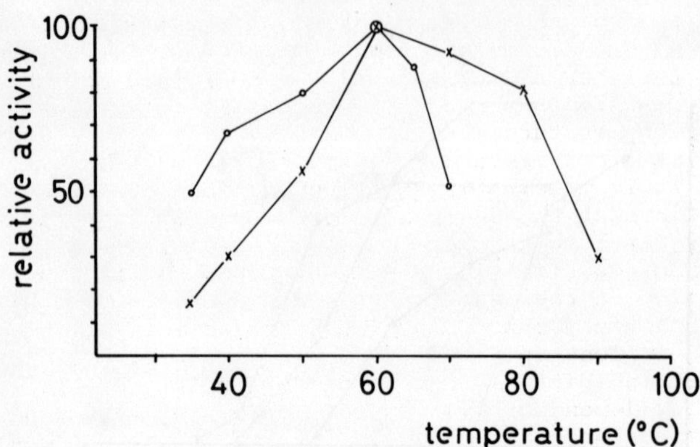

Fig. 4: Specific activity of thermophilic and mesophilic LDH from B. stearothermophilus at optimal pyruvate concentrations. thermophilic LDH (x —— x), mesophilic LDH (o —— o)

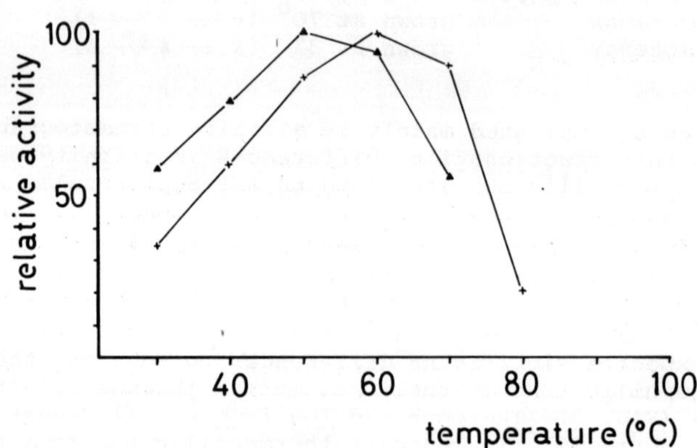

Fig. 5: Specific activity of thermophilic and mesophilic LDH from B. caldotenax at optimal pyruvate concentrations. thermophilic LDH (+ —— +), mesophilic LDH (▲ —— ▲)

As was expected from the different thermostabilities, the thermophilic LDHs are more active (as seen by comparison of the specific activities) at higher temperatures (70 - 80°) than are the mesophilic LDHs. On the other hand the mesophilic enzymes are more active at lower temperature (Fig. 4 and 5). The specific activities are measured at the different temperatures at optimal pyruvate concentrations: It is interesting that under these conditions during a 1 min. measurement the mesophilic LDHs from B. stearothermophilus and B. caldotenax show the same maximum activity at 60° as the thermophilic enzymes do, implying that they could function up to 60° which is already in the temperature range of the thermophilic enzyme. The temperature optimum of the specific activity of all LDHs is strongly dependent upon the pyruvate concentration, since saturation (V_{max}), inhibition and thermostabilization by the substrate are temperature dependent. The optimal pyruvate concentration at different temperatures is different for the thermophilic and mesophilic LDH from B. stearothermophilus: therm. = 30-60 mM, and mes. = 3-30 mM, but not for the thermophilic and mesophilic LDH from B. caldotenax (30-60 mM for both LDHs). As has also been established for B. caldolyticus[7], of both thermophilic and mesophilic LDH from B. stearothermophilus, the pyruvate requirement increases with temperature between 37° and 60° to about 10 fold in excess of the decrease in k_m (difference in k_m between 37° and 60° is only about 2 fold). It is reasonable to assume that pyruvate stabilizes the active enzyme.

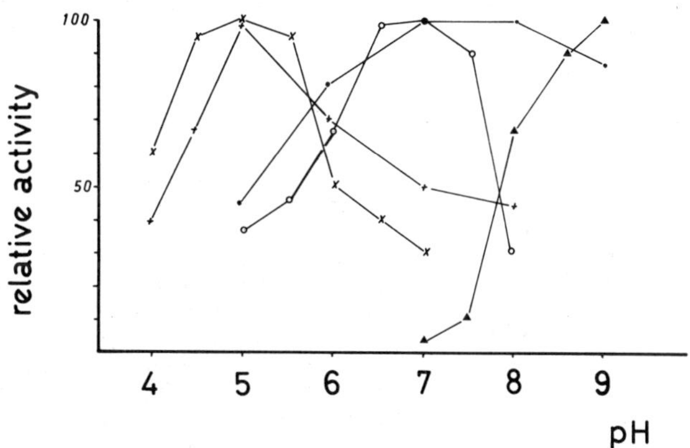

Fig. 6: pH-optima of thermophilic and mesophilic LDHs

B. stearothermophilus	(55°)	(x —— x —— x)
B. stearothermophilus	(37°)	(o —— o —— o)
B. caldotenax	(70°)	(+ —— + —— +)
B. caldotenax	(37°)	(▲ —— ▲ —— ▲)
Pig muscle		(● —— ● —— ●)

The thermophilic and mesophilic LDHs differ also in their pH-optimum (Fig. 6). Whereas the optimum pH for thermophilic LDH from B. caldotenax and B. stearothermophilus is about 5,it is about 7 for the mesophilic LDH from B. stearothermophilus, which is in a similar range to that of most of the other LDHs from higher organisms. In contrast, the pH-optimum for the mesophilic LDH from B. caldotenax is extremely high, at pH 9 (or higher). The significance of this finding cannot be discerned yet. The low pH-optimum of the thermophilic LDH is not necessarily restricted to thermophilic enzymes. Similar low optima were found for other LDHs of bacterial origin: for example pH 5.5 for that of Lactobacillus casei[8]. On the other hand, if the low pH-optimum is typical for bacterial LDH, the higher pH-optimum for the mesophilic LDH from B. stearothermophilus or B. caldotenax remains inexplicable.

b) Structures

In spite of their exhibiting, certainly very different, properties, the thermophilic and mesophilic LDH showed marked structural similarity. Since it has not yet been possible to isolate the very labile mesophilic LDH from B. caldotenax in pure form, it could not be compared structurally with the other bacterial LDHs. This is unfortunately a great restriction: in the following studies 2 thermophilic and only 1 mesophilic LDH can be compared. The molecular weights of all LDHs, determined by gel filtration on Sephadex G-150 are about 140'000. By SDS-gel electrophoresis the molecular weight of the identical subunits was determined to about 33'500 for the thermophilic LDH from B. stearothermophilus and 36'000 for both the thermophilic LDH from B. caldotenax and the mesophilic LDH from B. stearothermophilus. Therefore the bacterial LDHs investigated are tetramers like most of the L-LDH[9]. The isoelectric point of all LDH lies between 5.8 and 6.0.

The LDH investigated from B. stearothermophilus and B. caldotenax are very similar in their amino acid compositions, with the exception of the shorter polypeptide chain in LDH from B. stearothermophilus (molecular weight 33'500) which has only 303 instead of about 329-330 amino acid residues for the other LDHs (Table I). Comparing the amino acid composition it can be seen that the difference between the individual bacterial LDH species varies from about 6 to 25 amino acids (total) in contrast to great differences from e.g. the phylogenetically more distant dogfish LDH[10] which shows about 60 different amino acid residues. The LDHs from the thermophilic bacilli resemble the LDH from B. subtilis already described[11]. Noteworthy in the amino acid composition of the thermophilic and mesophilic LDHs is the difference in the ratio of lysine to arginine. The thermophilic LDHs always contain 10-11 arginine residues more than the mesophilic LDH and the mesophilic LDH, correspondingly, 8-10 lysine residues more than the thermophilic LDH. This might indicate a conservative exchange of these residues in the thermophilic and mesophilic polypeptide chain, whereby it is not known whether this ex-

Table I

COMPARISON OF AMINO ACID COMPOSITION OF LACTATE DEHYDROGENASES

	B. stearo-thermophilus 55°C	B. stearo-thermophilus 37°C	B. caldotenax 70°C	B. subtilis	Dogfish M-type	Rabbit H-type
Aspartic acid	35	40	38	39	37	35
Threonine	11	16	13	17	12	12
Serine	13	14	10	18	26	28
Glutamic acid	25	34	31	33	24	30
Proline	9	8	10	11	10	12
Glycine	27	31	33	31	24	23
Alanine	36	31	34	32	18	21
Cysteic acid	3	n.d.	n.d.	4	8	
Valine	26	30	29	29	34	34
Methionine	10	n.d.	n.d.	6	11	8
Isoleucine	22	23	25	20	20	21
Leucine	22	26	23	25	35	34
Tyrosine	11	11	12	13	7	7
Phenylalanine	13	11	15	13	7	6
Histidine	8	10	7	9	11	7
Lysine	12	24	15	23	29	24
Arginine	16	6	17	6	9	8
Tryptophane	4	5	n.d.	2	7	
TOTAL	303	(320)	(312)	331	329	

change of arginine for lysine (higher content of arginine) in the thermophilic enzyme is related to its higher thermostability. The possibility exists that comparison of the amino acid sequences and the 3-dimensional structures , where these exchanges have occurred, would yield more information. For example in the 3-dimensional structure, differences in the salt bridges from Lys or Arg, respectively, to the acid amino acids could be seen.

Comparison of N-terminal sequences with the amino acid sequencer demonstrates highly homologous sequences between the bacterial LDHs (Fig. 7). Of the first 41 amino acid residues at the N-terminus (12% of the total sequence) in the thermophilic LDH from B. caldotenax and B. stearothermophilus 86% are homologous, which indicates a close phylogenetic relationship of the two organisms. On the other hand 55% of the first 18 amino acid residues of the thermophilic and mesophilic LDH from B. stearothermophilus, respectively, are homologous. The possibility exists here that the homology within the chains may even be greater, as much as 80% since in the thermophilic LDHs from both bacteria only 67% of the first 18 amino acid residues are homologous. It is interesting to note that of the 7 exchanged amino acid residues (of the first 41 amino acid residues) 4-5 can be explained by a single base exchange.

In contrast to the homologous amino acid sequences in the bacterial LDH, dogfish LDH[10] shows only a sequence homology of 22 or 32% to the bacterial LDH in the first 18 or 41 amino acid residues, respectively (alignment see below).

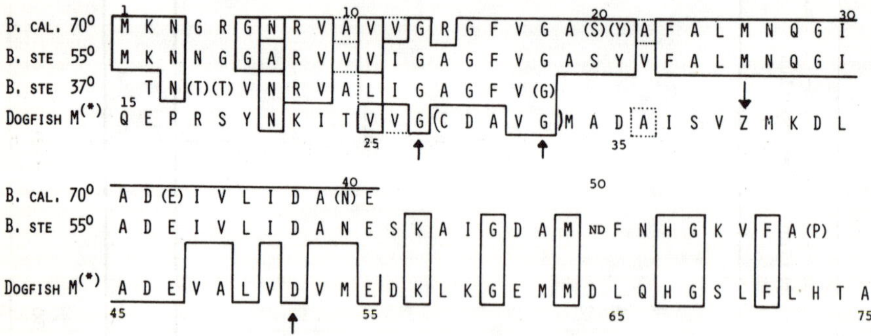

Fig. 7: Comparison of N-terminal amino acid sequences of lactate dehydrogenases
( ) amino acid residues not definitely proved
↑ important amino acid residues for alignment
(*) Taylor et al.[10]

It is noteworthy that the N-terminal sequences for the thermophilic LDH begin with methionine, but not those of the mesophilic bacterial LDH and dogfish. Since also the thermophilic alcohol dehydrogenase[12], triosephosphate isomerase[13], the β-chain of API[14], APII[15] have methionine at the N-terminus the questions arise as to whether this Met-N-terminus is typical of some thermophilic enzymes from B. stearothermophilus or B. caldotenax and as to the meaning of the Met at the N-terminus. Perhaps this could reflect a less exposed N-terminus in thermophilic enzymes and therefore a lesser susceptibility to degradation by aminopeptidases (API) after deformylation of the Form-Met-Polypeptide chain. In this connection a closer look at the N-terminus of thermophilic enzymes seems warranted. The closely homologous sequences in the thermophilic and mesophilic LDH are part of the coenzyme binding domains which have been studied by Rossmann et al.[16] and Brändén et al.[17] The sequence begins just before the region important for the nucleotide binding site (starting with βA) and are aligned with respect to the three invariant acid residues Gly (12 or 13), Gly (17 or 18) and Asp (38). The sequences of the bacterial LDHs are shorter than that of the dogfish by 15 or 16 amino acid residues, respectively. Thus, the greater part of the "extended arm" at the N-terminus, important for the interaction between the subunits (as postulated by Rossmann[16]) is missing in the bacterial LDHs.

SUMMARY AND CONCLUDING REMARKS

1) The adaptive system of thermophilic bacteria, as demonstrated with B. caldotenax, seems to be suitable to produce thermophilic and mesophilic enzymes for comparative studies.

2) If it may be assumed that the extensive homologies in the N-terminal sequences of the LDHs also extend over the entire polypeptide chain, comparison of these sequences together with investigation on the 3-dimensional structure offer the possibility of elucidating those structural details which may be responsible for thermostability and the other thermophilic properties. However, the difficulty still remains that the latter may be obscured by differences not related to thermostability etc. Nevertheless it may be hoped that comparison of the full sequences of not only the LDHs but also of a sufficient number of other enzymes of the same system will yield such details.

3) A further interesting goal with respect to the mechanism of enzyme adaptation would be reached if the differences in amino acid sequence of thermophilic and mesophilic LDH enzymes would throw light on the type of the amino acids always being exchanged. Here from the very hypothetical point of view the question arises as to whether the bacterial cell during the metabolic adaptation process or even by mutation/selection is able to modify just those few amino acid residues thermodynamically important for thermostability. Alternatively: does there exist a "rule" by which certain amino acid residues are invariably exchanged on a change from thermophilic to mesophilic enzymes?

4) Problems not mentionned here arise with B. stearothermophilus, which can be adapted poorly via the spores or on intermediate temperatures. Of great importance, but also a special problem in these studies on thermophilic and mesophilic enzymes produced by the same bacterium are a) the characterization of the thermophilic (70° or 55°) and mesophilic (37°) bacterial variants (in respect to type), b) the control of homogeneity of the bacterial culture (contamination, mixed population), c) proof of the genetic identity of the 70°- (55°-) and 37°-variant of B. caldotenax and B. stearothermophilus, which differ greatly in their phenotypes, for example in their metabolism, cell- or colony merphology. The taxonomical-biochemical identity or also the identity of morphology of the sporangia, since this should be an expression of the temperature dependent phenotype, cannot be used unconditionally as criteria of identity. Criteria such as the presence of identical enzymes in both variants or the identity of the genome (use of genetic markers, analysis of the DNA) are more reliable. Experiments with both variants of B. caldotenax demonstrated an identically high content of cytosine plus guanine in their DNA: 67.2% in the thermophilic DNA and 66.8% in the mesophilic DNA. In the thermophilic B. stearothermophilus the C+G content of the DNA was 56.5% and in the mesophilic variant 57.1%.

5) Despite the lack of further criteria for genetic identity and the presently not sufficient data concerning the entire amino acid sequences of the LDHs, this adaptive system with its thermophilic and mesophilic enzymes was presented to show the diverse possibilities in enzymology, protein chemistry, metabolisms and genetics that are immanent in such a system.

ACKNOWLEDGEMENT

This work was supported by the Schweizerischer Nationalfonds zur Förderung der wissenschaftlichen Forschung, Project 3.1640.73.

REFERENCES

1. H. Koffler, Bact.Rev. 21, 227 (1957)
2. L.L. Campbell and B. Pace, J.appl.Bact. 31, 24 (1968)
3. R. Singleton and R.E. Amelunxen, Bact.Rev. 37, 320 (1973)
4. H.U. Haberstich and H. Zuber, Arch.Microbiol. 98, 275 (1974)
5. L. Jung, R. Jost, E. Stoll and H. Zuber, Arch.Microbiol. 95, 125 (1974)
6. U.J. Heinen and W. Heinen, Arch.Microbiol. 82, 1 (1972)
   W. Heinen, Arch.Microbiol. 76, 2 (1971)
7. A. Weerkamp and R.D. MacElroy, Arch.Microbiol. 85, 113 (1972)
8. R. Holland and G.G. Pritchard, J.Bact. 121, 777 (1975)
9. J. Everse and N.O. Kaplan, Advances in Enzymol. Vol. 37, p. 65
10. S.S. Taylor, S.S. Oxley, W.S. Allison and N.O. Kaplan, Proc.Natl. Acad.Sci US 70, 1790 (1973)

11. A. Yoshida and E. Freese, Biochim.Biophys.Acta 99, 56 (1965)
12. J. Bridgen, E. Kolb and J.I. Harris, FEBS Letters 33, 1 (1973)
13. S. Artavanis and J.I. Harris, this volume
14. E. Stoll, L.H. Ericsson and H. Zuber, Proc.Natl.Acad.Sci US, 70, 3781 (1973)
15. E. Stoll, H.-G. Weder and H. Zuber, Biochim.Biophys.Acta, in press
16. M.G. Rossmann, M.J. Adams, M. Buehner, G.C. Ford, M.L. Hackert, P.J. Lentz, A. McPherson, R.W. Schevitz and J.E. Smiley, Cold Spring Harbor Symp. Quant. Biol. 36, 179 (1971)
17. J. Ohlson, B. Nordström and C.-J. Brändén, J.Mol.Biol. 89, 339 (1974)

STUDIES ON THE GENETICAL AND BIOCHEMICAL BASIS OF THERMOPHILY

JAMES A. LINDSAY and ERNEST H. CREASER
Research School of Biological Sciences, Australian National University, Canberra, Australia.

The mechanisms of adaption of living organisms to extremes of environment, particularly temperature, are poorly understood. Seemingly the most informative method of investigation would be to study the essential macromolecules necessary for the life process, from those organisms adapted to different temperature regimes. High temperatures are perhaps the most important, since they emphasise the stresses that must be placed on these molecules.

Previous work[1] has demonstrated little difference between the RNA, DNA and ribosomes from mesophiles and thermophiles. However, some differences have been noted in the molecular nature of proteins and enzymes, but they have been difficult to analyse since many of these comparative studies have been made on proteins and enzymes from different genera.

Our work with the enzyme histidinol dehydrogenase (HDH) was an attempt to reduce these comparative differences to a minimum, as we purified this enzyme from a range of species showing varied growth temperatures, all from the genus *Bacillus*. These were:-

B.psychrophilus	(15-28)
B.subtilis	(25-42)
B.stearothermophilus	(37-65)
B.caldolyticus	(37-87)

The problems of purifying one enzyme by normal methodology can be complex and often time consuming. We therefore tried to eliminate some problems by designing a series of affinity columns which could aid in purifying the HDH enzyme and yet function at various temperatures. The design of the columns was based on a series of inhibitor and substrate studies. In the final analysis four columns were made, their structures were:-

1. Sepharose-----Spacer C_6------Histidinol
2. Sepharose-----Spacer C_6------Histamine
3. Sepharose-----Spacer C_6------Histidine (attached at COOH)
4. Sepharose-----Spacer C_6------Histidine (attached at NH_2)

Briefly, purification is carried out by sorbing onto the column a crude cell extract which had been precipitated with 70% ammonium sulphate and dialysed in 0.05 M, pH 9.15 tris/HCI buffer. The column is then washed with tris buffer, to remove the majority of unwanted protein. Removal of the HDH is by elution with I$\underline{M}$ Imidazol pH 9.15 for the histamine column, and 0.I$\underline{M}$ Histidine pH 9.15 for the two histidine and histidinol columns. An example is shown in Fig 1:

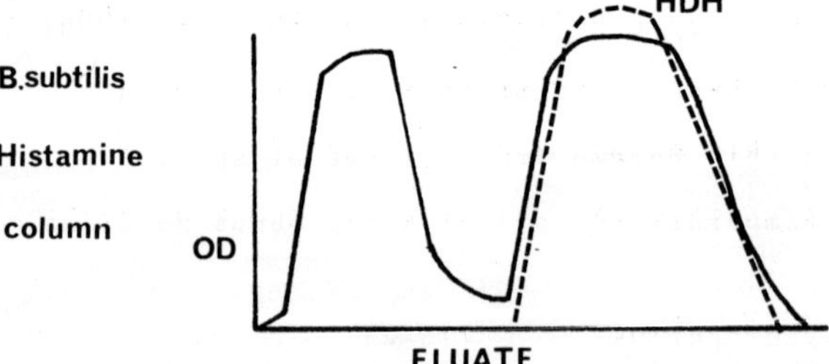

Most importantly for this purification method, all the columns may be used at various temperatures. For example *B.stearothermophilus* and *B.caldolyticus* HDH's were purified on the histidine COOH attached affinity column, at 55° and 65° respectively. The HDH's are subsequently completely purified by gradipore gel electrophoresis.

Enzyme Transformation.

Comparisons of the physical characteristics of the HDH's are shown in Tables 1 and 2:-

TABLE I

Km's (histidinol)	4.0 - 7.5 x 10^{-5}	
pH optima	9.8	
Temperature optima	B.psychrophilus	25
	B.subtilis	41
	B.stearothermophilus	74
	B.caldolyticus	85
SW_{20} value	4.2	
Molecular weight	59.000$\pm$ 1500	
Subunit number	2	

TABLE 2

INHIBITORS AND SUBSTRATES				% INHIBITION
B.psychr.	B.subtilis	B.stearo	B.caldo	Compound
0	0	0	0	L-histidinol - normal substrate
+45%	+60%	+26%	+ 6%	N-acetyl-L-histidinol
+30%	+40%	+29%	+11%	N-benzoyl-L-histidinol
70	70	54	64	Guanidine Carbonate
30	92	50	58	Imidazol
24	0	3	8	Adenine Sulphate
40	44	6	22	N-benzoyl histamine
30	71	76	22	Histamine
16	0	6	2	4-amino 5-imidazol-carboxyamide 2HCI
16	34	29	24	1, 2, 4 Triazole
24	93	83	44	5-imidazol, ethyl-amidinol valeric acid
42	88	63	47	Histidine

It would seem that basically the enzymes have similar structures although there are some noticable differences in the reaction to various inhibitors. Present work in our laboratory is directed towards elucidating the primary amino acid sequence of these various HDH enzymes to see if a molecular basis for thermophily can be found here.

This work although being informative, was not very satisfactory in answering the question "what controls thermostability". We therefore tried a totally new approach, that of genetical analysis, a preliminary account of which has appeared.(2)

Various scattered pieces of information showed that there may be a genetical component(s), McDonald[3] finding a single gene conferring growth at 55° in *B.subtilis* and Howard[4] a single gene allowing growth at 1° in Psuedomonas. Interestingly both these genes were found to be tightly linked to the streptomyocin resistance locus, perhaps indicating some association with the genes coding for rRNA, tRNA and ribosomal proteins, since the streptomyocin locus lies in a very conserved region of the *B.subtilis* genome, coding for the aforesaid properties.

Since *B.caldolyticus* does not have a transducing phage we were limited to DNA mediated transformation for genetic analysis. This method is however a problem since interspecific transformation is limited to a few closely related species of the *B.subtilis* group, and then to an even fewer number of loci.[7]

Two organisms were used, *B.subtilis* 25 *his*B (31) (str^S pur^{A-} (*ade* 16) leu^- his^- trp^- met^-; growth optimum 37°) as recipient and *B.caldolyticus* (str^r $pur A^+$ *ade*16) growth optimum 72°) as donor.

The standard phenol extraction method for DNA purification was found to be unsuitable for thermophiles.

Instead the DNA was extracted from *B.caldolyticus* and purified by treating the cells with 100 µg lysozyme in 5 ml 0.01 M EDTA (pH 8.0) for 15 min at 37°C; 10µg pronase was added and the cells stored at 4° for 16 hr; the resulting lysate was centrifuged to equilibrium in an ethidium bromide-caesium chloride gradient and the DNA band isolated.[5]

Transformation was carried out by the method of Spizizen.[6] Competent *B.subtilis* cells were treated with DNA (2.5 µg/ml) for two periods (40 min and 4 hr) after which 50µg/ml DNAase was added. Transformants for streptomycin resistance, and for adenine, histidine, leucine, tryptophan and methionine prototrophy were selected at 37° after growth of the treated cells in a rich broth for 4 hr to allow expression. Selection for high temperature growth ability was made by allowing samples of the 4 hr treated, 37° grown cells to grow further in a defined liquid medium for thermophiles at 70° for 18 hr and then plating for single colonies at 55°. The results are shown in Table 3.

TABLE 3

	*Thermophilic Plates 70°C growth.	
	-strept	+strept
40 min DNA treatment	714	11
4 hour treatment	1040	203
Controls		
(1) No DNA, no bacteria	no transformants	
(2) Bacteria, no DNA	no transformants	
(3) DNA, no bacteria	no transformants	
(4) Bacteria, DNA, DNase From Time 0	no transformants	

*Number of colonies per ml.

Selection at high temperature yielded transformants capable of growth at 70°, and transformants inheriting streptomyocin resistance or adenine prototrophy were obtained at 37°. Selection for inheritence of histidine, leucine, tryptophan or methionine prototrophy revealed no transformants.

Marmur *et al*[7] found a gradient of con servation in the DNA sequence of *B.subtilis* which extends from the origin (Pur A) to (Pur B), and since this sequence according to Shaw,[8] resists mutation it would seem possible that adaption to growth at high temperature may be a manifestation of the action of modified genes in this region. Further genetical analysis has given us more data, which allows us to map the gene(s) in this region.

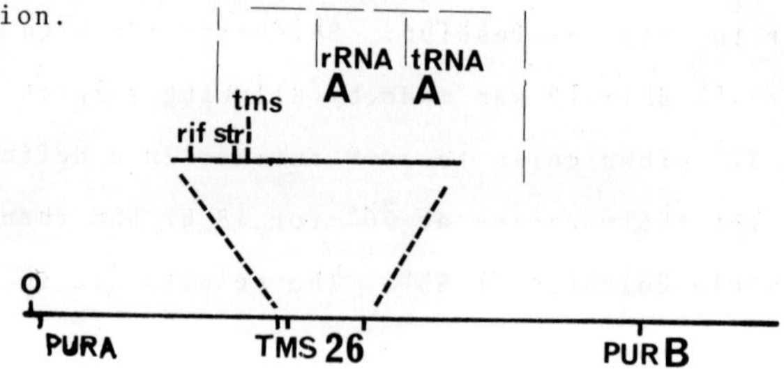

Growth curves of the transformants revealed markedly different rates at varying temperatures. At 70°, 20% of the transformants grew faster than the donor, and 10% much slower. At a lower temperature 37° these ratios are somewhat similar.

This data is suggestive of more than one gene conferring the ability to grow at high temperatures, and it would seem that for more complete thermostability ie. to grow at the faster rate, a majority of these genes must be inherited. The observed growth rates could be a function of the number of high temperature genes transferred. The rif^{r}, high temperature transformants showed only one growth pattern, corresponding to the highest rate shown in Figure 2.

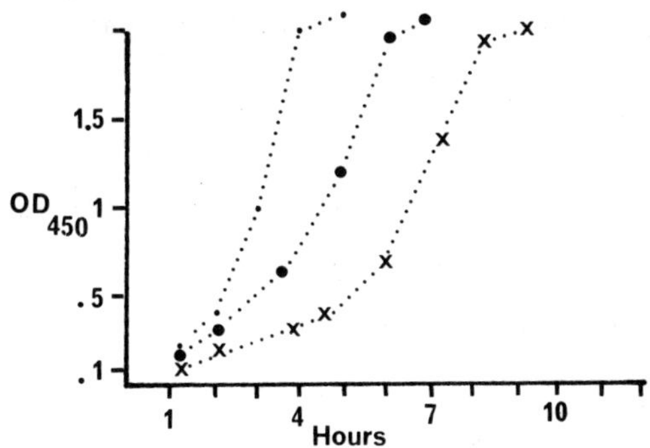

We selected one of the faster growing high temperature, *str*^{r} transformants, designated HT-I, and examined its HDH activity. Whereas the recipient *B.subtilis* (25 *his*B) HDH was inactivated by temperatures above 70°, the HDH enzyme from HT-I was thermostable even at 100°. Graph 2 shows the relative initial velocity curves for the various *Bacillus* HDH's.

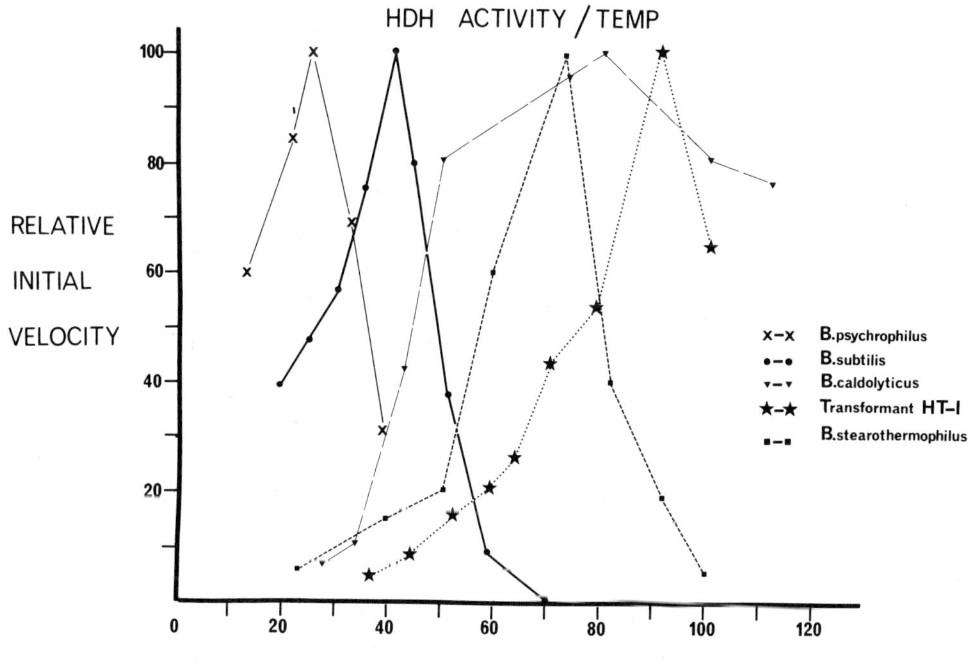

Identical results were obtained whether the enzyme preparation was in a crude extract or as a pure protein. There was however an unusual result with the HT-I HDH. The enzyme was not identical with that of either parent, inasmuch as it required 2-mercaptoethanol for full activity. Inhibitor studies revealed that in point of fact there may be some changes, possibly in the tertiary structure, to account for this.

TABLE 4

INHIBITORS AND SUBSTRATES % INHIBITION

B.subtilis	B.caldo	HT_1	Compound
0	0	0	L-histidinol
+60%	+ 6%	+90%	N-acetyl-L-histidinol
+40%	+11%	+25%	N-benzoyl-L-histidinol
70	64	50	Guanidine Carbonate
92	58	50	Imidazol
0	8	0	Adenine Sulphate
44	22	0	N-benzoyl histamine
71	22	60	Histamine
0	2	0	4-amino 5-imidazol-carboxyamide 2HCI
34	24	30	1, 2, 4 Triazole
93	44	77	5-imidazol, ethyl-amidinol valeric acid
88	47	51	Histidine

Interpretation of our results gives us two possible alternatives; either, that sufficient individual genes have been transferred to enable the transformants to grow at high temperatures or a small number of genes controlling wide pleiomorphic effects have been transferred.

We favour the latter, since it is highly improbable that multiple transformations of the needed frequency could occur. A working hypothesis is that there have been changes in the protein synthesising ability of the transformants at the ribosomal or tRNA level. These changes give rise to altered enzymes and proteins which can function at high temperatures. We feel that the answer may lie at the translational level of protein synthesis.

Various ideas, some of which have been suggested by Singleton and Ameluxen.[1]

(1) A change in the secondary structure of the tRNA molecule allowing binding of more than one amino acid.

(2) Variation in the types of tRNA's available for amino acid insertion.

(3) Conformational change in the synthetase which may allow for the enzyme acyl adenylate complex to bind a different tRNA.

(4) Non methylation of some tRNA's which allows for binding of other, not often used amino acids.

(5) Altered ribosomal proteins causing similar effects as str^{r} and str^{d} in *E. coli*.

These ideas will be tested shortly as we are setting up a cell free protein synthesising system, using the same genetic components as in the transformation experiments.

Interestingly, the gene(s) confering the thermostability do not affect other genetic properties of the transformants. We have used another strain of *B.subtilis* designated WB2802, which is derepressed for HDH levels. High temperature transformants of this strain also show derepressed levels of HDH activity when compared with the HDH from *B. subtilis* 25*his*B.

If the hypothesis that the transformation from meso to thermophily is caused by a few genes with very wide effects, is confirmed, one can speculate further. If we believe as Brock[9] suggests that survival at high temperatures is a very primitive adaption, then mesophiles and psychrophiles may have developed from thermophiles by losing some genetic material. All we have done is to put that property back.

(1) Singleton, R and Amelunxen, R.E. Bacterial Reviews 37 320-342 1973. (most recent review on subject)

(2) Lindsay,J.A. and Creaser,E.H. Nature 5510 650-652 1975

(3) Howard, J. 44th ANZAAS Conference, Perth 1973

(4) McDonald, W. C. Can J. Microbiol 15 1287-1291 1969

(5) Radloff, R. *et al* P.N.A.S. USA 57 1514-1521 1967

(6) Spizizen, J. P.N.A.S. USA 44 1072-1078 1958

(7) Marmur, J. *et al* Jou. Bact 85 461-467 1963

(8) Shaw, C.R. Biochem Genet. 4 275-283 1970

(9) Brock, T.D. Science 158 1012 1967

BACTERIAL DIVERSITY AT HIGH TEMPERATURE

Eric DEGRYSE

Laboratorium voor microbiologie en genetica
Vrije Universiteit van Brussel,
Opzoekingsinstituut van het C.O.O.V.I.,
Emile Grysonlaan, 1, B-1070 Brussel,Belgium.

INTRODUCTION

An increasing number of organisms which are able to grow at high temperature has been isolated during the last few years. Our purpose in this paper is to analyze their properties from a microbiological point of view. The discussion will only deal with thermophilic procaryotes to the exclusion of blue-green algae (Castenholz, 1969) and actinomycetes (Cross, 1968) which have been reviewed recently. Organisms will be considered as thermophilic if they grow above 50°C, as extreme thermophilic or caldo-active if they grow at temperatures in excess of 70°C.

As they have adapted to a single variable, temperature, thermophiles represent a choice material in the study of evolution. This type of adaptation, of which the results can be determined through a comparative analysis, appears correlated to the acquisition of thermostable proteins (Singleton and Amelunxen, 1973) and ribosomes (Pace and Campbell, 1967). Evolution towards high temperature has occured for a large number of bacteria with diverse morphologies and metabolic types. Clearly, thermophily is not the privilege of a morphologically and physiologically homogeneous group of organisms. This is confirmed by the G-C content of DNA isolated from thermophiles since it varies from 25 to 69 %. It should be asked now : "How are the metabolic and morphological properties clustered in the thermophilic bacteria"? Furthermore, since thermostability is enclosed in the primary sequence of the protein, one may expect that many mutational changes, particularly those affecting the structure of proteins and ribosomes, are necessary to acquire the capacity of growth at high temperature. Therefore it might be that adaptation to high temperature requires such fundamental changes that new types of bacteria emerge. It is necessary to know to what extent the type of clustering found in mesophiles is also found in thermophiles. An answer to these questions is of importance in evaluating the possible relationships between both groups of bacteria.

Since a recent review of the thermophilic organisms is not available it will be necessary to discuss them briefly, as divided in three broad classes : first the mycoplasmas, second the photosynthetic bacteria and forms related morphologically, third the eubacteria comprising gram-positive and gram-negative forms.

Thermophilic Organisms

1. Mycoplasmatales :

Thermoplasma acidophilum, isolated by Darland et al. (1970) is a bacterium resembling other mycoplasmas morphologically, both microscopically and macroscopically. However it is a free-living form adapted to high temperature and low pH. It is aerobic and does not require cholesterol. It stands somehow isolated from the other representatives of the Mycoplasmaceae. A closer relationship is perhaps to be found in mesophilic free living forms, if they are found.

2. Photosynthetic and Gliding Bacteria :

Chloroflexus aurantiacus (Pierson and Castenholz, 1974) is a filamentous gliding bacterium. Furthermore it can grow anaerobically in the light (with H_2S) and aerobically in the dark and the light. As suggested by Pierson and Castenholz, it might be the link between the green photosynthetic bacteria and the blue-green algae, of which some representatives also show gliding movements. A caldo-active gliding, but non-photosynthetic bacterium was isolated from hot springs (Lewin, 1970). Herpetosiphon geysericola is aerobic and forms filaments which may be sheated. It is cellulolytic.

3. Eubacteria. - A. Gram-positive :

Let us now turn to the gram-positive eubacteria.

a) Lactobacillacae.

Three species of the genus Lactobacillus can grow slightly above 50°C (Rogosa, 1974). They are Lactobacillus delbruckii, L. lactis and L. helvetici. It can perhaps be mentioned that "Lactobacillus thermophilus" is a sporeforming bacterium and might be Bacillus coagulans as suggested by Kitahara and Sizuki (1963).

b) Coryneform bacteria.

Among Coryneform bacteria there is a strictly anaerobic Eubacterium helwigiae, which is capable of growth at 55°C (see in Holdeman and Moore, 1972).

c) Actynomycetales.

The thermophilic actinomycetes have been reviewed (Cross,1968). It may be mentioned that no extreme thermophilic representatives are as yet described.

d) Endosporeforming bacteria.

Thermophilic organisms are known in the three genera of spore forming rod-like bacteria.

Let us consider the genus Clostridium (see Smith and Kobbs, 1974). In group III of the anaerobic Clostridia is classified Clostridium thermosaccharolyticum. This bacterium has 26 % G-C. Closely related and I think synonymous, is Cl. tartarivorum described by Mercer and Vaughn (1951).

Clostridium thermocellum belongs to group V. It is specialized in the fermentation of cellulose, cellobiose and xylose.

Morphologically and physiologically related to the mesophilic Desulfotomaculum ruminis, is the thermophilic Desulfotomaculum nigrificans. It is strictly anaerobic but contains cytochromes and uses sulfate as inorganic terminal electron acceptor. Its maximum temperature is 70°C (Campbell and Postgate, 1965).

In table I, we have compiled the properties of the aerobic thermophilic sporeforming bacteria. Apart from the well-known Bacillus coagulans and B. stearothermophilus, it may be interesting to note that thermophilic strains of B. subtilis, B. licheniformis and B.brevis also exist (see Gibson and Gordon, 1974). Bacillus acidocaldarius was isolated by Darland and Brock (1974). Both its thermolabile spores and its high G-C content (63 %) are atypical for a Bacillus. In addition to the organisms presented in table I it may be mentioned that Heinen and Heinen (1972) isolated three caldo-active Bacillus strains. There is also a sporeforming bacterium growing strictly autotrophically on reduced sulfur compounds. It was called "Thiobacillus thermophilica" (Egorova and Deryugine, 1963). However since it is sporeforming it might be a new genus related to Bacillus.

Table I : the aerobic sporeforming bacteria of the genus Bacillus (Gibson and Gordon, 1974).

Species	subtilis	licheniformis	coagulans	stearothermophilus	brevis	acidocaldarius
morphological group	1	1	1 - 2	2	2	1 - 2
sporangium	c	c	st - t	t	ct	st - t
maximum temperature	55	55	60	75	60	70
optimum pH	6-8	6-8	5-7	6-8	6-8	3-5
stability of the spores at 100°C	+	+	+	+	+	-
motility	+	+	+	+	+	
fermentation	-	+	+	+	-	-
denitrification	-	+	-	-	-	-
G-C content[1]	(42-43)	(43-47)	47	53	(45)	62

abbreviations : c : central - ct : central to terminal - t : terminal st : subterminal

[1] in brackets : if determined on mesophilic strains of the same species.

e) Methane producing bacteria.

Finally a Methanebacterium thermoautotrophicus has been isolated (Zeikus and Wolfe, 1972). It is an extreme thermophile which is metanogenic in the presence of CO_2 and H_2 , the only substrates used for growth. It is morphologically similar to the mesophilic M. formicicum.

- B. Gram-negative bacteria :

Let us now analyse the gram-negative bacteria.

a) Chemolitotrophic bacteria

Among the chemolitotrophic bacteria two morphological groups might be distinguished : first the rod-shaped "Thiobacillus and second the spherical extreme thermophilic forms which lack a peptidoglycan wall.

To the first group belongs a facultatively autotrophic Thiobacillus described by Williams and Hoare (1972). The properties in common with mesophilic facultatively autotrophic bacteria are compared in table II. Note the extensive similarities of the unnamed thermophile with Thiobacillus novellus.

Table II : The aerobic facultatively autotrophic sulfur oxidising bacteria of the genus Thiobacillus (Vishniac, 1974)

Species	novellus	unnamed[1]	intermedius
optimum temperature	30	50	30
optimum pH	7.8-9	5.6	6 - 7
growth on single organic compounds	-	-	+
thiosulfate oxidation in autotrophic medium	1/3	2/3	3/3
sulfite oxidase[2]	+	+	
repression of thiosulfate oxidation by organic compounds	+	-	-
growth on sulfur	-	-	+
motility	-	-	+
G-C content	66	66	

1 : described by Williams and Hoare (1972).

2 : AMP independent sulfite oxidase.

The second group unites morphologically similar organisms growing at very high temperature and low pH. Nevertheless wide variations in G-C % and some metabolic differences became apparent. These organisms might be the result of a convergent evolution imposed by their extreme habitat (De Rosa et al. 1975). It should stimulate microbiologists to apply, in addition to the classical tests, new techniques in order to obtain phenotypical data reflecting the genotypical differences.
The properties of these organisms have been compiled in table III, where they can be compared with those of rod-shaped counterparts of the genus Thiobacillus. It should be noticed that the Italian sulfur oxidisers are somewhat unrelated and might be a new genus.

Table III : The spherical shaped thermophilic sulfur oxidisers

Genus	Thiobacillus[1]	Sulfolobus[2]	Unnamed[3]	Unnamed[4]	Thiobacillus[5]
Species	acidophilus	acido-caldarius	idem	idem	ferrooxidans
optimum temperature	30	70 - 75	75 - 85	65	35
optimum pH	3	2 - 3	3 - 4		2.5-5.5
substrate	sulfur	sulfur	sulfur	sulfur	sulfur
[a]stimulation of growth on sulfur		+	+	-	-
[a]repression of sulfur oxidation		-			
growth on sulfur	facultative	idem	idem	obligate	idem
growth on Fe^{++}	-	-	facul-tative	obligate	idem
motility	+	-	-	-	+
peptidoglycan	+	-	-	-	+
isoprenoid ether lipid	-	+	+		-
G-C	63	60 - 70	40 - 47	54 - 60	56 - 57

[a] on addition of yeast extract

References : 1 : Guay and Silver, 1975 - 2 : Brock et al., 1972 -
3 : De Rosa et al., 1975 - 4 : Brierly and Brierly, 1973.
5 . Temple and Colmor, 1951.

b) Pleomorphic bacteria

A spherical, pleomorphic bacterium was isolated from hot springs in Tiberias (Kahan, 1961). It reproduces by budding and can give rise to polarly flagellated cells. It is aerobic and does not grow at 30°C. It might be considered as belonging to the genus Hyphomicrobium.

Another pleomorphic bacterium was described by Jackson et al. (1973). Thermomicrobium roseum is a slow growing (generation time under optimal conditions is 5 hrs) aerobic caldoactive strain with 64 % G-C. Its affiliation is at present obscure.

A completely different organism is capable at 50°C of heterotrophic and autotrophic growth with H_2 . It was called Hydrogenomonas (McGee et al., 1967). However the genus Hydrogenomonas has been given up. Therefore, on the basis of its polar flagellation, this organism is to be considered as a Pseudomonas species.

Finally Brock and Freeze described a widely distributed group of extreme thermophiles. Thermus aquaticus was found to be aerobic and displayed a rod-filament dimorphism. It gives also rise to special spherical forms, called rotund bodies and it has a marked penicillin sensitivity.

Strain HB8 isolated from a thermal spa in Japan (Oshima and Imahori, 1971) does not form the filaments nor the rotund bodies of Thermus aquaticus. Because of higher temperature optimum and salt tolerance it was called Thermus thermophilus (previously Flavobacterium thermophilus) (Oshima and Imahori, 1974). Carotenoid-less strains resembling Thermus were isolated in the U.S.A. by Ramaley and Hixon (1970) and in Great Britain by Williams and Pask-Hughes (1973).

We have analysed the type strains of Thermus aquaticus (strain YTI) and of Thermus thermophilus (strain HB8), the only strains described so far, and have compared them with strains NH and DI (kindly provided by Dr. Williams) and with 3 new belgian isolates (Degryse and Glansdorff, in preparation). We have extended the original observations and found that all strains are morphologically and physiologically similar. Indeed the filaments and rotund bodies occur in all strains but are probably involution forms occuring under adverse conditions. Nevertheless, the general appearance of these rotund bodies and of the penicillin sensitivity among these strains favours their taxonomic importance, already suggested by Brock and Freeze (1969).

Except for flagellation, it seems that Thermus aquaticus is related to the mesophilic genera Pseudomonas and Alcaligenes. Ascribing this organism to one of these genus would require the demonstration of a relationship with the relevant species.

CONCLUSIONS

What conclusions may be drawn from this survey ? It seems that although a great number of properties have adapted to high temperature, these properties are not clustered at random. Almost all these properties are found in mesophiles and, most important, the same kind of clustering is observed for thermophiles as well as for mesophiles. Indeed thermophiles do resemble mesophiles usually at the intra- and interspecies level. Clearly thermophily does not involve, in the majority of cases, extensive changes. Because of the observed similarities it follows that adaptation to a high temperature habitat was rather conservative, that is through small and limited modifications of existing properties. Therefore, differences in growth temperature should not be considered sufficient for the creation of new species for strains with otherwise identical or nearly identical properties.

We think that at least some species exist which are related to mesophiles to the same extent as mesophiles are related among themselves, the expression of the difference being more spectacular for the thermophiles.

Nevertheless the constraints imposed by very high temperatures have influenced bacterial life. Only 7 genera are represented above 70°C. They are Herpetosiphon, Bacillus, Methanobacterium, Sulfolobus, the Italian sulfur oxidisers, Thermomicrobium and Thermus.

The relationship with mesophilic counterparts is obscured for at least some of these caldoactive organisms. This seems particularly true for organisms growing at high temperature and low pH.

These" microbiological " observations obtained in fact on the organisms as a whole, seem to be supported by the studies of its components. In a recent review of proteins of thermophilic organisms, Singleton and Amelunxen (1973) concluded that these proteins appear similar to their mesophilic counterparts. They write (Singleton and Amelunxen, 1973) : "Their points of similarity include : 1 molecular weight, 2 subunit composition, 3 allosteric effectors, 4 amino acid composition, 5 primary sequences".

As noted on several occasions the G-C content of the DNA of thermophiles has the same value as that of mesophiles. Indeed Pace and Campbell (1967) could not find a relationship between growth temperature and G-C %. This lack of correlation is not observed at the level of the ribosomes, where indeed a higher G-C content is observed, accounting in part for the higher melting temperatures.

Nevertheless, with an increase in temperature there is a decrease in the expression of life (see also Brock, 1969) : together with a decrease in genus diversity there is also a decrease in the number of metabolic activities adapted. Indeed none of the extreme thermophiles, mentioned above, ferments. Nitrogen fixation is absent above 60°C (Postgate, 1971 ; Stewart, 1973). The use of light as energy source does not occur above 70°C (Brock, 1969). Furthermore the growth rate of the high temperature strains of thermophilic blue-green algae is slower (Castenholz, 1969).

Organisms growing at high temperature and low pH show the results of a convergent evolution : absence of motility, replacement of an ester linkage by an isoprenoid ether lipid, absence of a peptidoglycan wall. Weiss has proposed (1974) that the absence of the peptidoglycan wall might be the consequence of its lability. As an alternative we suggest that the peptidoglycan-forming enzymes have not evolved simultaneously with the rest of the organism. Likewise we might

attribute the filaments formed by Thermus aquaticus to the sensitivity of the septum forming enzymes to superoptimal conditions.
It has also been reported by various workers that a number of enzymes isolated from Bacillus stearothermophilus are thermolabile at the temperature of synthesis. We have also found in Thermus aquaticus strain Z05 that even with the supplementation of certain vitamins, growth on glucose and on pyruvate ammonium is slow as compared to mesophiles, whereas growth on glucose, glutamate and ammonium is fast. (see table IV). It appears that specific enzymes are limiting for glucolysis on glucose and gluconeogenesis on pyruvate ammonium.
We think that in thermophiles and in extreme thermophiles in particular, a number of enzymes might not be completely adapted.

Table IV : Growth rates of Thermus aquaticus Z05 compared to those of mesophiles

Organisms	Escherichia coli[1]	Bacillus subtilis[2]	Pseudomonas aeruginosa[3]	Ps. putida[4]	Thermus aquaticus Z05[4]
Substrates					
glucose and glutamate					65 '
glucose	62'	90'	120'	120'	
malate	90'	102'			
glutamate	240'		110'	84'	120'
acetate	240'			110'	120'
pyruvate	105'			105'	270'

References : 1 : Kornberg, 1965.
2 : Diesterhaft and Freeze, 1973.
3 : Isaac and Holloway, 1972.
4 : Degryse, in preparation.

What are the reasons for the absence of whole families at high temperatures ? No single explanation exists but rather a combination of several factors might come into play. It is quite likely that all organisms have not been in contact with a high temperature habitat for the same era. Indeed the scarcity of a selective medium (see for example Castenholz, 1969) could have restricted the number of species, because only widely distributed bacterial groups could evolve or have evolved.
It has been noticed that few strictly and preferential fermentative thermophiles exist . This might be attributed to the inefficiency of

fermentation compared to respiration, as an energy yielding metabolism. Therefore, to remain competitive, a fermentative cell has to catabolize more substrate per time unit than the respiring cell.
Finally the genetic background of the adapting organism may have played an imperative role. Many of the existing thermophiles might represent a selection of organisms whose genome had not to be modified too much in order to yield thermostable cell components. The only actual indication that this may be true, is the observation that between a series of closely related meso- and thermophilic species the G-C content is conserved. In order to prove this statement more directly one has to perform DNA-DNA and DNA-RNA hybridizations.

ACKNOWLEDGEMENTS

I wish to express my sincere thanks to Drs. Glansdorff and Piérard for many stimulating discussions and for their help during the preparation of the paper.

REFERENCES

Brierly, C.L., Brierly, J.A., 1973. Can. J. Microbiol. 19, p.183-188.
Brock, T.D., 1969. 19th Symp. Soc. Gen. Microbiol. p. 15-41.
Brock, T.D., Brock, K.M., Belly, R.T., Weiss, R.L., 1972. Arch. Microbiol. 84, p. 54-68.
Brock, T.D., Freeze, H., 1969. J. Bact. 98, p. 289-297.
Campbell, L.L., Postgate, J.R., 1965. Bact. Revs. 29, p. 359-363.
Castenholz, R.W., 1969. Bact. Revs. 33, p. 476-504.
Cross, T., 1968. J. Appl. Bact. 31, p. 36-53.
Darland, G., Brock T.D., Samsonoff, W., Conti, S.F., 1970. Science 170, p. 1416-1418.
De Rosa, M., Gambacorta, A., Bulock, J.D., 1975. J. Gen. Microbiol. 86, p. 156-164.
Diesterhaft, M.D., Freeze, E., 1973. J. Biol. Chem. 248, p. 6062-6070.
Egorova, A.A., Deryugina, Z.P., 1963. Mikrobiologiya 32, p. 439-440.
Gibson, T., Gordon, R.E., 1974. In "Bergey's Manual of Determinative Bacteriology" 8th ed., p.529-550. Buchanan, R.E., and Gibbons, N.E. eds., the Williams Wilkins Company, Baltimore.
Guay, R., Silver, M., 1975. Can.J.Microbiol. 21, p. 281-288.
Heinen, V.J., Heinen, W., 1972. Arch. Microbiol. 82, p. 1-23.
Holdeman, L.V., Moore, W.E.C., 1972. In "Eubacterium" p. 23-30. Holdeman, L.V., Moore, W.E.C. eds., Anaerobe Laboratory Manual, Virginia Polytechnic Anaerobe Laboratory, Blaksburgh, Va.
Isaac, J.H., Holloway, B.W., 1972. J. Gen. Microbiol. 73, p. 427-438.

Jackson, T.J., Ramaley, R.F., Meinshein, W.G., 1973. Int. J. Syst. Bact. 23, p. 28.
Kahan, D., 1961. Nature 192, p. 1212-1213.
Kitahara, K., Suzuki, J., 1963. J. Gen. Appl. Microbiol. 9, p.59-71.
Kornberg, H.L., 1965. In "Colloques du C.N.R.S. : Régulation chez les microorganismes", p. 193-207.
Lewin, R.A., 1970. Can.J. Microbiol. 16, p. 517-520.
Mercer, W.A., Vaughn, R.M., 1951. J. Bact. 62, p. 27-37.
Mc Gee, J.M., Brown, L.R., Tisher, R.G., 1967. Nature 214, p.715-716.
Oshima, T., Imahori, K., 1971. J. Gen. Appl. Microbiol. 17, p. 513-518.
Oshima, T., Imahori, K., 1974. Int. J. Syst. Bact. 24, p. 102-112.
Pace, B., Campbell, L.L., 1967. Proc. nat. Acad. Sci.(USA) 57, p. 1110-1116.
Pierson, B.K., Castenholz, R.W., 1974. Arch. Microbiol. 100, p.5-24.
Postgate, J.R., 1971. In "The Chemistry and Biochemistry of Nitrogen Fixation, p.161-190. Postgate J.R. ed. Plenum Press, London, New York.
Ramaley, T.J., Hixon , 1970. J.Bact. 103, p. 527.
Rogosa, M., 1974. In "Bergey's Manual of Determinative Bacteriology" 8th ed. p. 576-593. Buchanan, R.E., Gibbons, N.E. eds. The Williams Wilkins Company, Baltimore.
Singleton, R., Amelunxen, R.E., 1973. Bact. Revs.37, p. 320-342.
Smith, L., Hobbs, G., 1974. In "Bergey's Manual of Determinative Bacteriology" 8th ed. p. 551-572. Buchanan, R.E., Gibbons, N.E. eds. The Williams Wilkins Company, Baltimore.
Stewart, W.D.P.,1973. Ann. Rev. Microbiol. 27, p. 283-316.
Temple, K.L., Colmer, A.R., 1951. J. Bact. 19, p. 605-611.
Vishniac, W.V., 1974. In "Bergey's Manual of Determinative Bacteriology" 8th ed. p. 456-461. Buchanan, R.E., Gibbons, N.E. eds. The Williams Wilkins Company, Baltimore.
Weiss, R.L., 1974. J. Bact. 118, p. 275-284.
Williams, R.A.D., Hoare, D.S., 1972. J. Gen. Microbiol. 70, p.555-566.
Williams, R.A.D., Pask-Hughes, R.A., 1973. J. Gen. Microbiol. 77, p. X.
Zeikus, J.G., Wolfe, R.S., 1972. J. Bact. 109, p. 707-713.

ROUND TABLE AND GENERAL DISCUSSION ON STRUCTURE AND FUNCTION OF THERMOPHILIC ENZYMES AND PROTEINS

CHAIRMAN (H. NEURATH, Seattle):
The members of the panel whom you see here had a briefing session yesterday and decided to try to highlight those aspects of the discussion of the last two and a half days which represent recurrent themes or which require further clarification. The general topic of the session this afternoon will be protein structure and thermophily. We are ascending from the primary structure of proteins in terms of composition and partial sequences to the three-dimensional structure as it relates to homologous amino-acid sequences. We shall next proceed from the monomeric to the sub-unit enzymes or, in other words, consider the tertiary and the quaternary structure of thermophilic proteins. Many of the enzymes which are of interest in connection with this symposium are metallo-enzymes and we therefore feel that a consideration of the role of metals in the stability and functions as they relate to thermostability should be included in this discussion.

As we have heard during the last two days, there is a question how the temperature optimum of enzymes, as measured by activity, is related to conformational stability, and the question of the reversible denaturation needs to be considered in this connection. Some confusion in this area has been generated in the past by the inconsistency with which melting temperatures and melting profiles have been expressed. We feel that some clarification would be in order. It is also clear from the discussion of the last two days that in many cases thermophilic and mesophilic enzymes of different phylogenetic origins have been compared with each other, thus raising the question of how to dissociate the background of phylogeny from the intrinsic thermal stability.

As a general procedure, I would like to suggest that we first hear a discussion among the members of the panel and then invite audience participation. The time that we spend on each topic will depend on the nature of the discussion and the clarifications that we can achieve, and I think it would be appropriate to start with the primary structure of protein, the relation of the amino acid composition to thermal stability and the extent to which partial sequences shed some light on this question. Dr. Harris, would you like to open the discussion of this topic?

J.I. HARRIS, Cambridge:
Attempts have been made to relate protein structure to thermal stability by comparing amino acid compositions of mesophile and thermophile proteins. However, it seems obvious from what has been said here this week that there is no generally valid correlation between the content of any particular amino acid, or group of amino acids (e.g. hydrophobic or charged residues), and thermal stability. Clearly what matters is not the overall composition but the location of particular residues in the three-dimensional structure.

I said in my lecture that thermophilic enzymes tend to have fewer SH groups than their counterparts in mesophiles. For example, lobster and pig muscle GPDHs contain five and four cysteines, respectively. Of these one (Cys-149 in the sequence) is essential and cannot be deleted or substituted. Perhaps it is significant that GPDH from an extreme thermophile contains just one cysteine, Cys-149. Furthermore, whereas PFK from rabbit muscle possesses 16 cysteines per subunit, the B. stearothermophilus enzyme contains only two, while T. aquaticus enzyme has no cysteine. Presumably SH groups are not essential for PFK activity and one might argue that the cysteine residues may have been deleted so as to prevent oxidative inactivation of the enzyme under aerobic conditions at 75°C. There is also the interesting example of alcohol dehydrogenase: subunits of liver, yeast and B. stearothermophilus enzymes contain 14, 9, and 5 cysteines, respectively. The overall trend is thus again apparent. Nevertheless in the B. stearothermophilus enzyme a cysteine has been inserted in a position homologous to Thr-43 in liver and Ser-40 in yeast. Moreover cysteines homologous to Cys-46 in liver have been conserved in the other two, which is reassuring since this particular cysteine has been shown to be a ligand to the essential Zn in liver ADH. It is also reassuring that the insertion of a cysteine (Cys-43) in the B. stearothermophilus enzyme has occurred at a position that is known to be buried in the interior of the protein and thus less likely to be a potential source of instability. It is therefore entirely possible that a thermophile protein will be found to have as many or even more SH groups than a mesophile counterpart - that in itself would not necessarily invalidate my hypothesis. One would need to know the exact locations of the cysteines in the 3-D structure, and in this connection I might also point out that although the cysteine in B. stearo.-ADH that is homologous to the essential Cys-46 in liver has been conserved, it is located in a more protected environment in the native structure since, unlike its counterparts in liver and yeast, it is not reactive towards thiol reagents.

Such considerations as these could of course also apply to other residues with potentially reactive side chains so that it is not altogether surprising that there is no obvious correlation between amino acid composition per se and thermal stability, and I doubt whether such comparisons serve any useful purpose.

J. BREWER, Georgia:
Dr. Ljungdahl and I have been looking at amino acid compositions for quite some time and aside from bloodshot eyes we found that, including the data presented in this symposium, there seems to be a trend toward more arginine in maybe three cases out of four in proteins from thermophiles. Possibly the arginine is bulkier and more hydrophilic than whatever it replaces. It might be that if we could look at the three-dimensional structure for some of the few proteins whose three-dimensional structure is known and look particularly for any substitutions of arginine for something else, the arginine should then be at the surfaces of the proteins, the ones exposed to the solvent so that they can interact with the solvent and hence impart some additional stability to the protein.

E. STELLWAGEN, Iowa:
John, could I ask a question of this correlation with arginine and that is: isn't it just the lysine-arginine ratio that changes?

J. BREWER:
Not necessarily. We see no change in the ratio in the case of the synthetases or the dehydrogenases. We looked at a series of about five or six different sets of enzymes whose compositions were known, that is a series of one enzyme from a thermophile and one to six or seven from a mesophile. There was one instance in which there was no correlation of ar-

ginine with thermostability (the α-amylases). There was one instance of a possible correlation, the class one, aldolases. The other four or five cases showed a correlation with arginine content with thermostability, that is thermophily. And we did not necessarily see a change in the lysine-arginine ratio. In other words, it seems to be more a trend toward increased arginine than toward increased arginine and decreased lysine.

H. ZUBER, Zürich:
As I pointed out in my talk the thermophilic and mesophilic lactate dehydrogenases from B. stearothermophilus and B. caldotenax show remarkable differences in the ratio of argine to lysine. The thermophilic LDH contains about 10-11 Arg residues more than the mesophilic LDH, and the mesophilic LDH, correspondingly, about 8-10 Lys residues more than the thermophilic LDH.

J.I. HARRIS:
In GPDHs there is also a positive numerical correlation between the total number of arginines and thermal stability. Thus arginine contents range between 9 in lobster, 15 in B. stearothermophilus and 16 in T. aquaticus. Close inspection of the 3-D structure does not, however, reveal any positive interactions of arginines that might contribute towards the stability of either the tertiary or quaternary structure of the enzyme. There is one interesting example of a Lys to Arg replacement. I refer to Lys→183 in the sequence which in all known mesophile GPDHs is the site of an SH NH_2 acyl transfer reaction with the thermophilic enzyme but only when the sequence was done did we realise why. The Lys→Arg substitution does however maintain a basic side chain in this position in the protein.

R. SINGLETON, Newark:
A couple of years ago, when I was at NASA Ames Research Center, I set up the two classes of proteins I mentioned the other day. I treated all of the thermophilic proteins as one class and a collection of corresponding mesophilic proteins as another class.

	REECK & FISHER (207 PROTEINS)		NON THERMOPHILIC (56 PROTEINS)		THERMOPHILIC (15 PROTEINS)		t_c/t_o
	MEAN	sd	MEAN	sd	MEAN	sd	
Asx	10.7	2.6	12.3	2.2	10.2	2.1	1.76
Thr	5.7	1.9	6.0	2.2	6.1	1.4	.14
Ser	6.3	2.5	6.3	1.6	4.4	1.4	1.62
Glx	10.6	3.3	9.0	2.9	10.2	2.6	.98
Pro	4.8	2.1	4.4	1.2	4.9	3.2	.42
Gly	8.1	3.1	8.7	1.8	9.3	1.4	.77
Ala	8.5	2.8	9.8	2.3	10.2	3.1	.33
Cys	2.3	2.7	2.9	4.3	1.6	4.0	.56
Val	6.8	2.0	7.6	2.1	8.0	2.0	.28
Met	1.9	1.1	1.6	.9	1.7	.8	.12
Ile	5.0	1.7	5.9	1.7	6.3	1.7	.25
Leu	8.1	2.5	6.4	2.6	7.1	2.6	.61
Tyr	3.3	1.6	3.3	1.6	3.5	2.4	.06
Phe	3.7	1.4	3.5	1.2	3.2	1.5	.54
Lys	6.5	2.7	5.9	2.6	5.9	1.5	.03
His	2.2	1.2	1.9	.9	2.3	.6	.12
Arg	4.4	2.0	3.1	1.4	3.8	1.4	1.07
Trp	1.3	1.0	1.2	1.4	1.2	1.5	.03

The mean amino acid composition was calculated for the two classes and a students t-test performed to detect any difference between thermophilic and non-thermophilic means. At P=0.01, significant differences were observed for aspartic acid/asparagine, serine, and possibly arginine. Thus, there is some slight statistical evidence to support the idea that perhaps small changes may exist in amino acid compositions of thermophilic protein. (NASA Conference on Microbial Adaptation to Extreme Environments III. 1974. Academic Press, in press).

B.W. MATTHEWS, Eugene:
I wonder if there is another quite different fact that we ought to consider, particularly in case of the extracellular enzymes. Most of the thermophilic bacteria are isolated from environments which are more salty than would be the case for bacteria at large and this might cause differences in the surface residues which have nothing whatever to do with thermostability.

F.C. WEDLER, Rensselaer:
I think it's worth pointing out that halophiles are often thermophiles. It seems to me you have to worry about those amino acids which are perhaps in the core versus those which are on the outside. We don't yet either know about aspartic versus asparagine in many cases, and the analyses, as far as cysteine is concerned, need to be done very carefully as well. So the composition is perhaps not as meaningful, probably the least meaningful thing that we have to deal with. It's very hard to make any correlations, it seems to me.

L.G. LJUNGDAHL, Georgia:
I would like to make a comment on cysteine. Like Dr. Harris, I have noticed that there generally is a trend to find less cysteine in thermophilic proteins, but this is not true for the thermophilic proteins from Clostridium thermoaceticum when compared with corresponding proteins from Clostridium formiaceticum. Methylene tetrahydrofolate dehydrogenase from both organisms has 8 cysteine residues and formyltetrahydrofolate synthetase, also from both organisms, has 24 residues. And, if I remember correctly from Dr. Himes' work, formyltetrahydrofolate synthetase from Clostridium cylindrosporum has 24 cysteine residues per molecule and they exist as sulfhydryl groups. I assume it is so in C. thermoaceticum enzyme too.

J.I. HARRIS:
I would of course agree that there need not necessarily be a direct relationship between the total number of cysteines and thermal stability. My point is that one needs to know where they are located in the 3-D structure. Cysteines can occur on the "inside" or on the "outside" and it is my contention that "outside" cysteines are a potential source of instability (e.g. due to oxidation or disulphide exchange reactions), and that they are deleted or, alternatively, protected from the "outside" by changes in the secondary or tertiary structure.

A. FONTANA, Padova:
The observations of Dr. Ljungdahl point out that general answers to the problem of thermostability of thermophilic enzymes do not seem to hold. Certainly, thiol groups are very reactive, they easily oxidize to the disulfide stage and would determine intra- and intermolecular crosslinks causing denaturation of enzymes. The enhanced thermal stability of 6-phosphogluconate dehydrogenase from B. stearothermophilus in respect to the enzyme from E. coli does not seem to be related to differences in content and behaviour of thiol groups since the enzyme from both sources has the same number of cysteine residues and in both cases inactivation occurs upon selective modification by the SH-reagents p-chloromercuribenzoate and 5,5'-dithiobis-(2-nitrobenzoic acid), the activity being fully restored by addition of β-mercaptoethanol. In addition, in both cases it has been observed similar protection against inactivation by the substrate 6-phosphogluconate and the coenzyme NADP.

CHAIRMAN:
Is there any non-panelist who would like to contribute to this discussion? I hope you don't feel intimidated by the fact that you are sitting up there.

R. SINGLETON:
I would like to ask Dr. Harris: Is he proposing that heat denaturation as a general process involves an oxidation of SH groups and that this is why thermophilic proteins have a decreased SH content?

J.I. HARRIS:
Yes, I think it is one of the mechanisms of thermal denaturation in thermophile as well as in mesophile proteins. SH groups tend to oxidise to disulphides, and GPDH is for example heat-inactivated by formation of an intrachain disulphide bond between Cys-149 and Cys-153 in the sequence (unless of course it is kept in the presence of mercaptoethanol or DTT). Formation of this bond seems to trigger off a conformational change that results in precipitation and irreversible inactivation of the enzyme. I would have thought that such reactions are even more likely to occur at the growth temperatures of thermophiles. There are of course other mechanisms of denaturation and I am not saying that proteins lacking cysteine are necessarily heat stable.

R. SINGLETON:
That's what I'm saying.

J. FEDER, St. Louis:
The neutral proteases from the Bacilli are an example of a group of enzymes in which disulfide bonds cannot be implicated in stability. A number of these enzymes do not have disulfide bonds including both the thermophilic thermolysin and the mesophilic enzymes from B. megaterium and various strains of B. subtilis. In the case of the Bacilli neutral proteases, one might conclude that the lack of disulfides is more characteristic of a group rather than of thermophile or mesophile.

It should also be pointed out that relative to Dr. Matthews' earlier comments, all of these are extracellular enzymes.

O.H.W. KAO, Albany:
Relating to Dr. Harris' comment on -SH groups, we have observed that most phycocyanins from thermophilic organisms contain more 1/2 cystine groups than the ones from mesophilic organisms. However, we do not know whether that is one of the factors associated with thermostability.

J.I. HARRIS:
If all the cysteic acid was really derived from cysteine (SH) (and not from cystine (-S-S-)) residues then they must be buried in the interior of your proteins!

CHAIRMAN:
If there are no further comments, we can go to the next topic, namely how homologous sequences relate to the three-dimensional structure of these enzymes. I'd like to ask Dr. Matthews to start the discussion.

B.W. MATTHEWS:
I think the discussion immediately preceding and through the conference made it fairly evident that, knowing the composition of a protein does not necessarily give a clue to its thermostability or lack of thermostability. Knowing its amino acid sequence also need not give any really reliable information on this point. Also I think it fair to say that given the three-dimensional structure of a protein, it is not at all clear whether it will be thermostable. Prof. Neurath, in his opening remarks, referred to the problem of separating thermostability from "a background of philogeny" which complicates the vital issue of getting to the point of what it is that makes a protein thermostable. Now we have recently engaged in a study which does relate in a rather direct way to this problem. We've been studying the structure of the lysozyme from Bacteriophage T4 [PNAS 71, 4178 (1974)]. The lysozyme is not particularly thermostable, however there are available many mutants of the enzyme which have different stability and activity to that of the wild-type enzyme. One of the aims in carrying out this project was that we would be able to look at an enzyme and compare its structure, its stability, its activity with that of mutant enzymes in which only one, for example, amino acid substitution was made, and in this way we would eliminate the background of confusion that normally exists in comparing different species, or, for that matter, enzymes from the same source. The diagram below illustrates the conformation of the protein.

Those mutants which do not materially affect the activity of the enzyme are indicated "N" (non-essential). Those mutants which drastically modify the activity of the enzyme are indicated "E" (essential) and you see that all the "E"'s are clustered around the opening through the center of the molecule which we believe to be the active site of the enzyme. Mutants which do not affect the activity are in general far from the active site , so that there is a rather good correlation between the genetic evidence and the three-dimensional structure.

There is a mutant that I wanted to mention which bears on the problem of stability. The mutant is the following: at position 88 a tyrosine in the wild-type enzyme is replaced by histidine. The consequence of this change is that the enzyme with the histidine is a cold sensitive enzyme. At 37°C the activity of this mutant is about 40% of wild-type enzyme,

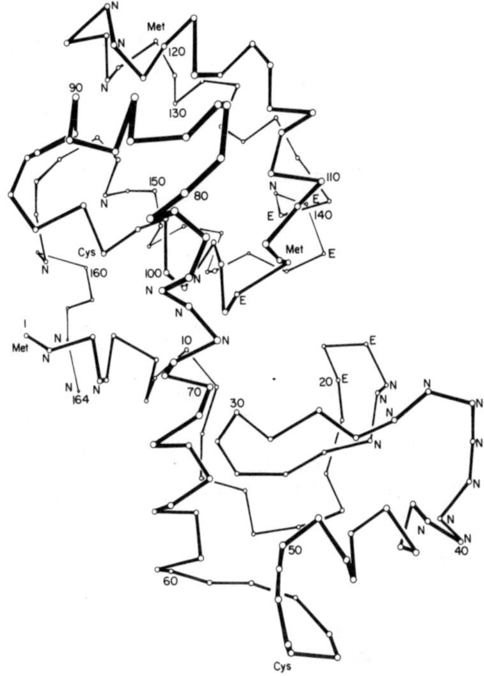

but at 20°C the activity of the mutant is less than 0.01% of the activity of the wild-type. So the simple substitution of a tyrosine for a histidine at 20°C totally inactivates the protein. At 40°C the enzyme has an activity which is comparable with that of the wild-type enzyme. I think this is perhaps the most convincing evidence that one could ask for to support the view point that the changes in the stability of enzymes can in fact be due to very subtle changes in their structure, and their composition.

I want to make a different comment on the use of amino acid sequences to predict secondary structure. At the time we were about to determine the lysozyme structure we wrote to all the people who were involved in predicting secondary structures from amino acid sequence and invited them to tell us what the secondary structure of this protein would be (Biochim. et Biophys. Acta, in press). We calculated the correlation between the prediction and the observation and the best prediction that we received for the helices had a correlation coefficient of about 40%. All the other predictions were less than 40%. In the case of the beta structure and the bends the correlations ranged from negative values up to about 10 or 20%. So I think one ought to be rather skeptical in using these methods to predict secondary structure in proteins.

E. STELLWAGEN:
Wasn't this very same experiment done with a dfferent protein?

B.W. MATTHEWS:
Yes, that's true.

E. STELLWAGEN:
And weren't the results more encouraging?

B.W. MATTHEWS:
Somewhat. A similar comparison was made with adenylate kinase, and in that case the agreement was somewhat better, the correlations ranging up to about 50 or 60% [Nature 250, 140 (1974)]. But one has to ask whether one is going to be content with a correlation of 50 or 60%, at best.

F.C. WEDLER:
Two things: First of all, you have to ask where the mutation occurs and you don't really know how important that is unless you know the tertiary structure, so even a comparison of homologous sequences doesn't tell you much. The other thing is, if you only had a sequence or an analysis for thermolysin you might have no idea that it could be thermostable. And indeed, the apoprotein is certainly not a thermophilic protein. In glutamine synthetase for example, the one from E. coli has a structural site for divalent cations and it's a thermophilic protein: it will stand 65 degrees for 15 minutes without denaturation. The one from B. subtilis falls apart at 40 degrees and dissociates upon dilution, so surely from what we've talked about so far, it's very difficult to predict whether it's going to be thermophilic or which way it's going to go. There are special co-factors obviously that contribute, and the composition of the protein itself is only one of many number of factors that can play into the picture. I think you really have to specify whether you've got a metal site in the enzyme, because these co-factors (particularly with the monomeric system) obviously can be very important.

CHAIRMAN:
No, we'll come back to the metals on the point no. 4. Do you have any comments on Dr. Matthews' statements?

From the view point of the cold sensitive mutant, the wild type is a thermophilic enzyme?

B.W. MATTHEWS:
That is right.

L.L. CAMPBELL, Newark:
I want to ask Brian about the labeling of his first slide that showed essential and non-essential substitutions. In the particular mutation shown in position 88, tyrosine would be essential at 20°C. You see it's a definitional problem we deal with when we talk about thermostability. When you substitute for tyrosine in that particular location, you have an enzyme that is cold labile; therefore tyrosine would be essential for activity at that temperature. I believe you have to expand your "red dots" depending on what temperature you are talking about when assessing substitutions or deletions as being essential or non-essential for enzyme activity.

B.W. MATTHEWS:
Your point is well taken. My point in showing the slide was simply to emphasize the fact that very small changes in the structure of a protein can demolish the protein structure and that this illustrates I think in a very dramatic way the problem that we've been faced in trying to separate out what it is that causes one protein to be more stable than another.

CHAIRMAN:
Could I ask whether you have considered the possibility that this particular amino acid substitution causes an aggregation of monomers and the inactivation is due an association involving the active site of one molecule and the surface of another molecule?

B.W. MATTHEWS:
I don't think that aggregation has been completely ruled out.

F.C. WEDLER:
Could I raise a question to the panel? There's been some talk about a greater content of polar amino acids which presumably reside on the surface of thermophilic proteins. It's my feeling that perhaps this has something to do with disrupting or restructuring the water near the surface of the protein, and I'd like to raise the idea that perhaps this has something to do with the rate at which the water reaches the interior and contributes to the rate of denaturation of the protein. What do the panel members think about that?

CHAIRMAN:
Who would like to comment on this?

J. BREWER:
I think it was Lakowicz and Weber who examined the rate of diffusion of oxygen into a number of proteins by fluorescence techniques and found that the oxygen could get right into the middle of these proteins, whose structure was known from X-ray studies. The diffusion coefficients for oxygen for tryptophans inside proteins was nearly the same as it was for tryptophans which were known to be exposed. So, this has worked a seeping revolution in peoples' ideas about protein structure: while people still tend to think in terms of the inside and outside of proteins, they may not be quite as sharply distinct as we've been thinking. If oxygen can get right through, diffuse right through a protein, perhaps water can slowly diffuse into a protein also. Perhaps it's just a question of the hydrophobicity of the residues around a particular residue that restricts water diffusion to some extent.

F.C. WEDLER:
Can I respond to that very quickly? Obviously oxygen is not water, in terms of polarity it's a very different molecule. Certainly the holes are big enough to let it in, but the question is whether with a very polar molecule (water) it really will.

B.W. MATTHEWS:
I think it's true that the average protein has more hydrophobic residues on a surface than we tend to believe. We once compared the surface residues of thermolysin, carboxypeptidase and α-chymotrypsin and didn't at that time see any obvious differences in the surface residue composition of these three proteins.

E. STELLWAGEN:
In order to extract any general rules of thermophily with respect to protein structure, the picture seems very pessimistic if I understand your comments. It is clear that amino acid compositions are not very informative. Unfortunately, comparative sequences within a homologous series of proteins also do not appear to give any obvious indications of the basis of thermophily. It would seem then that there are no general mechanisms for enhancing thermostability of proteins. This suggests that attempts to develop a general view of enhanced thermostability will be futile.

S.M. FRIEDMAN, New York:
I have a more optimistic outlook. Furthermore, there is some unpublished data available which sheds light on a mechanism to explain thermostability. Yarbrough and Koffler have evidence to indicate that a few critical tyrosine residues present in flagellin from thermophiles are responsible for the thermostable property of this particular protein.

R. SINGLETON:
I've seen these Koffler and Yarbrough data (Abstracts, NASA Conference on Microbial Adaptation to Extreme Environments, II 1972), but it seems to fit in more or less with the theme that's been developing at this meeting, i.e., that there may be one or two residues involved in protein stabilization but there are no major differences. However, I don't believe that there is any generalized mechanism of thermophily there. It's interesting data, but there's no generalization to be obtained from it.

S.M. FRIEDMAN:
But then again there might not be any generalization to be made for the property of thermostability from one protein to another. But at least in answer to Dr. Stellwagen's query, all is not hopeless maybe.

CHAIRMAN:
If I may respond to Dr. Stellwagen also, it seems to me one doesn't have to be too pessimistic because given the sequence and the three-dimensional structure of insulin one would never guess that it has a physiological function. But once you do know the physiological function there is hope that you may be able to relate it to the sequence and the three-dimensional structure. And the same may be true of thermophilic enzymes.

K. RAO, London:
I think Dr. Stellwagen should wait till tomorrow because tomorrow is the concluding day and I do have some data in showing thermostability and sequence on ferredoxins. I am going to present it tomorrow and we may have something there. So, wait till 6 o'clock Thursday.

J.I. HARRIS:
Perhaps one could summarise the discussion by saying that there are many different ways in which a protein could acquire thermal stability. As I pointed out in my lecture, the additional free energy of stabilisation that is required to confer thermal stability can be very small - of the order of 5 kcals/mole. Bond energies of this magnitude can be gained in a variety of different ways and through the same types of bonds that are commonly involved in stabilising proteins at mesophilic temperatures. It follows that different proteins even from the same thermophilic organism will be stabilised in different ways, so that we are unlikely to uncover any universally applicable recipe for thermophily.

H. GROSJEAN, Brussels:
The sequence and X-ray crystallographic data give us a static view of a protein. May be some essential characteristics of a thermostable protein should be found in some other parameters, such as in the potential flexibility or in some other dynamic aspects of protein structure for which, of course, both the sequence and the crystallographic data do not give us any information.

CHAIRMAN:
May I ask Dr. Matthews whether it is practically feasible to carry out X-ray structure analysis of thermostable enzymes at elevated temperatures in order to compare them with the structure of the mesophilic enzymes at a lower temperature?

B.W. MATTHEWS:
To a point it's possible, in fact we tried with crystals of thermolysin to see to what extent we could measure data at high temperatures. The highest temperature at which we could maintain a crystal for long enough to get reliable measurements was about 60 or 65 degrees and these data suggested that the structure in the crystal was somewhat more mobile but that on the average it was the same structure that we were seeing at a lower temperature. Technically it is difficult for a number of reasons to collect data beyond about 60 to 70 degrees even for a very stable protein.

R. SINGLETON:
Did you do the crystallization at a higher temperature or did you simply take the crystals and heat them?

B.W. MATTHEWS:
No, in the experiment we used pre-grown crystals. I think at high temperatures we would have problems with autolysis.

J.I. HARRIS:
There are of course methods other than X-ray crystallography for looking at the 3-D structure of proteins; for example, spectrophotometric methods such as NMR, ESR and CD, which have the additional advantage of being applicable to protein solutions. In the course of this meeting we have been given examples of several proteins in which no conformational changes could be detected at temperatures ranging between 20° and 60° to 70°C, and for this purpose I feel that solution (e.g. spectrophotometric) methods are of more general application than X-ray crystallography.

J.N. JANSONIUS, Basel:
I agree with Dr. Harris that this probably is the case. One of the technical reasons that Dr. Matthews was mentioning is of course the unavoidable increase in atomic vibration with temperature, leading to a blurring of detail in the electron density map that would result from such high temperature X-ray work. Interpretation of such a map would be correspondingly more ambiguous.

J. LINDSAY, Canberra:
I wonder though, Dr. Matthews, if the crystal structure studies are meaningful in terms of looking at the enzyme or the protein in the cell, in terms of "how, does it actually work in the cell?" You're not really going to get anything from your 3D studies on crystal forms, at any temperature; you've got to look at the enzyme or protein in the cell and you can't do that.

CHAIRMAN:
Those are fighting words.

B.W. MATTHEWS (from Australia):
From a fellowship-countryman, what's more. Well, you've obviously touched a nerve. It's a point that crystallographers, particularly, have been worried about. I think it's fair to say that all of the evidence which is available, and there's now a lot of data, leads one to conclude that the results of protein structures as determined in the crystals are very relevant to their structures in solution. The evidence is of two forms: Firstly, one can take the same enzyme grown in different crystal forms under very different conditions, for example in the presence of alcohol or in the presence of ammonium sulfate, and can independently determine the structures of those crystals. In every case, without exception, the structure that one determines is the same within experimental error. This suggests that the formation of the crystal does not radically alter the structure of the protein because if changes occurred then it would be likely that the two structures would be different. That's the first general line of evidence. The second general line of evidence is that one can show, and it has been shown for many proteins, that the protein in the crystal is active. Crystals of proteins are different to crystals of small molecules and salts. They are rather like a box of ping-pong balls, where the ping-pong balls are the molecules and the open spaces are occupied by free solvent through which one can diffuse in substrates and one can diffuse out products. I think it's fair to say that if an enzyme in a crystal can carry out its biological function, then the structure that one has in the crystal is relevant to what one has in solution. One may have certain reservations, perhaps, that the protein isn't as free to move in the crystal as it is in solution, or perhaps that the next molecule in the crystal blocks the active site. These are complications that one clearly has to be concerned with for every crystal structure, however I think it is unfair to say that the results of protein crystal structures determined crystallographically are not relevant to the structure in solution or in the cell.

J. LINDSAY:
Nobody has yet done a temperature inside the cell, we can talk about temperature of the media it's grown in and we can talk about the conditions, the components of the media. We can't really talk about what is in the cell at the time the enzyme is functioning and that is really what the whole thing is about.

J.I. HARRIS:
Thermolysin is after all an extra-cellular enzyme. In so far as it functions outside the cell it probably is possible to correlate solution and crystal studies. On the other hand results obtained with thermolysin may not be particularly relevant to intracellular multi-subunit thermophile enzymes in which subunit interactions also play a role in thermal stability. The thermolysin (apo)-protein as such is not heat stable - it needs Ca^{++} to stabilise it.

J. FEDER:
It is important to distinguish between the intrinsic stability of a protein and the stability due to extrinsic factors that might be present within the cell. I would suggest that there are systems in which one can estimate these intracellular effects on stability. For example, in some cases an enzyme can be assayed within the cell without disrupting the cell. Substrate diffuses into the cell is cleaved and products diffuse out. The stability of the system can be compared to that of the isolated enzyme.

CHAIRMAN:
Unless anybody has some burning issues to raise relating to homologous sequences and their relation to 3 D, we shall turn to the next discussion point.

A. FONTANA:
I would like to recall our observations related to the findings of Dr. Matthews that crystals of thermolysin are stable up to about 60-70°C. We have carried out a detailed study on the conformation and conformational stability of thermolysin by circular dichroism and fluorescence emission measurements. It has been found that up to about 75°C the circular dichroism spectrum of thermolysin in Tris-HCl buffer, pH 7.0, containing 10 mM $CaCl_2$ does not change significantly in shape and intensity of ellipticity. Analogously, upon heating up to about 80°C the enzyme in a calcium-containing buffer only a monotonic decrease in fluoresecence emission intensity at 333 nm was observed, due to thermal quenching [J.A. Gally and G.H. Edelman, Biochim.Biophys.Acta 60, 499 (1962)]. The temperature effect on the fluorescence properties of thermolysin was very similar to that found with a mixture of tryptophan and tyrosine in the same ratio as contained in the enzyme. In conclusion, these physico-chemical measurements would indicate that the conformation of thermolysin in a solution containing calcium ions is maintained up to 75-80°C, similarly to the enzyme heated in the crystal state. On the other hand, as I reported in detail yesterday, the apoenzyme, i.e. the EDTA treated enzyme, is rather unstable being denatured at about 48°C.

K. IMAHORI, Tokyo:
Coming back to the fluctuation of a three-dimensional structure, we have studied the hydrogen-deuterium exchange for both mesophilic and thermophilic enzymes. The thermophile enzyme has a larger number of slowly exchangeable hydrogen than the mesophilic enzyme.

CHAIRMAN:
Well, the problems are difficult enough when we consider the stability of the folded polypeptide chain within a monomeric protein or enzyme molecule. But when we come to subunit enzymes, we have to consider the interaction between the various subunits or between the various polypeptide chains and the areas of contact between the subunits. And so I think it is appropriate that we discuss the question of the thermal stability of enzymes which form a quaternary assembly and I'd like to ask which member of the panel would like to open the discussion.

E. STELLWAGEN:
I don't think there is any correlation. The interactions between subunits are the same interactions which are within subunits and it seems to me we've just translated the problem from a tertiary level to a quaternary level. While thermophilic enolases are hexamers and thermolabile enolases are dimers, most thermophilic enzymes are not larger than their mesophilic counterparts.

J.N. JANSONIUS:
In part Dr. Stellwagen is right. However, there are differences. For example, charge interactions will usually not occur inside a monomer (althrough they may occur in its surface). On the other hand, in stabilization of quaternary structure this type of interaction often plays an important role.

L.G. LJUNGDAHL:
Actually, the question is if the quaternary structure can be responsible for thermostability. I think it can and I believe we have evidence for that in formyltetrahydrofolate synthetase from Clostridium thermoaceticum. This enzyme is a tetramere as are the mesophilic enzymes from other Clostridia. The mesophilic enzymes dissociate to subunits when one takes potassium or ammonium ions out of the solution. The C. thermoaceticum enzyme does not do that indicating a stronger interaction between the subunits. The enzyme from C. cylindrosporum dissociates also to subunits around pH 4 as was found by Himes and Rabinowitz and their coworkers. They found also that the monomers of the C. cylindrosporum enzyme are quite stable and that the enzyme can be reconstituted. The C. thermoaceticum enzyme also dissociates at pH 4, but the monomers are not stable and the enzyme cannot be reconstituted. It seems to me that this may indicate that the monomers of C. thermoaceticum are less stable than the monomers from C.cylindrosporum enzyme. The higher stability of the C. thermoaceticum enzyme may thus depend on the quaternary structure. Of course, I am not sure about this, but I believe our results indicate that.

J.I. HARRIS:
We should perhaps recall that multi-subunit enzymes are initially synthesised in the cell as monomers. Presumably therefore the synthesized polypeptide chain must be thermodynamically stable at the growing temperature to enable it to fold correctly and to form a stable quaternary structure. It follows that a thermophile monomer needs to be more stable, thermodynamically as well as kinetically, than its mesophile counterpart to enable it to acquire a stable tertiary and quaternary structure at the elevated temperature. In addition one imagines that the inter-subunit bonds also need to be strengthened to ensure the stability of the active multimeric form at high temperatures. However one would need to know how much of the tertiary structure has to be formed to enable the embryonic monomer to interact with other monomers in order to arrive at any quantitative estimates for the relative stabilisation of the tertiary and quaternary structures.

CHAIRMAN:
It seems to me that what we've been talking about is the effect of temperature as one of the forces which are responsible for the maintenance of the three-dimensional structure of the monomer on the one hand, and of the quaternary assembly on the other. While fundamentally, there may be no difference between them, I think there could be a significant difference in degree. If we accept the fact that the surface of a protein molecule is largely composed of polar residues, whereas the inside of a monomer is largely occupied by hydrophobic residues, it seems to me that the forces which are responsible for the maintenance of the quaternary structure are not necessarily the same as those which are responsible primarily for the maintenance of the internal structure of the monomer. If that is true, one could argue that the temperature effect on the maintenance of the conformation of a subunit enzyme doesn't follow necessarily the same rules as those which describe the maintenance of the structure of the monomer. I notice that Dr. Matthews is shaking his head.

B.W. MATTHEWS:
I will adopt the opposite point of view. It's true that a monomeric enzyme may have a largely polar surface but an oligomeric enzyme will have subunits constructed so that part of the surface is not polar and that's what causes them to aggregate. So, I would think that it's likely that the sorts of forces which stabilize oligomers will be of the same type as those which stabilize individual subunits.

F.C. WEDLER:
We have a small theory floating in our laboratory which says that "more is more" instead of "less is more", i.e. more structure gives more stability. Certainly one is forming bonds and secondly one is excluding solvent water when one has an aggregate. Now with the dodecameric glutamine synthetase there are obviously hydrophobic patches of some kind or at least structure-forming areas above and below the subunits, so that this aggregate can stack up. Also, this is a heat-driven process, which does suggest hydrophobic interactions. Another small theory which we are looking at is that indeed, if the thermophiles are primitive ancestors of the mesophiles then one of the first events following the synthesis of protein monomer was aggregation with other monomers. Indeed, the aggregation process as a thermostabilizing mechanism, for which we have some

evidence, may have occurred first, and then some of the more complex intersubunit interactions may have evolved, once lower temperatures were achieved and a greater flexibility could be allowed.

J. FEDER:
I would like to ask if anyone has information relating to the effect of temperature on quaternary structure stability. Is there an optimum temperature for tetramer formation?

CHAIRMAN:
Good question, but nobody is willing to give an answer. Maybe nobody has the answer.

O.H.W. KAO:
We have been studying the aggregation properties of phycocyanin from either mesophilic or thermophilic organisms. So far, we were unable to find a positive correlation between the quaternary structure and the thermostability of the protein.

CHAIRMAN:
Do we have any hybridization experiments of subunits of mesophilic and thermophilic enzymes which would shed some light on this problem?

J.I. HARRIS:
Dr. Artavanis in my laboratory has succeeded in hybridising triosephosphate isomerase (TIM) from a thermophile (B. stearothermophilus) and a mesophile (chicken muscle). The hybrid dimer comprising one thermophile and one mesophile subunit was isolated in close to theoretical yield, implying that the tertiary and quaternary structures of the two forms are closely related. A study of the stability of the hybrid to heat, urea and SDS, showed that the rate-limiting step in the inactivation of TIM by urea and SDS is the dissociation of the dimer into subunits, whereas its thermal stability was shown to be related to the stability of the individual subunits.

Dr. Artavanis has also determined the complete amino acid sequence of B. stearothermophilus TIM. (It has a sequence homology of about 38% with the chicken muscle enzyme.) This enabled us to relate the sequence of the thermophile enzyme to the 3-D structure of the chicken muscle enzyme which has been determined by Professor Phillips and his colleagues at Oxford. This comparison showed that residues implicated in the catalytic activity and in subunit contacts were highly conserved in the thermophilic enzyme. Moreover two important subunit contacts were changed in the thermophile enzyme. These changes of serine-(OH)-(OH)-serine and tyrosine-(OH)-(COO^-)-aspartic acid contacts in the chicken enzyme to phenylalanine-(-⬡-)-(-⬡-)-phenylalanine and histidine (N)-(COO^-)-aspartic acid contacts in the thermophile enzyme could well provide enough additional stabilisation energy to account for the greater thermal stability of the dimeric structure of B. stearothermophilus TIM.

G. CACACE, Naples:
I would like to make a comment about the subunit structure of an RNA polymerase we have isolated from a highly thermophilic microorganism, Caldariella acidophila. The subunit composition of this RNA polymerase is very similar to that of the E. coli enzyme, but the molecular weight of the larger subunits, β and β', is smaller than that of the other bacterial RNA polymerases so far described. The reduction in molecular weight of the two larger subunits could be related to the stability of the quaternary structure of this enzyme which is optimally active at 80°C.

Y. NOSOH, Tokyo:
Glutamine synthetase from B. stearothermophilus we're working with was not dissociated into subunits, when treated with up to 3 M urea in the presence of Mg^{2+}. The activity and the hydrophobic property of the enzyme, however, decreased remarkably. So I don't think the quaternary structure seems to be closely related to thermostability, although Dr. Harris suggested the importance of the relation.

J.I. HARRIS:
In the case of TIM we found that, to a first approximation, inactivation in urea or SDS could be correlated with dissociation of the dimer. On the other hand thermal inactivation of the hybrid (C-S) dimer at 60° appeared to be related to the stability of the monomers. At 60° the hybrid dimer dissociates - the chicken monomer is unstable at this temperature and denatures, while the thermophile monomer is stable and is able to reform active dimers. This shows quite convincingly that the tertiary structure of B. stearothermophilus TIM is stable at 60°C (the growth temperature of the organism). On the other hand it is not stable in urea or SDS (the enzyme is irreversibly inactivated in these media as soon as the dimer dissociates) - but then it doesn't need to be since there is no urea or SDS inside the cell!

K. IMAHORI, Tokyo:
As I told you yesterday, Dr. Susuki and I are trying to have a hybrid of a glyceraldehyde-3-p-dehydrogenase obtained from several species. Although we have not yet a general condition in which we can make hybridization up to the present stage we feel that some salt such as borate is critically useful. A minimum of salt concentration which is necessary to prepare the hybridization reflects the stability of the quaternary structure of the enzyme.

H. ZUBER, Zürich:
In this connection I would like to remind you of the important role of metal ions in the association - or the aggregation - process of oligomeric enzymes. For example, some metalloenzymes possess a second metal atom in addition to the one of the active site. The former seems to be important firstly for the tertiary structure of the subunits, and secondly for the association of the subunits to the whole enzyme. In the case of the thermophilic aminopeptidase T (API) we found two metal atoms per subunit, one in the active site, and the other is perhaps important for the tertiary and quaternary structure of this 12-subunit enzyme. Furthermore, we have observed that the stability of the API depends strongly on the type of metal bound; the cobalt enzyme (although more active) is not as thermostable as the zinc enzyme.

CHAIRMAN:
This is as nice a transition as I could have thought of to come to the next topic of our agenda but before we proceed, I think we'll break here for a few minutes in order to allow a change of tapes and as soon as this has been accomplished, we'll discuss the role of metals in thermophily.

J.I. HARRIS:
Dr. Matthews showed earlier how the stability of the protein to heat could be influenced by a single amino acid substitution in the primary structure. My slide (illustrating the work of Dr. R. Jack in my laboratory) is concerned with the thermal stability of B. stearothermophilus aldolase. This is a dimeric Zn-containing enzyme with a half-life of about

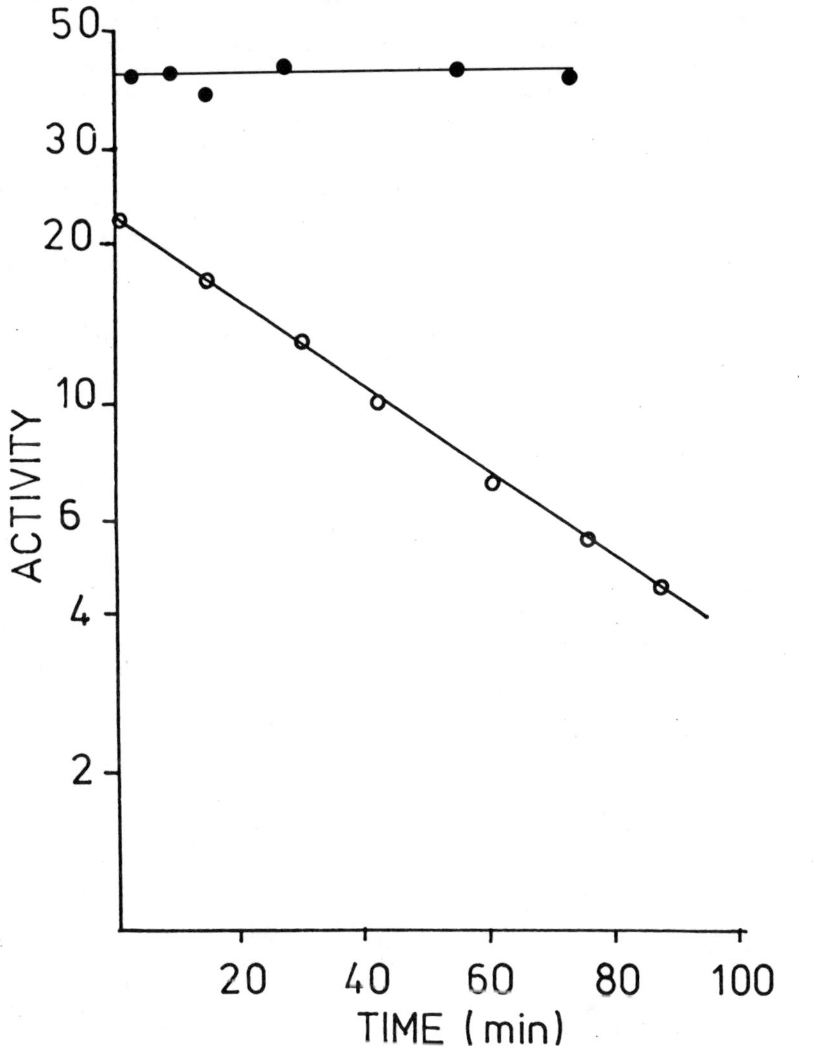

Thermal inactivation of Aldolase at 64°

○ Zn^{++}
● Co^{++}

30 min at 64°C. The apo(Zn-free)enzyme is appreciably less stable and we believe that the serious loss of activity that occurs during isolation of the enzyme is due to a gradual loss of Zn^{++}. We found, however, that loss of activity could be largely prevented by adding Co^{++} to the buffers. Presumably the enzyme has a greater affinity for Co++ and enzyme prepared in this manner contained two atoms Co^{++} instead of two atoms of Zn^{++}. Note that the Co^{++} enzyme is not inactivated to any detectable extent at 64°! This experiment shows that the thermal stability of a protein can be profoundly influenced by merely changing one divalent metal for another. One imagines that Co^{++} binds to the same sites as Zn^{++} and that the tertiary and quaternary structures of the Zn and Co aldolases are largely unchanged. There are no amino acid substitutes, no differences in side-chain interactions within monomers or of inter-subunit contacts in dimers; there's not even a difference in SH content! (B. stearothermophilus aldolase contains three SH groups per subunit but these are not reactive in the native dimer.) I doubt whether even the combined skills of protein chemists and X-ray crystallographers will provide an explanation for this result - but what about NMR or ESR?

A. FONTANA:
During this meeting we had occasion to refer quite often to the role of calcium on protein stability. In particular, calcium has been shown to play a major role in the thermostability of thermolysin and α-amylase. However, the stabilizing effect of calcium is not restricted to thermophilic enzymes, since mesophilic enzymes like trypsin, desoxyribonuclease A, glutamine synthetase and others show enhanced stability in presence of calcium ions, although the effect of these ions is not so dramatic as, for example, in the case of thermolysin. The mechanism by which calcium ions exert their effects on macromolecular conformation and conformational stability, at low concentration, is related to the possibility of a bridging function within the polypeptide chain, thus reducing the polypeptide backbone flexibility and thus enhancing stability. Alternatively, stabilization of a polypeptide structure of net negative charge can be achieved by removal of intra-molecular repulsion from anionic groups, calcium ions being bound to carboxylate groups [P.H. von Hippel and T. Schleich, Acc.Chem.Res. 2, 257 (1969)]. However, the reverse effect, i.e. thermal destabilization, has been observed at higher concentrations of calcium ions. The melting temperature of ribonuclease is reduced by about 20°C in 2M calcium ion concentration, from 61°C to near 40°C [P.H. von Hippel and K.Y. Wong, J.Biol.Chem. 240, 3909 (1965)]. In conclusion, it is important to differentiate between specific ion effects at lower concentration, i.e. at mM concentration, or at higher ionic concentrations where the effects characteristic for the Hofmeister series become dominant. And just to extend the subject under discussion, I like to recall that not only metals exert their effects on protein conformation and conformational stability, but anions as well. Unfortunately, in studying protein thermostability of thermophilic enzymes, investigations on the effects of low salt concentrations on protein stability are rare. The influence of different buffer systems at low ionic strength on melting temperature of proteins can be definitely significant, as in ribonuclease, where, relative to a HCl/KCl buffer, a phosphate buffer produces a ΔT_m of 12°C and a sulfate buffer of 14°C [A. Ginsburg and W.R. Carrol, Biochemistry 4, 2159 (1965)].

E. STELLWAGEN:
It seems to me that anything that binds preferentially to the native form of a protein will raise its melting temperature and anything that binds preferentially to the denatured form will lower the melting temperature simply by coupling binding equilibria with the N→D transition. If calcium preferentially binds to the native conformation of thermolysin, it will increase the thermostability of the complex relative to that of the apoprotein. However, if calcium preferentially binds to the denatured form of ribonuclease, it will decrease the thermostability of ribonuclease.

CHAIRMAN:
Does this presuppose that the metal combines with the polypeptide chain after it has assumed its specific conformation?

E. STELLWAGEN:
I am suggesting that the metal combines with a specific conformation in order to enhance thermostability.

J. BREWER:
This brings up another aspect. Any preferential interaction, any binding at all to the native form would impart at least a kinetic stability to a protein. Until it denatured the protein would be stabilized by any interaction with any ligand.

R. SINGLETON:
I would like to address a question to Dr. Harris about the aldolase he mentioned previously. The aldolase situation is a little bit different from what Dr. Stellwagen has pointed out in that the zinc is necessary for the catalytic activity of the enzyme. Is zinc the natural metal for this particular aldolase or is cobalt? Is the cobalt enzyme catalytically active and if so, is it as active as other aldolases?

J.I. HARRIS:
Yes, Zn^{++} is the natural metal. It can be removed with EDTA (the metal-free enzyme stays as the dimer but is totally inactive) and replaced by Co^{++}. There is an absolute requirement for a divalent metal ion, and addition of two atoms of Co^{++} to apoenzyme restores activity fully. The only change detected so far is the increase in thermal stability.

R. SINGLETON:
I wonder what would happen if you would grow the organism on a zinc free medium and feed it cobalt to see if it would preferentially be incorporated.

J. FEDER:
One also observes increased activity by substituting cobalt for zinc in thermolysin. However, the binding constant for cobalt is less than that for zinc. A few years ago we determined the binding constants for zinc and cobalt to apo-thermolysin by kinetic analyses of phenanthroline inhibition curves. The dissociation constant for zinc-thermolysin was about $2x10^{-13}$M and that for cobalt-thermolysin was about $3x10^{-10}$M. It is interesting to note that cobalt is considerably less inhibitory to the enzyme than zinc at higher concentrations.

I would like to comment on the role of calcium in the thermostability of thermolysin. Obviously there is more than one role for divalent cations in stability. However, the role of the four calciums in thermolysin, particularly those between the two lobes, might be described as a bridging moiety. It seems quite obvious that they have similar function to a disulfide bond. The first such example, I believe, was for the calcium in Anfinsen's nuclease. Dr. Matthews has shown this to be the case for thermolysin. Our studies certainly demonstrated that this was necessary for thermostability. However, I believe that the intrinsic thermostability of thermolysin cannot be related to the calcium alone. Certainly if one reduces a disulfide, one loses activity due to the resultant loss of conformational integrity. However, since a covalent bond has been disrupted, and this is common to both mesophiles and thermophiles, we do not speak of it being the stabilizing factor. The calcium in thermolysin, I believe, has an analogous role. Both the mesophilic and thermophilic neutral proteases require calcium for stability. However in the presence of excess calcium, one enzyme has considerable thermostability while the other does not.

Earlier today there was a study presented on a thermophilic and mesophilic α-amylase. Both require calcium for stability. The stability of the calcium-free enzymes in denaturants such as urea was studied and the thermostable enzyme still was more stable. This says that in the rest of the tertiary structure there is built-in thermostability above any role for calcium.

I think most of what has been presented to-date here is concerned with the sum total of all of those non-covalent interactions which result in a more stable conformation; and calcium does not fit in this class if we consider it to be another kind of ionic interchain bond having a role not unlike a disulfide bond.

L.G. LJUNGDAHL:
I would like to comment about metalloproteins like ferredoxin and the formate dehydrogenase, which I discussed yesterday. These proteins are more thermostable when isolated from thermophilic organisms and yet they have the same metal content. It seems that the thermostability has nothing to do with the metals and with ferredoxin it is possible to take the iron out, but the protein part is still stable at high temperatures.

O. ANDRESEN, Denmark:
There's one thing I've been missing during most of the lectures concerning the role of metals in the stability of enzymes and that investigations as to what importance will the pH of the reaction be. It might be that calcium adds stability on both sides of an optimal pH at which calcium is not necessary and that might have something to do with the isoelectric point.

CHAIRMAN:
Dr. Matthews, would you like to comment on this calcium binding problem?

B.W. MATTHEWS:
I will comment on metal stabilization, at least as we have thought about it looking at the thermolysin structure. Ionic interactions are thought to be entropy driven and in that respect are very similar to hydrophobic interactions. In both cases the energy of interaction will actually increase with temperature, so one can say, from a thermodynamic point of view, there is really nothing to choose between an ionic

interaction in the presence of water and a hydrophobic interaction in the presence of water. In the case of thermolysin, the enzyme binds calcium and if you remove the calcium the enzyme is unstable. On the other hand, as has been pointed out, there are plenty of stable enzymes which do not bind metals. One can ask the question: Why is it that in the case of thermolysin the enzyme happens to bind calcium when one could make the case that there's no need for metal binding? The only suggestion that I can make is that perhaps ionic interactions or interactions through a metal are in some sense more directional than a hydrophobic interaction and perhaps this is important either in maintaining the three-dimensional structure or perhaps, as Dr. Neurath suggested, this is important in promoting the correct folding of at least part of the structure. I don't feel we understand why it is that some proteins happen to bind metals for a structural reason and others do not.

F.C. WEDLER:
I'd like to make a comment in regard to what Dr. Feder suggested, I think I can cite at least one example where what he says seems to be true. First of all if one compares mesophilic glutamine synthetases (the E. coli and B. subtilis enzymes) in the E. coli case there is a structural metal site which is the tightest site of the several and the effect of filling that site is to completely wipe out any cooperative interactions between subunits. The second thing that might be said is in the Bacillus line enzymes that we've studied (mesophilic versus thermophilic) where we appear to get a greater degree of disulfide crosslinking, especially in the B. caldolyticus system then we also lose cooperative ATP binding.

CHAIRMAN:
Since we have been talking about thermophily, I think it is quite important that we clarify our thoughts as to what we mean in terms of temperature, temperature optimum, conformational stability and how do we best express melting temperatures and melting profiles. I think these points are related to each other and are really crucial to any discussion of the problem, and so again I'd like to ask for volunteers to start the ball rolling. How about it, Dr. Stellwagen, you raised the point so many times that you may just as well help us to understand the problem.

E. STELLWAGEN:
I think there need be some standardization as to how one measures thermostability. First, one needs to establish whether a given protein exhibits one or more thermal transitions. If multiple transitions exist, then different experimental probes such as enzymic activity,CD, or $\Delta pH/\Delta T$ measurements may sense different transitions for the same protein. Thus a standard technique is needed to compare the thermostability of a variety of proteins. I would hope that such a standard technique would measure gross structural stability as opposed to functional stability which may be localized and perturbed by the enzyme assay itself owing to preferential binding of substrates to the functional form of the protein.

CHAIRMAN:
Maybe we could break this problem down into two parts: one is the question of the conformational stability and the other is the question of enzyme inactivation. These are not necessarily the same and, as you in-

dicated, the first one is a kinetic problem and the second one is a thermodynamic problem.

J.I. HARRIS:
There is perhaps a fundamental dilemma here. How do we measure conformational changes under conditions comparable to those that exist in the cell? The "melting temperature" of an enzyme can vary with protein concentration, pH, nature of buffer ions, etc., and I think it's going to be very difficult to arrive at a set of standard conditions that would be relevant to enzyme stability in vivo.

K. IMAHORI:
In discussing the melting temperature, we should be aware of at least two or three points. First, if this denaturation is reversible or irreversible. In the case of reversible, you can obtain completely the same melting temperature irrespective of what kind of method you may take. For example, by preincubation method as Dr. Stellwagen has pointed out or by the continuous heating method. These two methods give completely the same temperature. However, in the case of irreversible inactivation this is not an equilibrium problem but a kinetic problem. Thus melting temperature depends on incubation time at that temperature or heating rate in the case of the continuous heating. We have experienced in continuous heating melting temperature depends remarkably on the rate of heating, so in that occasion you should be very careful about in what conditions you are measuring. When you measure by remaining activity, you should be aware of how long you've been heating at each temperature and this is a very important thing for discussing the melting temperature.

H. GROSJEAN:
Thermophilic proteins might have a special ability to renature more easily than other proteins after heat treatment. Their particular thermostability may simply reflect this essential property.

J. FEDER:
We are attempting to establish a standard procedure for describing stability. The underlying problem, however, is that in order to quantitatively describe a process, one must write some mechanistic model and we do not have one at this moment. We have heard two experimental approaches for describing stability measurements. One involved monitoring conformational changes by some physical parameter such as CD or spectrophotometry. The other involved following loss of activity of an enzyme as a function of temperature or a denaturant. In the latter case, most of these studies involved incubation for a given period of time and an enzyme activity measurement. It then is difficult to correlate these data with those obtained for a different time period under slightly different conditions. If one observes that these are generally first order decay processes then it would be important to have a kinetic description of the process yielding a rate constant for decay. The temperature dependence of this process then could be used for quantitating the stability of the enzyme.

It should be reiterated that the proteases present particular problems for describing stability. For all other enzymes, assuming they are free of contaminating proteases, one can speak of denaturation. The proteases, however, present a complicated picture due to the autolytic process.

In addition to denaturation, one has autolysis of unfolded enzyme. It is therefore, necessary to carry out these studies in the presence of inhibitors which will prevent autolysis.

CHAIRMAN:
May I confess my ignorance and raise the question whether there's any correlation between temperature optimum and the energy of activation of the enzyme inactivation process. In other words if you make Arrhenius plots do the thermophilic and the mesophilic enzymes differ from each other?

J. BREWER:
I don't think too many people looked at this, just from a casual look at the slides that have been presented. It would appear they are not consistently different. In some cases you have very sharp falloffs of activity versus temperatures. This is more generally the case, but in others you have rather gentle declines, getting a relatively low enthalpy of activation of denaturation. I think Dr. Feder's remark about people attempting to define a mechanism from their data is particularly à propos; we're dealing,first, with a kinetic process and so we should include time in the description of it. In the second place, it's not always - indeed may seldom be - a strictly first order process, so we should consider the extent of inactivation to be one of the parameters. Perhaps we should just propose something like 90% inactivation over a ten-minute period as a flat criterion for expressing relative thermostabilities.

T.K. SUNDARAM, Manchester:
Can I ask the experts a technical question? When reporting the thermostability of proteins - and here I am not thinking of proteases but non-protease enzymes - how important is it to define the protein concentration, in other words does this depend on the protein concentration?

R. SINGLETON:
Because protein stabilization can depend on protein concentration it is vitally important that the concentration be specified. Furthermore, if a dependence upon concentration is observed, I believe this should be reported, and if possible the stabilization studied at a point where the concentration effect is minimal.

T.K. SUNDARAM:
What's the experience in general? Is there a dependence on protein concentration in general?

R. SINGLETON:
Initially, I would say no, if one considers thermophilic proteins as a class, however, somebody else may have a comment on this point?

J. BREWER:
You also should consider the possibility of dissociation, presumably the monomers have a different thermostability than the native form. I suppose it would be best to attempt to describe the thermostability of what you think is the native form and pick a protein concentration at which that is the predominant species.

F.C. WEDLER:
We've talked about kinetic and thermodynamic processes and maybe I can restate the problem if that is needed. The question is whether these enzymes are more stable because there is a higher kinetic barrier to denaturation or whether the ground state of the native protein is lower. I don't know quite how to find out what the difference in the ground states of the native versus the denatured form would be, because the equilibrium is probably unmeasurable. But certainly one can look at the kinetic barrier. It's a tough problem but the answer will probably be that both things are true.

Y. NOSOH:
We have been talking about heat inactivation of thermophile enzymes, but I am thinking it is also important to consider heat denaturation of the proteins.

E. STELLWAGEN:
Yes, I think that's the point I was attempting to make this morning and one I hoped I had raised earlier in this discussion. My own feeling is is that it's the conformational stability that would be most useful as opposed to the functional stability. In order to compare the thermostability of various proteins, it would be useful to employ a standard set of conditions for measurement of thermostability. For example, all measurements could be made in 0.1 M NaCl at pH 7. The melting temperature could be defined as a selected magnitude for the first order rate constant for denaturation observed in this solvent to account for variations in the kinetics of thermal denaturation from protein to protein.

F.C. WEDLER:
I would like to mention something related to the question of the effect of protein concentration. There's obviously the problem of aggregation or disaggregation of the pure homogeneous enzyme. In addition, a couple of enzymes we have looked at are intrinsically more stable in the crude cell extract than they are when purified, even at comparable total protein concentration. If we take a purified protein and put in bovine serum albumin, then it's more stable than if we don't have it. So one wonders, when one deals with a purified protein and looks at the denaturation curve, if one is looking at anything like what might be going on in the cell.

J. LINDSAY:
That was exactly the question that I was wanting to put to Dr. Matthews before, that perhaps there was something else involved in that what you really see is not the true thing that's happening in the cell. If you can add bovine serum albumin you get different results, then what's really happening in the cell?

CHAIRMAN:
I think there's enough substance for further discussions but the time has run out. So, let me take the opportunity to thank the panel and all other participants for their contributions and declare the session as adjourned.

AUTHOR INDEX

SUBJECT INDEX